PRINCETON
LECTURES ON
BIOPHYSICS

PRINCETON LECTURES ON BIOPHYSICS/

NEC Research Institute, Princeton, USA
23 – 29 June 1991

Edited by

W Bialek
NEC Research Institute

World Scientific
Singapore • New Jersey • London • Hong Kong

Published by

World Scientific Publishing Co. Pte. Ltd.
P O Box 128, Farrer Road, Singapore 9128
USA office: Suite 1B, 1060 Main Street, River Edge, NJ 07661
UK office: 73 Lynton Mead, Totteridge, London N20 8DH

PRINCETON LECTURES ON BIOPHYSICS

ISBN 981-02-1325-5

Printed in Singapore by Utopia Press.

Preface

A Little bit of History

Physics and biology were not always separate subjects. In the 19th century, Helmholtz, Maxwell, Ohm and Rayleigh all made great contributions to what we would now call the biological sciences. To these physicists, understanding the mechanics of the ear and the perception of sound were parts of acoustics, muscles constituted a perfectly sensible testing ground for the laws of thermodynamics, and so on. In the intervening century, physicists have often been drawn back to biology.

One historical thread tying physicists to biology is an interest in uncovering the molecular basis of life. Bohr inherited an interest in the subject from his physiologist father, and passed his thoughts on to a young colleague named Delbruck, who became one of the founders of molecular biology. A whole generation of physicists were influenced by Schrodinger's little book *What is Life?*. The early work on the structure of biological molecules --- Crick and Watson on DNA, Perutz and Kendrew on proteins --- was carried out in a small shed outside the Cavendish laboratory, under the watchful eye of Bragg himself. Ideas on the feedback mechanisms which regulate gene expression were developed by Monod in conversations with Szilard.

A second thread begins with the experiments of Galvani and Volta on "animal electricity," and continues through Helmholtz' work on the propagation of the nerve impulse and the quantitative investigation of the sense organs. This thread is picked up in this century with the British physiologists Hill, Hodgkin, Huxley, and Katz, who provided a quantitative description of the dynamics of nerves, muscles and synapses. Attempts to test Helmholtz' ideas about the mechanics of the inner ear have driven the development of beautiful experimental techniques for the measurement of small motions, beginning with the work of von Bekesy.

Despite these frequent interactions, physics and biology have developed separate intellectual traditions. Biology is taught largely as a celebration of the diversity and complexity of life on earth, a complexity which seems to exist even at the molecular level. Physics is taught as a search for simple and universal principles. Biology is divided in many different slices: molecular, cellular, organismal and population, neurobiology, immunology, developmental biology and genetics. A budding Delbruck would have a hard time learning about biology from modern texbooks, which are often written in a "headline" style more familiar to medical students than to physics students. In addition, the systems which provide the greatest inspiration for a physicist may be off the beaten track of the biologist.

As physics has matured, boundaries between sub-disciplines have tended to dissolve. We know that the nuclear reactions which make the stars shine are the same ones which we can reproduce here on earth, that broken symmetries in many-body systems can be understood in the same framework whether we are talking about electrons in solid state physics or the Higgs field in elementary particle physics. The ideas of scaling unify the equilibrium statistical mechanics of phase transitions both with the high-energy behavior of quarks and gluons and with the chaotic dynamics of systems far from equilibrium. In this context, the exclusion of living things from the provenance of physics would seem arbitrary.

About the Lectures

The Princeton Lectures on Biophysics are meant to be a place where physicists can come and learn about the phenomena of biology and about the attempts of other physicists to understand these phenomena. The ground rules are that, no matter how abstract we may end up, the discussion has to be grounded in the (sometimes harsh, but often inspiring) realities of life. You will notice that all of the chapters in this book have at least one figure with real data. Part of what makes this an exciting time to work at the interface of physics and biology is that so many biological systems have been characterized to the point where quantitative, "physics-style" experiments are possible. This is good news for both theorists and experimentalists, and the topics for the first set of lectures were chosen to highlight some of these particularly ripe areas, although obviously not in an exhaustive fashion.

The lectures themselves were held at the NEC Research Institute for one week in June 1991. The students were primarily advanced Ph.D. students and postdocs from physics departments, although there were exceptions. A relatively small fraction were already working in some area of biophysics. Many came unsure about whether one could reconcile an interest in biology with the desire to remain a part of the tradition and community of physicists; some left committed to follow precisely this path. It was an intense an lively week, and perhaps the most remarkable fact was that the activities seemed to peak on Thursday, which is very unusual for a one week event.

The format of the lectures was intended to maximize discussion, with long breaks and strong coffee. Each lecture was two hours long, and some lecturers gave two talks. Many of the students had taken their last biology course sometime in high school, yet I think that in most cases the lecturers managed to avoid jargon and fill in enough background to make things comprehensible. Because we covered a very broad range of topics, everyone (including the lecturers) was ignorant of the background for some of the talks, and in an important way this helped the communication process.

About this Book

This volume is not a precise record of the lectures. Several of the lecturers were unable to write up their notes, and in one case a lecturer from the 1992 installment agreed to pinch hit. The result is a book (coincidentally) evenly divided, with four articles following the molecular thread and four following the sensory systems. We begin with Hong *et al.*, who describe a heroic experiment in which they pump excitations into a protein molecule at relatively low energies and observe the resulting changes in the rate at which the molecule makes transitions between different states, states which are actually part of the cycle through which the protein carries out its biological function. This is a sort of resonant activation spectroscopy which provides a window into a little-studied (but much speculated) energy range for protein dynamics. The chapter captures some of the wit and irreverence of Austin's lecture, and the experiments have become much clearer with time. They challenge us to think about the "elementary excitations" of these complex molecules, and the way in which these excitations couple to the biologically significant dynamics.

The spectroscopic theme continues with Loppnow, who describes the dynamics of the visual pigments. The response of these molecules to the absorption of a photon is ultimately what triggers our perception, and this response turns out to be astoundingly fast; time-resolved studies of the visual pigments have been among the first experiments done with each new generation of fast pulse lasers. Loppnow explains how Raman spectroscopy allows the direct measurement of the forces of the molecule in the instant after the photon is absorbed, and develops a scenario for the molecular dynamics in the crucial first 50 femtoseconds of vision.

Proteins are large, complex and heterogeneous molecules. They are not random, displaying several repeating structural motifs, but they are not crystalline either. Understanding how these structures arise --- and ultimately predicting the structures of particular proteins --- is the protein folding problem. For biologists this is a central issue, since they would like to know in advance how new genetically engineered molecules will behave. For physicists, as Chan *et al.* describe, there are wonderful statsitical mechanics problems which can be attacked with both analytical and numerical techniques. This chapter, pieced together from two recent articles, provides as authoritative and up-to-date a view of this exciting area as I have seen anywhere.

One of the simplest of chemical reactions is electron transfer between donors and acceptors fixed in space. Biology uses these reactions in a variety of energy converting systems, notably photosynthesis. Here the simplicity hides tremendous subtlety --- how do you control the electron's path, to insure that the photon's energy is really trapped in a state sufficiently long-lived to do useful chemistry? Betts *et al.* tackle the problem of electron pathways through proteins, approximating it down to a managable combination of quantum mechanics and graph search. The simple theory is sufficient to organize and predict the results of a long series of experiments on specially modified proteins. These arguments point the way toward rigorous tests of the theory in which one attempts to design new proteins which have certain specified electron transfer reactions.

The study of unicellular organisms provides a bridge between the molecular and the macroscopic. Here the problems being solved are familiar from our own experience --- how do we find our way to a desirable location --- but bacteria have to answer these questions by examining only a handful of molecules in their environment, and their behavior is in turn controlled by only a handful of molecules inside the cell. Kruglyak gives an introduction to these issues in the context of bacterial chemotaxis, the system whereby bacteria can swim toward higher concentrations of useful molecules (or away from aversive ones). There is the tantalizing possibility that much of bacterial behavior will be understandable as a strategy for dealing for the physics of the bacterial environment.

In the 1991 lectures the subject of chemotaxis was covered by Steve Block, who gave two lectures on the mechanism and control of biological motors, including the remarkable rotary engine of bateria. One of the important emerging techniques for studying biological motors is the optical trap, which allows the experimenter to apply foces comparable to those produced by single motor molecules. This work points the way toward quantitative physics experiments on force generation by single molecules, and many groups are now working toward this goal. For a recent example see the article by Block *et al.* in *Nature* **348,** 348-352 (1990).

The nervous system is, in part, a machine for processing incoming information. Information theory tells us that even the simpler problem of moving information from one place to another is not trivial; there are limits, analagous to the limits on heat engines in thermodynamics, on the ability of systems, including nerve cells, to transmit information. Atick provides a self-contained introduction to these deep ideas, and then proceeds to explore their application in the early stages of visual processing. The focus is on making optimal use of the available capacity through the construction of efficient representations, where efficiency is given a precise mathematical definiton. This leads to predictions about what different pieces of the visual system, from flies to goldfish to monkeys, should be doing if they are to provide the maximally efficient represenations of the world as it varies in space, time and color. These predictions are in striking agreement with experiment, suggesting that the brain may indeed have found the optimal solution to its information transmission problems.

One of the problems in testing the information-theoretic ideas is that the efficiency with which the nervous system represents information is not measurable by experiments on any single neuron. Instead one needs methods which allow simultaneous recording of the activity of many individual cells. In the 1991 Lectures Jon Art discussed high resolution optical imaging techniques which can be used with dyes that are sensitive to the electrical activity of an array of cells. In the 1992 installment, Markus Meister returned to the problem, describing multi-electrode arrays which have be used to record activity from up to one hundred cells at once in the retina. Both of these approaches are developing quickly. For review of optical recording methods see the article by Grinvald in *Ann. Rev. Neurosci.* **8**, 263-305 (1985) or the recent contribution by Chien and Pine in *Biophys. J.* **60**, 697-711 (1991). The methods of Meister *et al.* are described (in part) in *Science* **252**, 939-943 (1991).

Our experience of the world is dominated by our visual system, but of course this is not the case for all animals. A dramatically different view of the world is provided by the bat's echolocation or sonar system, described in Simmons' contribution. Rather than scanning the world with a narrow beam, as in man-made systems, the bat broadcasts widely and uses his computational power to synthesize an image of the three-dimensional world. Simmons reviews the latest in a long series of experiments which aim to characterize this image and the remarkable way in which the bat constructs it out of the different clues in the echo waveform. I cannot resist pointing out that Jim held his audience at full attention for a solid three hours straight, alternately pulling out some remarkable experimental result and telling amusing stories about bats. I hope that some of the excitement we all experienced comes through in the text.

Part of what made Thursday such a remarkable day was that Simmons' lecture was followed by Nicolas Franceschini's. Nicolas descibed a series of beautiful experiments which have gradually elucidated various parts of the fly's visual system, from the optics and and pigment molecules [see N. F. in *Photoreceptors*, Borsellino and Cervetto, eds., pp. 319-350 (Plenum Press 1984)] to the neurons which compute motion across the visual field and provide a signal to guide the bug's flight [see N. F. *et al.* in *Facets of Vision*, Hardie and Stavenga, eds., pp. 360-390 (Springer-Verlag 1989)]. At the end he described a robot which was designed using some of the principles of visual processing which had been learned from the fly [Pichon *et al.*, *SPIE* **1195**, 44-53 (1989)], and showed how the attempt to build this robot focused attention on new questions about the fly itself.

The chapters by Atick and Kruglyak introduced the idea that sensory systems operate at or near fundamental limits imposed by physics and information theory. I take up this theme in the last chapter, reviewing the evidence for optimal performance in a wide variety of different systems, and developing the theory of optimal signal processing to the point where we can make successful predictions about the dynamics of real neural circuits on the hypothesis that these circuits are optimized for particular tasks.

Acknowledgements

Many of the participants in the Lectures kindly pointed out that it was probably the best organized workshop they had attended. Starting with the tentative list of speakers and vague ideas about where to house the students, Cynthia Woodhull created a real organization. Angela Cramer joined her in the month or so before the Lectures, and together they had responsiblity for most everything. Getting more than sixty people together, keeping them fed, housed, transported, and generally happy is an amazing accomplishment. Cynthia and Angela also hounded the lecturers into producing enough notes and reprints to fill a thick binder, which we all keep as the real record of what happened that week.

Seemingly all of the adminstrative staff at NECI joined at one time or another, helping with everything from contracts to xeroxing. As a scientist at NEC I have an easy life in large part because the vice-president for physical science, Joe Giordmaine, believes that the administration exists to help the scientists. This is an uncommon perspective, and it extends to the sponsorship of events like the Lectures. Joe took an interest in every aspect of the project, providing much needed advice and support.

Producing this book was a large effort. I am writing this quite late at night, and Mary Anne Rich has just gone home, having put in overtime for yet another day. She has not only worked hard on the manuscript, but kept all the other things running so that I was free to think about the editing. Most of the figures were scanned and then incorporated into the compuscript, and this was made possible by Brad Gianulis and the facilities of the graphics lab. The difference between the 'almost final' and the true final version of the manuscript was substantial, and Allan Schweitzer and Nick Socci deserve much of the credit for this tranformation.

When we started I didn't realize that the funds for this event were coming directly from the president of the Research Institute, Dawon Kahng. Dawon was very excited about bringing so many new faces to the Institute, and he attended almost all the lectures. He attended the banquet, but only after I had promised that there would be no speeches. Dawon's style as president could be annoying. He didn't sit in his office and administer. He once confided to me that adminstration wasn't very interesting; he wished he could be back in the lab. He would wander the halls, bothering us with his ideas and telling us what was wrong with our latest papers. During the week of the lectures, and for months after, we had many enjoyable conversations about what we both had learned. Shortly before the 1992 installment of the lectures, Dawon died. As a small token of my appreciation, this book is dedicated to him.

W. B.

Contributors

- *J. Atick,* School of Natural Sciences, Institute for Advanced Study. Present address: Rockefeller University.

- *R. Austin,* Department of Physics, Princeton University.

- *J. N. Betts,* Department of Chemistry, University of Pittsburgh.

- *D. N. Beratan,* Department of Chemistry, University of Pittsburgh.

- *W. Bialek,* NEC Research Institute.

- *R. Callendar,* Department of Physics, City University of New York.

- *H. S. Chan,* Department of Pharmecutical Chemistry, University of California at San Francisco.

- *K. A. Dill,* Department of Pharmecutical Chemistry, University of California at San Francisco.

- *M. Hong,* Department of Physics, Princeton University.

- *L. Kruglyak,* School of Natural Sciences, Institute for Advanced Study. Present address: Department of Theoretical Physics, Oxford University.

- *J. Plombon,* Department of Physics, University of California at Santa Barbara.

- *D. Shortle,* Department of Biological Chemistry, John Hopkins University School of Medicine.

- *G. R. Loppnow,* Department of Chemistry, Princeton University.

- *J. N. Onuchic,* Department of Physics, University of California at San Diego.

- *J. J. Regan,* Department of Physics, University of California at San Diego.

- *J. A. Simmons,* Department of Psychology and the Neuroscience Program, Brown University.

Participants

- *Ajay*, University of North Carolina at Chapel Hill.

- *Dmittrii Averin*, State University of New York at Stony Brook.

- *Marc Baldus*, University of Florida.

- *Aniket Bhattacharya*, University of Maryland.

- *William Bruno*, Los Alamos National Laboratory.

- *Michel Carreau*, Massachusetts Institute of Technology.

- *Jack Chan*, Cornell University.

- *Yih-Yuh Chen*, California Institute of Technology.

- *Kelvin Chu*, University of Illinois at Urbana-Champaign.

- *Craig Danko*, University of Pittsburgh.

- *Johannes Dapprich*, University of Florida.

- *Lei Du*, University of Delaware.

- *Michael Fasolka*, University of Pittsburgh.

- *Marla Feller*, University of California at Berkeley.

- *Francisco Figueirido*, Rutgers University.

- *Gyongyi Gaal*, University of Pennsylvania.

- *Neil Gershenfeld*, Harvard University.

- *Leh-Huh Gwa*, Rutgers University.

- *Yiwu He*, Boston University.

- *Robert Jinks*, Syracuse Unversity.

- *Michael Kara-Ivanov*, Weizmann Institute.

- *Peter Leopold*, University of California at San Diego.

- *Albert Libchaber*, NEC Research Institute.

- *Robert Lingle, Jr.*, Louisiana State University.

- *Kenji Okajima*, NEC Fundamental Research Labs, Tsukuba.

- *Gendi Pang*, Clarkson University.

- *Matteo Pelligrini*, Stanford University.

- *Vitor Pereira Leite*, University of California at San Diego.

- *Maria Person*, University of Chicago.

- *Zongan Qiu*, University of Florida.

- *Fred Rieke*, NEC Research Institute.

- *Daniel Ruderman*, University of California at Berkeley.

- *Christoph F. Schmidt*, Harvard University.

- *Allan Schweitzer*, NEC Research Institute.

- *John Sharpe*, Unversity of Colorado.

- *Norberto Silva*, University of Illinois at Urbana-Champaign.

- *Adam Simon*, NEC Research Institute.

- *Dorine Starace*, Syracuse University.

- *Joseph Strzalka*, Princeton University.

- *Nilgun Sungar*, Calpoly San Luis Obispo.

- *Arjun Surya*, Syracuse University.

- *Karel Svoboda*, Harvard University.

- *Michel Wall*, Princeton University.

- *David Warland*, University of California at Berkeley.

- *Dong Wang*, University of Michigan.

- *Gerard Weisbuch*, Laboratoire de Physique Statistique de l'ENS, Paris.

- *Jin Wu*, University of Rochester.

- *Scott Yost*, University of Florida.

- *Zhou Zou*, Stanford University.

Contents

Collective Modes in Bacteriorhodopsin
Mi Hong, Robert Austin, John Plombon, and Robert Callendar   1

The First Picosecond of Vision: Raman Studies of Rhodopsin
Glen R. Loppnow 27

Statistical Mechanics and Protein Folding
Hue Sun Chan, Ken A. Dill, and David Shortle 69

Finding Electron Transfer Pathways
J. J. Regan, J. N. Betts, D. N. Beratan, and J. N. Onuchic 175

Physical Constraints and Optimal Signal Processing in Bacterial Chemotaxis
Leonid Kruglyak 197

Could Information Theory Provide An Ecological Theory of Sensory Processing?
Joseph J. Atick 223

Time-Frequency Transforms and Images of Targets in the Sonar of Bats
James A. Simmons 291

Optimal Signal Processing in the Nervous System
William Bialek 321

Collective Modes in Bacteriorhodopsin

Mi Hong, Robert Austin, John Plombon, and Robert Callender

Introduction

This document is in essence an addendum to the original work that we presented at the NEC 1991 Summer School. At that time we had just completed a series of measurements using far-infrared radiation to perturb electron transfer rates in photosynthetic reaction centers. A great deal of the data presented in that talk (we don't think it went over very well) has been presented in a paper that we published in a special issue of *Chemical Physics* [3]. Writing that paper was interesting since we decided to write a document detailing the uncertainties in our minds concerning the far-infrared experiments we were doing. We had a *tough time* getting that paper published in a relatively uncensored form. Much of the spirit of the original talk at NEC is still there (no pictures of the interior of men's and women's washrooms, however) and if you glance at the paper you will find a good amount of our original NEC talk. However, in the year since we gave that talk we have carried out further experiments concerning the role of collective modes in protein dynamics, and it is these further experiments that we would like to discuss in this paper. We hope that the editors of this series will allow us to maintain the note of skeptical but enthusiastic enquiry that we need to propagate in physics.

Let's examine the issues at hand. Proteins are large, highly condensed polymers that are roughly spherical in shape. Their mission, if we ignore their important structural roles, is to catalyze chemical reactions in living organisms. The reactions that they catalyze often run up-hill in free energy and hence require an external source of energy such as ATP. Usually these reactions are *extraordinarily* slow to proceed in the absence of the protein. Thus the proteins act as quite marvelous highly specific mesoscopic reactors. If we were chemists, which we most assuredly are not (we are far too stupid) then our approach might be to treat each protein as a wonderful puzzle, complex and quite unique, to be carefully unraveled and explained based upon detailed chemical mechanisms. But, we have this training as physicists where we are taught to look for global mechanisms of unifying importance.

In the experiments discussed at the NEC summer school we explored the use of far-infrared (FIR) photoexcitation to search for functionally important collective modes in proteins. Lots of people define FIR in different ways: we will take the semi-arbitrary cut-off of FIR as those excitations lying below 400 cm^{-1}. Looking back at these experiments with the experience (and lack of imagination and physical drive) of advanced age, we would say that the experiments definitely put the horse before the cart: we tried to see IF excitations in the FIR could cause a change in a reaction rate. One rationale for doing this reverse order experimentation is simply to see if there is anything going on: if not, then forget it!

In our case we have seen that FIR pumping at 50 cm^{-1} to 80 cm^{-1} does seem to influence reaction rates in an *athermal* way. Now just what does that catch word *athermal* mean anyway? It means that if you pump energy into some particular state that the energy resides in that state for a "substantially" long period of time. While the energy is in that state and not in other states the system is not in thermal equilibrium. Hence, athermal. A thermal change in the rate would occur if the energy of the absorbed FIR photon *very rapidly*, on the order of picoseconds, thermalizes into all the degrees of freedom of the system. Thus, while a specific mode may heated to an effective "temperature" of 100K in the instantaneous absorption of a 100 cm^{-1} photon, after relaxation the amount of energy/mode increase is on the order of 10^{-2} K. This is what one of our skeptical colleagues at Bell Labs means when he characterizes the FIR excitation as a blow-torch, quickly heating all modes. However, there are experiments one can do to find out how long these FIR modes live, and that is one of the main purposes of this paper. In this addendum to the original NEC lecture, we will discuss a simple experimental test of relaxation rates and a recent theoretical result by Angel Garcia which also bears on the issues at hand.

The Bacteriorhodopsin Puzzle

Bacteriorhodopsin (bR) is a very interesting protein that has many important properties. The molecule has a molecular weight of 26,534 daltons and consists of a single polypeptide chain consisting mostly of 7 α-helical stretches which repeatedly span the membrane in which the protein is found [16]. Figure 1.1 shows a schematic picture of the protein. bR has a chromophore (retinal), a linear conjugated polyene, which is covalently linked to lysine-216 in the backbone of the protein. The retinal undergoes conformational cis-trans transitions when photo-excited by a visible photon [25] and ultimately the pumping of a proton (a positive charge, that is) across the membrane, "uphill" against the chemical potential gradient.

The trans-cis conformational transition of the chromophore in bR leads to conformational transitions in the protein itself [6, 13, 21]. Usually, these conformational transitions are monitored *indirectly* by observing the perturbations that the protein structure makes upon the chromophore

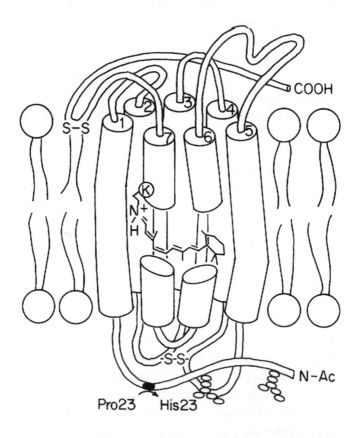

Figure 1.1: A schematic view of the polypeptide chain of the closely related protein rhodopsin, taken from the note by Applebury [2]. The site of the retinal attachment is shown, and the α-helix segments are denoted by large cylinders. Lipid molecules composing the membrane are also shown roughly to scale.

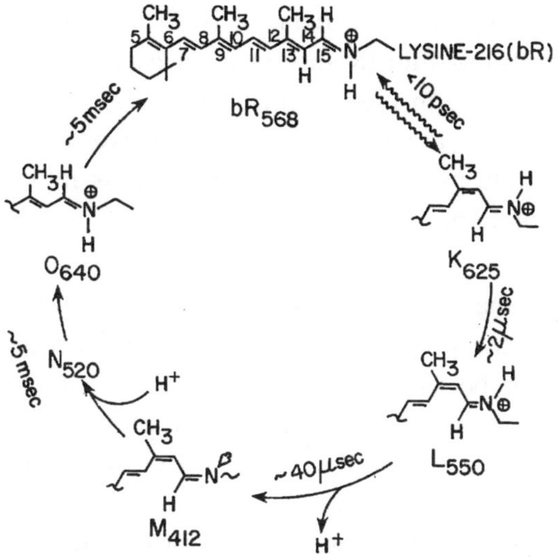

Figure 1.2: The photochemical cycle of bR. The absorption maximum of the chromophore are denoted by the subscript. Lifetimes of the states at room temperature are given. From the review by Khorana [19].

absorbance in the visible range of the spectrum. This work has been very important: it has shown (a) that the trans-cis isomerization occurs very rapidly, in less than a picosecond [22] and (b) that the protein cycles through a series of metastable states. Figure 1.2 shows a plot of the various states that have been identified via optical spectroscopy, and their lifetime at room temperature [11]. The lifetimes of these states are highly temperature dependent and show freeze-outs over a range of temperatures, some as low as approximately 100 K [24]. Since this temperature corresponds to energies on the order of 100 cm^{-1}, it has been speculated that in fact collective protein motions are involved in this kinetic freeze out [23]. It is interesting that at low temperatures the K state (bR$_{630}$) and the light-adapted bR *trans* ground state (bR$_{568}$) can be created reversibly by absorption of the appropriate color photon [26].

The presence of these intermediate states tells us something interesting: in these large structures there exist metastable conformational states which are separated by energetic barriers. It is impossible to tell from either the optical spectra or mid-IR spectra what exactly is the conformation of the macromolecule in these metastable states, but the relatively low energy barrier of the state suggests that the state is a "soft" state consisting of small-amplitude perturbations to many atoms rather than a highly localized deformation.

As interesting as these optical measurements may be, ultimately one really wants to probe the protein conformation as a function of time. One approach is to probe directly in the infrared where the local vibrational transitions of the molecular elements of the protein are evident. There is in fact a substantial amount of work in the *mid-IR* (3000 cm^{-1} to 500 cm^{-1}) on structural changes in bR. Figure 1.3, taken from the work of Hess and his group [15] shows some of the characteristic changes seen using time-resolved FTIR. Although many of the features seen are dominated by chromophore changes, and in fact can be compared to the resonance Raman results [20], there are also protein features that can be ascertained. In particular, the backbone amide stretch from 1671-1650 cm^{-1}, which is a sensitive indicator of the α-helix conformation, shows the same time-dependent changes as the chromophore bands do. This seems to indicate, as Gerwert *et al.* stress, that "all reactions in various parts of the protein are synchronized to each other and no independent cycles exist for different parts."

Although the amide stretch band is a good place to look for semi-local conformational changes, it would be better if indicators could be found for truly large scale motions in the protein. It would be nice if we had information about the FIR portion of the spectrum where one would hope that the collective modes might have some oscillator strength variations with conformation.

The most interesting and experimentally accessible states are the light adapted state bR$_{568}$ and the first excited state K$_{630}$, easily isolated at low

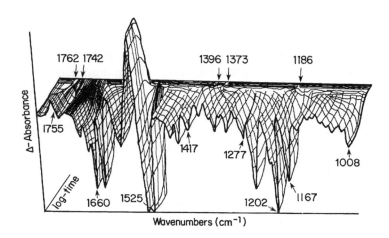

Figure 1.3: A time-resolved image of the difference FTIR spectra of hydrated bR after optical excitation. This Figure was taken from the work of Gerwert et al. [15]. The FTIR had a time resolution of $\sim$ 7 ms and a spectral resolution of 4 cm^{-1}. The temperature was 278K. The spectra correspond to M $\rightarrow$ bR changes.

temperatures (below 70K). So-called light adapted bR$_{568}$, bR which has been continuously exposed to light levels on the order of milliwatts/cm^2, will have its chromophore in the *trans* configuration at low temperatures if cooled quickly in the dark. As we mentioned above, in response to green light the chromophore makes a trans-cis isomerization [25] and the protein is now in the K state, with chromophore absorbance maximum at 630 nm. This trans-cis isomerization is actually photoreversible at low temperatures: illumination with red light drives the system back to the trans state. Both states seem to be very stable at temperatures below 40 K: the thermal relaxation rate between the two states seems to be very slow. However, at 70 K the relaxation rate is quite fast, on the order of milliseconds.

It is known that the trans-cis photo-driven energy storage is highly effective: of the 2 eV carried in by the visible photon, about 1 eV is stored as chemical energy [17]. There is something important and puzzling in the last fact: somehow the protein absorbs 2 eV and stores 1 eV by transferring a charge effectively over at least 5 nm. One would guess that maybe energy could be stored locally by the trans-cis isomerization, but how can energy be transferred over a large distance without losing it?

One possibility is that the energy is stored "mechanically" in a strained conformation of the protein. In fact, this is the essence (we think) of John Hopfield's obvious but subtle idea of how the R-T free energy difference is stored in hemoglobin [18].

Such an idea still has problems, however. A strained conformation implies that many atoms over a large volume of the protein are subtly moved relative to the conformation before the "event" to a metastable configuration. At least initially, this conformational shape change must occur at a time scale fast enough to compete with thermal and viscous relaxation, since the pre-event state is a local minimum in free energy by definition. It is easy to see how 2 eV photons could use the Franck-Condon principle [8] to snap a protein into a metastable state, but how can soft energies on the order of 100's of cm^{-1} do such a thing? Remember that the protein doesn't stay in the initial metastable state after the "event", it moves to other metastable states with high retention of the initial energy deposition. It stays cocked.

There is one answer to this puzzle with which not many would agree, namely that the energy is transferred in the same way water waves transfer energy over large scales in a highly viscous medium: by excitation of large scale collective motions whose group velocity carries energy faster than the relaxation process can remove it. The key is to make the wavelength sufficiently long that the diffusive relaxation time becomes very large compared to the wave period. Of course, the longest the wavelength can be in a protein is the diameter of the protein, about 5 nm typically. If the speed of sound c in a protein is about 10^5 cm/sec (typical for liquids) the frequency of the wave is approximately 10^{11} Hz. Can such a wave travel across a protein without attenuation?

What sets the time scale for that relaxation? We can roughly guess this criterion in a protein. Let's imagine that we have a 5 nm diameter protein again. We assume again that the group velocity of sound c in a protein is roughly the same as water, 1.4×10^5 cm/s [1]. Attenuation (relaxation) of acoustic waves occurs through two basic mechanisms: viscous damping and thermal diffusion. It isn't of course clear that the macroscopic equations that come from the Navier-Stokes equation are applicable at the length and frequency scale we wish to discuss: we are at the murky line between mesoscopic and atomic scale phenomena. However, it is interesting to see what the predictions of continuum mechanics are.

The attenuation coefficient $\alpha_{thermal}$ for an acoustic wave due to thermal diffusion is:

$$\alpha_{thermal} = \frac{\gamma - 1}{\gamma} \times \frac{\omega^2 \kappa}{2\rho c^3 c_v} \qquad (1.1)$$

where γ is ratio of specific heat at constant pressure to constant temperature, ω is the frequency of the wave, κ is the thermal conductivity, ρ is the density of the fluid and c_v is the specific heat/gram at constant volume. If the attenuation length is set to the diameter of the protein, and we assume that a protein macromolecule is approximately like water in its physical properties, we get that the maximum frequency ω_{max} at which we can expect acoustic transmission of energy is roughly 2×10^{12} Hz. The corresponding phonon energy, incorrectly assuming a linear restoring potential is equal to $\hbar \omega_{max} = 1.1 \times 10^{-3}$ eV or 10 cm^{-1}. This corresponds to FIR frequencies if these oscillations are excited by absorbed photons, and we would expect that 10 cm^{-1} photons could excite acoustic modes which could propagate across the molecule. The attenuation coefficient $\alpha_{viscous}$ for viscous damping is given by:

$$\alpha_{viscous} = \frac{2\omega^2 \eta}{3\rho c^3} \qquad (1.2)$$

where η is the viscosity of the medium. For illustration, we will do the calculation for water at 20 C. Our experiments are mostly done at cryogenic temperatures where the protein is undoubtedly a solid, in which case this expression is basically meaningless. In any event, we find that again if we want transmission across 5 nm that the maximum frequency ω_{max} is 3×10^{11} Hz. This would indicate that viscous damping would strongly attenuate FIR modes, in fact this may be why we observed no signal with FIR excitation at 50 cm^{-1} above approximately 180 K in myoglobin [5]. However, the question of the decoupling frequency of the solvent from the molecule (mid-IR transitions in solution are of course quite sharp) and the internal viscosity of the protein leave many questions unanswered.

In sum, what we have in mind here is that some sort of collective modes, in this frequency range, can have a long enough attenuation length and can live long enough to move the protein to a collectively strained state. These waves that carry stress should propagate in picoseconds across the 5 nm protein and move it to a strained configuration. They are NOT solitons, Davydov or otherwise [12], they carry very small amounts of energy. Nor do our ideas agree with a model proposed by Bialek and Goldstein [10] in which the functional dynamics of the molecule are dominated by quasi-harmonic vibrations. The strained metastable states are the source of reaction rate heterogeneity at low temperatures. As long as we are at it, since two of us were lucky students of Hans Frauenfelder we wish to point out that what we are proposing here is in some sense a version of Hans' brilliant idea of conformational substates in proteins which have a corresponding distribution in reaction rates [4]. We simply claim here that we can pump between Hans' states with FIR radiation. Can it be done?

We suspect that the reader is skeptical about such a scenario. There is a famous quote by A. Einstein about the violent opposition of mediocre minds, but the people that invoke it angrily on their office doors always seem to be people with mediocre ideas, so we'll leave it alone. For now. Wait till we get older.

Effects of FIR Irradiation on the Photocycle of bR

The first experiment we did was simply to find if pulsed FIR radiation had any effect on the bR photocycle bR_{568}-K_{630} at low temperatures. We refer the reader back to our papers on FIR effects in electron transport [3] and ligand recombination [5] for details about the technique. In the case of the bR experiments, thin films of hydrated bR were deposited on a plastic disk (TPX, a common plastic, has excellent transmission in the optical and FIR regions) via careful deposition of a solution containing bR vesicles. The final optical density of the films was quite high, on the order of 10 optical density units (OD) at the 568 nm absorbance maximum of light adapted bR (OD is the base-ten logarithm of the absorptions; a sample with OD=10 transmits 10^{-10} of the incident light). The bR thin films were held in place in a copper sample holder mounted in a Janis flow cryostat with Z-cut quartz optical windows. A quartz-halogen lamp with a 3 inch water filter to remove IR was used to maintain the bR in a cycling state: the green portion of the spectrum drove the bR to the K state, and the red portion of the spectrum drove the K to bR cycle. The basic idea is that the system will settle down into an equilibrium concentration of K and bR. Pulsed FIR, if it does allow some of the protein to leak out of the K state will leave a perturbed amount of bR and K states immediately after the FIR pulse. If the ratio of the bR/K is monitored at an appropriate wavelength sensitive to the amount of (in our case) the K state present then we would expect a *prompt* change in the absorbance of the signal. *Prompt* means that since we expect that any sort of reasonable lifetime for FIR excitations will be less than a nanosecond, the FIR sets up a quasi-equilibrium new set of rate constants during the laser pulse. When the FIR laser pulse is over, the rates relax back to the normal rates in nanoseconds. In our case, we used a 650 nm filter of 20 nm bandwidth to monitor the amount of K state absorbance.

Since we have length constraints, we won't go into a detailed description of

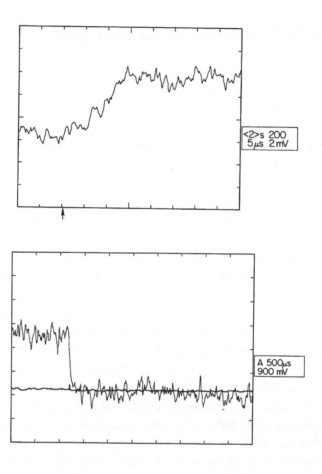

Figure 1.4: (a) The transient absorbance change seen at 650 nm due to a pulse of FIR radiation at 80 cm^{-1}, sample temperature was 7 K. The sign of the signal is such that the induced signal corresponds to a increase in transmitted light, or less K state during FIR irradiation. The time scale is 5 μsec/division. Since the laser pulse is 10 μsec in duration, we identify this signal as a prompt signal. (b) The same signal observed at 7 K displayed at 500 μsec per division. The flat baseline indicates that the signal was prompt and no additional temperature induced changes occurred.

the experiment but instead present some "highlights." Figure 1.4 displays some of the results from these experiments at 7 K. Figure 1.4a shows the observed transmission changes seen at 5 μsec/div, while Fig. 1.4b shows the observed changes seen at a longer time scale. These results show that the absorbtion changes occur during the the 5 μsec duration of the FEL pulse and in fact that the change rises linearly with time during the pulse. The only reasonable explanation outside of FIR induced reaction rate changes that we have for such a signal would be a change in the absorption spectrum due to sample heating. However, unlike the case of the reaction centers where the 860 special pair band turned out to be surprisingly temperature dependent [3], the temperature dependence of the absorption spectra of bR is far less than the 860 band [7] and the same analysis we applied to the reaction center system will serve to show that we cannot explain the observed signals seen in bR. The FIR induced absorbance change seen at 70 K has a changed sign and now indicates a an absorbance increase with FIR irradiation. In fact, we find that the FIR induced bR signal for a particular hydrated sample has two temperatures where no signal is seen, at approximately 50 K and 150 K. It is unlikely that any simple temperature induced broadening could give rise to such a complex scenario, but it is possible that the thermal occupation of further states of the bR photocycle and subsequent FIR perturbation of the rates could give rise to a complex signal $vs.$ temperature. Because of this complexity, we will restrict our comments to the signal observed at temperatures below 50 K, where we presumably know what we are dealing with in terms of the states of the protein. We also have to point out that the sign of the signal seen at room temperature (data not shown) is a function of the hydration of the sample: highly dried samples show an absorbance decrease with FIR, while highly hydrated samples show an absorbance increase. Again, a simple thermal broadening is probably not responsible for this behavior.

A last and highly informative piece of data can be found in Fig. 1.5. We show a double-time resolved signal. In this experiment only steady state red light (650 nm) illuminated the sample and no green light, hence the sample should be predominantly in the bR_{568} state. A 532 nm doubled YAG pulsed laser (10 nsec) was used to transiently drive the system to the

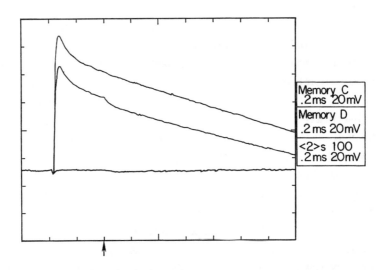

Figure 1.5: A double time-resolved experiment. As we describe in the text, a 630 nm monitoring light created a predominantly bR_{568} state. A 10 nsec 532 nm doubled Nd:YAG laser pulse was used to create a transient excess K_{630} population. After a 15 msec time delay the FIR laser was fired (the arrow shows the firing position in time of the FEL). A 2 msec filtering time constant was used to enhance signal/noise in this single-shot experiment. The temperature was 7 K. The top curve in the Figure is the response with the FEL radiation blocked to show that there is no electronic pickup from the FEL, while the bottom curve shows the response when the Nd:YAG laser is blocked but the FEL illuminates the sample, showing that virtually the entire signal comes from the K state generated by the Nd:YAG laser pulse at 532 nm.

K_{630} state, increasing the absorbance. Then, following a fixed delay after the 532 nm pulse the FIR FEL was fired and resultant absorbance change observed. If our analysis of the signal at low temperatures is correct, an *increase* in transmission should be seen....which we do see. If the Nd:YAG laser is not fired then no signal is seen, consistent with the claim that at least at low temperatures the FIR signal comes exclusively from the K state. Both the K state signal induced by the 532 nm laser pulse and the FIR induced change are eventually lost due to optical pumping of the red light of the monitoring beam.

It is informative from the Nd:YAG experiment to estimate the quantitative influence of the FIR pulse. Since no signal is seen when the Nd:YAG laser does not fire, there is effectively no base of K state molecules present in the absence of the Nd:YAG laser pulse. At the time of the FEL pulse, the "amount" of K state is approximately 50 mV in terms of ΔI. The FIR shot removes about 4 mV of K state in 12 μsec. Since we have confirmed, in data not shown, that the signal is linear with FEL energy, we get simply that the rate of K state loss is $(4/50) \times 1/12 \times 10^{-6}$ sec^{-1}, or 7×10^3sec^{-1}. If we know the lifetime of the collective modes, and the pumping rate of the FIR photons, we can convert this number to a true transition rate in the excited state.

An Experiment to Measure Collective Mode Lifetimes

Let's suppose that collective modes can be excited and that in the excited states the reaction rates are different. The most important question that has to be answered is: what are the lifetimes of collective modes in proteins? If the lifetime is on the order of 1 period, about 10 picoseconds from our calculation above, then there is a possibility of significant kinetic steerage of a reaction. Lifetimes on the order of 100 ps to 1 nanosecond would represent considerably less damping and greater efficiency. A one nanosecond lifetime would be fantastic. What are the relaxation rates?

The brute force way to measure the lifetime of an excited state is to do a pump-probe experiment: put in enough energy to equalize the populations in the ground and excited states, then probe with a weak beam the resulting loss of absorbance with time after the pump. The relaxation rate $\mathcal{K}_r$ of the excited state, also known as the longitudinal relaxation rate or $1/T_1$ in analogy to NMR jargon, is obtained by measuring the recovery of the initial ground state population. Unfortunately, the UCSB FEL has a 10 μsec FIR pulsewidth T, much too long to do direct time-resolved work. The next best thing is to make a quasi-equilibrium pump-probe experiment: assume that the relaxation rate is fast enough (that is, $1/\mathcal{K}_r \ll T$) that a steady state saturation is obtained during the FIR pump pulse. Measurement of this bleach can then be somewhat indirectly back-calculated to obtain the rate.

There is a clever way to make such an experiment that doesn't require the separation of the FIR beam into a separate pump and probe beam, and doesn't require great linearity in the FIR detectors. Let the FIR beam be focussed onto a spot of some diameter D in the material of interest. The material is assumed to have an absorbance A_s (the absorbance is the same as OD, as we mentioned above). Let there be an absorbing filter of absorbance A at the frequency of interest. Let the incoming FIR pulse have total energy E and pulse duration T. A FIR detector is placed after the sample. A beam splitter in front of the entire apparatus is used to measure the (possibly variable) beam energy in a separate detector. Figure 1.6 should make the configuration clear.

If the filter is put in FRONT of the sample, then the energy incident on the sample is $E \times 10^{-A}$, and the energy incident on the detector is $E \times 10^{-(A+A_s)}$. Let the filter now be put in BACK of the sample. The sample now sees the full pulse energy E, but the detector still has incident on it the same attenuated energy $E \times 10^{-(A+A_s)}$. Clearly, if the incident power $P = E/T$ is insufficient to cause appreciable saturation the ratio of the incident energy to detected energy will be the same independent of te filter position. However, if the FIR power is sufficiently high to saturate the system then the detector will record greater energy transmission when

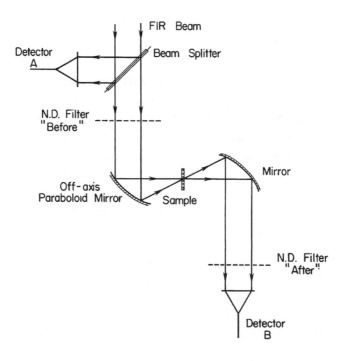

Figure 1.6: The schematic outline of the saturation experiment performed on the UCSB FEL. In these experiments a 50 nsec FIR pulse from a CO_2 laser shifted by an alcohol vapor column was used to produce the FIR pulse. The sample was essentially the same bR thin film as was used in the optical experiments discussed in Figure 1.5. A beam splitter was used to divert a small amount of the FIR beam overt to Detector A. An off-axis aluminum parabolic reflector was used to focus the FIR onto a 0.2×0.2 cm spot. A second off-axis parabolic reflector collimated the beam and directed the beam onto a second Detector B. The ratio of B/A was averaged by a LeCroy 9400A transient recorder for 50 laser shots. The laser was run at 0.3 Hz. The absorbance filter mentioned in the text could be placed either "before" or "after" the sample.

the filter is placed AFTER the sample then when it is placed BEFORE the sample. Let's call this ratio R:

$$\mathcal{R} = E_{after}/E_{before} \tag{1.3}$$

Since the OD of the sample is known at the illuminated FIR wavelength, the number of molecules of the absorbing molecule is known from the visible OD, the energy/photon is known, the energy of the FIR laser pulse is known, and the pulse width, we know the rate at which FIR photons are hitting the protein molecules, call that rate $\mathcal{K}_p$. Further, we assume that the relaxation rate of the excited state is some unknown number $\mathcal{K}_r$. If the laser pulse is much longer in duration than the mean excited state lifetime then the sample will come into a steady state value of ground N_g and excited state population N_e. In equilibrium we have:

$$
\begin{aligned}
\frac{dN_g}{dt} &= -\mathcal{K}_p N_g + \mathcal{K}_p N_e + \mathcal{K}_r N_e = 0 \\
\frac{dN_e}{dt} &= +\mathcal{K}_p N_g - \mathcal{K}_p N_e - \mathcal{K}_r N_e = 0
\end{aligned}
\tag{1.4}
$$

This yields:

$$\mathcal{K}_r = \mathcal{K}_p \left(\frac{N_g}{N_e} - 1 \right) \tag{1.5}$$

Finally, we note that:

$$\mathcal{R} = \frac{N_g}{N_g - N_e} \tag{1.6}$$

in the limit where $N_e \ll N_g$. Then we have:

$$\mathcal{K}_r = \mathcal{K}_p \frac{1}{1 - \mathcal{R}} \tag{1.7}$$

The experiment to determine $\mathcal{K}_r$ was carried out recently at the UCSB FEL. The protein sample we made studied was essentially the same bR thin film as used in the experiments cited above. The bR sample was light adapted before cooling to 7 K. A FIR absorption spectrum of the BR sample was measured to have an absorbance at 100 cm^{-1} of 1.0 OD. The sample was mounted in a cryostat so that the temperature could be easily varied. Since we expect that the T_1's for proteins will be rather short, we need the highest possible intensities for the FIR pump beam to see appreciable saturation. The highest flux FIR source at the UCSB CFELS is a CO_2 pulsed laser which drives an alcohol vapor column to produce stimulated Raman emission off the rotational levels of alcohol molecules. The FIR output from this laser is quite intense: typically 1.0 millijoules in a 50 nanosecond pulse width, photon energy 100 cm^{-1}. The FIR beam was focussed to a $4 \times 10^{-2} \text{ cm}^2$ spot size on the sample. This corresponds to about 10^{19} photons/cm^2. The sample OD at 563 nm was 10, and using a ϵ of $60 \text{ mM}^{-1}\text{cm}^{-1}$ gives about 4×10^{15} molecules in the illuminated spot. We then find that the pump rate $\mathcal{K}_p$ is $2 \times 10^9 \text{ sec}^{-1}$. That is, every 0.5 nsec a bR molecule absorbs a 100 cm^{-1} photon.

After some false starts due the nasty ability of FIR radiation to bounce all over a lab, we found that the ratio of the signal detected on the pyroelectric detector with a 1.0 OD filter in back of the sample (I_b) and in front (I_f) of the sample was 1.2 ± 0.05, sample at 7 K. Note that the ratio of I_b/I_f is greater than 1, as is expected for a saturation of the absorbing levels in the FIR. Simple tests such as using just the TPX plastic sample holders yielded nulls with R = 1.00 ± 0.01. The saturation was only evident when the sample BR film was present.

A value of $\mathcal{R} = 1.2$ then yields our desired result, namely that *the relaxation rate $\mathcal{K}_r \sim 10^{10} \text{ sec}^{-1}$ at 100 cm^{-1}!* Now you know. You may recall that this is roughly, within an order of magnitude, of what we guessed would be the relaxation rate from simple Navier-Stokes equation noodling.

The reader should be aware that a *100 ps lifetime for a 100 cm^{-1} mode* is pretty heretical. That is a reasonably long time for energy to bounce around the protein before it becomes thermalized.

The Excited State Transition Rate

Now that we have K_r, we can continue our calculation to get the excited state reaction rate. The pump rate K_p in the Nd:YAG laser experiment has to be modified since the CO_2 laser was not used, but instead the CFELS FEL. The pump rate is modified by the increased pulse energy (16 mJ vs. 1 mJ) and increased pulse width (12 μsec vs. 0.05 μsec). The net effect is that the pump rate in the Nd:YAG experiment should have been approximately 10^8 sec^{-1}. Thus, FIR photons excited the bR molecules only every 10 nsec in that experiment, and we believe the excitation rattled around the protein for about 0.1 nsec after excitation. The effective rate then has to be multiplied by that "duty factor" of 100. *We finally get that the rate of K to "something" when excited by 100 cm^{-1} photons is approximately 7×10^5 sec^{-1}.* You read it here first. This number isn't totally crazy: the rate of decay of the K state at room temperature is approximately 10^6 sec^{-1}.

A Recent Theoretical Paper by A. Garcia

A weird experiment like this one has little value if there is no theoretical back-up behind it. Fortunately, after this experiment was done there have been a few provocative theoretical *simulations* of protein dynamics in aqueous environments. We want to address here at the end the question of how one could expect an excited state to have a lifetime of 100 picoseconds, when most people would expect that any collective modes would have at most a few picosecond lifetimes. One possibility is that it is totally wrong to think of these excited states as some sort of normal modes of a harmonic system.

In principle, computers might be able to model the dynamics of proteins. The basic problem in any computer dynamics simulation is to somehow get a view of the dynamics of the process. If you look at each atom in the biomolecule individually it would appear to be oscillating in some random manner with very little correlation with atoms some distance away. Basically, the whole object seems to a quivering chaotic mass of atoms. The question is, as we have discussed above, if there is any sort of collective aspect to the motions in the molecule. Since Bill Bialek is an enthusiastic promoter of neural processing in organisms, we could make the analogy to studying 1 neuron firing versus looking for some sort of a collective response in the entire neural network.

Angel Garcia has pointed out in a provocative paper [14] that collective motions in condensed polymers such as proteins cannot be viewed as a normal mode analysis problem: the interactions between the different amino acids are highly non-linear with the distance between the groups and the polymer is held together very weakly so that the non-linearities are strongly present. The result is that the complex motions are highly anharmonic and metastable in nature: the protein can be expected to reside in some particular conformation for some time τ and then will switch to another conformation. Note that because of the metastability expected of these conformations that even if the system is highly overdamped and dissipative it is still possible for the system to display anomalously long "excited state" lifetimes.

Figure 1.7 shows a plot of the neighboring dihedral angular displacements of a series of amino acids in the protein crambin as a function of time. The protein is simulated in an aqueous environment by 1315 water molecules. The temperature of this simulation was 300 K. The most striking thing to notice in this simulation is that the motions of the amino acids are not distributed in a simple gaussian manner around an average position, nor is the motion a random walk in phase space. Rather, the motions are metastable in nature, characterized by confinement to a particular angular range for an extended time followed by a very rapid jump to another metastable minimum. The motions are also correlated, as you would

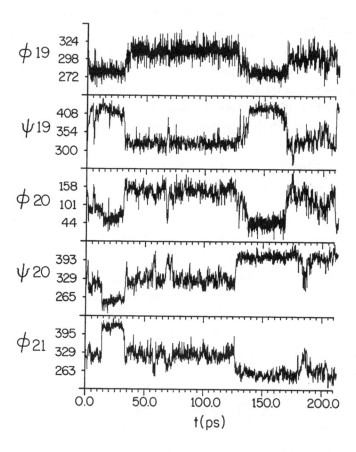

Figure 1.7: A plot of the dihedral angle ψ and ϕ displacements of the 19th, 20th and 21st amino acid in the protein crambin plotted vs. time. Note the completely anharmonic nature of the motions and the correlations between the different amino acids as they move. This Figure is taken from the article by Garcia [14].

expect in a highly condensed molecule: as one amino acid moves one way, others must move the other way to make room. Interestingly, the average lifetime of these metastable states is on the order of 100 picoseconds, although the transition to a particular state takes about a picosecond, indicating that the actual structural transition time is quite rapid.

Now, such a simulation is done in a fluid solvent, and there has been no attempt to try and calculate the oscillator strength connecting the states, so it may be that these collective modes are in no way correlated with the broad continuum of absorbance observed in protein thin films. However, it is intriguing to notice the correlation between the metastable lifetimes and the saturation recovery times observed in the experiment reported on here. Stay tuned for details, as a famous member of our faculty has noted it isn't over until the Fat Lady sings and we don't see her backstage yet. One of you doubting experimentalists will have to do the job.

Acknowledgements

This work was generously supported by the Office of Naval Research. We are particularly grateful the staff of the CSFEL for support of this work Chris Felix provided the instrumental idea of using the before-after filter technique. We hope that the UCSB FEL remains committed to allowing outside users to use the marvelous facility that the ONR has created.

References

[1] H. Anderson, ed.(1989). *Physics Vade Mecum* (American Institute of Physics, New York).

[2] M. Applebury (1990). *Nature* **343**, 316-317.

[3] R. H. Austin, M. K. Hong, C. Moser, and J. Plombon (1991) *Chem. Phys.* **158,** 473-486.

[4] R. H. Austin, K. W. Beeson, L. Eisenstein, and H. Frauenfelder (1975). *Biochemistry* **14,** 5355-5373.

[5] R. H. Austin, M. W. Roberson and P. Mansky (1989). *Phys. Rev. Lett.* **62,** 1912-1915.

[6] K. Bagley, V. Balogh-Nair, A. A. Croteau, G. Dollinger, T. G. Ebrey, L. Eisenstein, M. K. Hong, K. Nakanishi, and J. Vittitow (1985). *Biochemistry* **24,** 6055-6071.

[7] B. Becher, F. Tokunaga and T. G. Ebrey (1980). *Biochemistry* **17,** 2293-2300.

[8] R. S. Becker (1969). *Theory and Interpretation of Fluorescence and Phosphorescence* (Wiley Interscience, New York).

[9] D. Beece, S. F. Bowne, J. Czege, L. Eisenstein, H. Frauenfelder, D. Good, M. C. Marden, P. Ormos, L. Reinisch, and K. T. Yue (1981) *Biochem. Photobiol.* **33,** 517-522

[10] W. Bialek and R. Goldstein (1987). In *Protein Structure, Molecular and Electronic Reactivity,* R. Austin *et al.,* eds. (Springer-Verlag, Berlin).

[11] R. Callender and B. Honig (1977). *Ann. Rev. Biophys. Bioeng.* **6,** 33-55.

[12] P. L. Christiansen and A. C. Scott, eds. (1990). *Davydov's Soliton Revisited* (Plenum Press, New York).

[13] J. E. Draheim and J. Y. Cassim (1985). *Biophys. J.* **47,** 497-507.

[14] A. Garcia (1992). *Phys. Rev. Lett.* **68,** 2696-2699.

[15] K. Gerwert, G. Souvignier, and B. Hess (1990). *Proc. Natl. Acad. Sci. USA* **87,** 9774-9778.

[16] R. Henderson, J. M. Baldwin, T. A. Ceska, F. Zemlin, E. Beckman and K. H. Downing (1990). *J. Mol. Biol.* **213,** 899-929.

[17] B. Honig, T. G. Ebrey, R. H. Callender, U. Dinur, and M. Ottolenghi (1979). *Proc. Natl. Acad. Sci. USA* **76,** 2503-2507.

[18] J. J. Hopfield (1973). *J. Mol. Biol.* **77**, 207.

[19] H. G. Khorana (1988). *J. Biol. Chem.* **263**, 7439-7442.

[20] T. Kitagawa and A. Maeda (1989). *Photochem. Photobiol.* **50**, 883-894.

[21] H. Marrero and K. J. Rothschild (1987). *Biophys. J.* **52**, 629-635.

[22] R. A. Mathies, S. W. Lin, J. B. Ames, and W. T. Pollard (1991). *Ann. Rev. Biophys. Chem.* **20**, 491-518.

[23] S. J. Milder and D. S. Kliger (1986). *Biophys. J.* **49**, 567-570.

[24] P. Ormos, D. Braunstein, M. K. Hong, S. Lin, and J. Vittitow (1987). In *Biophysical Studies of Retinal Proteins*, T.G. Ebrey *et al.*, eds. (University of Illinois Press, Urbana).

[25] W. Stoeckenius and R. A. Bogomolni (1982). *Ann. Rev. Biochem.* **51**, 587-616.

[26] L. Zimanyi, P. Ormos, and J. K. Lanyi (1989). *Biochemistry* **28**, 1656-1661.

The First Picosecond of Vision: Raman Studies of Rhodopsin

Glen R. Loppnow

Introduction

Vision begins when a pigment molecule absorbs a photon and changes its structure. In this lecture we will examine in detail the forces which cause this structural change. The pigment molecules have a remarkably fast dynamics, and we will see at least in outline how this fast dynamics comes about. It turns out that Resonant Raman spetroscopy is ideally suited to this task, since the Raman cross-section is directly related to the forces which the molecule experiences just after absorbing a photon. To make sense out of all this I will review the theory of Raman scattering in some detail, and I will try to give a clear picture of how the experiments get done. Let's begin by setting the more macroscopic stage, with the eye itself.

The Eye

In higher animals, image formation occurs in a specialized organ called the eye. The gross structure of the eye varies from organism to organism, but can be roughly divided into three types: pinhole, compound, and refracting. The pinhole eye is the most primitive, consisting of an aperture and a semicircular layer of photoreceptor cells, and is found only in lower invertebrates and vertebrates such as the *Nautilus* mollusc [20]. This type of eye suffers from a trade-off between low light levels and high definition related to the size of the pinhole; the collection angle varies directly while resolution varies inversely with the size of the aperture. The compound eye of invertebrates retains much of the high resolution of the pinhole eye, but improves the light-collection efficiency by essentially making each photoreceptor cell a pinhole eye and then distributing them on a spherical surface. The refracting eye is the most sophisticated eye, retaining both high-resolution and high light-collecting efficiency.

Vertebrates and a few invertebrates, such as the squid and octopus, have refracting eyes (Fig. 2.1). The advantage of the refracting eye is that all the rays originating from one point on the surface of the object are focused onto one point in the retina, resulting in both a high light-collection efficiency and a high degree of resolution. In this eye, the light rays are focused onto the photoreceptors in the retina by a lens system. The lens system in most refracting eyes consists primarily of the cornea (coarse focus) and lens (fine focus). A resting eye has a total focal length of ~ 17 mm while an eye focused on the nearest object it can resolve has a focal length of $\sim 13.5mm$ The difference in focal lengths is the limit of accommodation of the lens. In humans, accommodation is achieved by changing the radius of curvature of the lens with the ciliary muscle. In some fish, who lack most of the focusing power at the cornea/water interface due to the similarity of refractive indices, accommodation is achieved by moving the lens back and forth along the optical axis until the correct image distance is achieved.

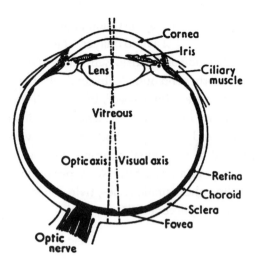

Figure 2.1: Horizontal cross-section of a human eye [20]. Light is focused by the cornea and lens onto the photosensitive retina. The colored iris is a sphincter muscle that controls the amount of light reaching the retina. The choroid or pigment epithelium is black to absorb scattered light.

The biggest disadvantage of the refracting eye is the large number of different media that light must pass through to reach the retina. These substances scatter the light and may also be dispersive or absorbing. For example, the human eye contains a yellow filter, the macular pigment, which probably acts as an ultraviolet absorber. The filter also absorbs wavelengths ≤ 360 nm which reduces the sensitivity in the blue region of the visible spectrum [81]. Such filters predominate in the retinas of turtles, fishes, and birds [9, 42]. Diseases and aging can also affect the ability of the refracting eye to function properly. Cataracts, a clouding of the lens, can cause blindness because too much light is scattered or reflected and not enough reaches the retina. Aging can harden the lens, resulting in a loss of accommodation. It has also been documented that yellowing of the lens with age changes the perception of color by humans [5, 55].

In all types of eyes, the actual detection of light occurs in the photoreceptors of the retina. Humans have two types of photoreceptors, rods and cones. Rod cells have long, cylindrical outer segments consisting of

$\sim 2000 - 10,000$ disks containing the visual pigment. The hollow, pancake-shaped disks are composed of a lipid bilayer. The visual pigment spans the lipid bilayer on either side of the disk, with one end of the molecule inside the pancake and the other end outside. In cone cells the plasma membrane invaginates into the intracellular space, but doesn't forms distinct disks. Both cells also have an inner segment, which contains most of the other cellular organelles, and a synaptic body through which the excitation signal is transmitted to the retinal neurons. Fish and other vertebrates have evolved certain variations of the two types of photoreceptors found in humans, including short single cones, twin cones, double cones, triple cones, and even quadruple cones [72].

The visual process

All known visual pigments are composed of a transmembrane protein (opsin) with a conjugated polyene (vitamin A derivative) covalently attached to it. On the basis of its similarity to bacteriorhodopsin, the structure of the visual pigment is thought to be seven alpha helical cylinders encircling the vitamin A chromophore [16]. The conjugated polyene chromophore is responsible for initiating the visual event by absorbing a photon with energies in the visible spectrum. The visual pigments of most terrestrial vertebrates contain a vitamin A_1 derivative commonly called retinal [19]. Visual pigments of most fish, some birds, and some amphibians contain a vitamin A_2 derivative, 3,4-didehydroretinal [19]. Most invertebrates have visual pigments which contain a 3-hydroxyretinal prosthetic group [71]. This difference in chromophore types is partly responsible for shifting the sensitivity of the fish pigments to the red and insect pigments into the ultraviolet. Bovine rhodopsin has been extensively studied as a model for other visual pigments due to its stability [20], ease of preparation [3], and close analogy to other visual pigments. Bovine rhodopsin contains an 11-*cis* A_1 retinal chromophore connected to the 41 kD opsin protein via a protonated Schiff base linkage to lysine-296 [10]. The chromophore in all known visual pigments is in the 11-*cis* configuration (see Fig. 2.2).

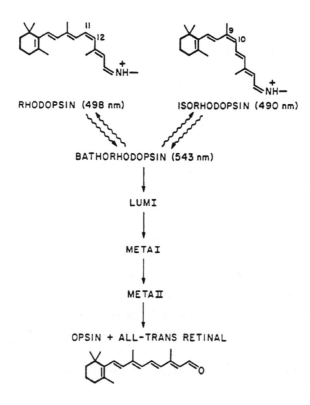

Figure 2.2: The bleaching sequence of rhodopsin. Absorption maxima in nm are shown in parentheses. Upon absorption of a photon by rhodopsin, the retinal chromophore undergoes 11-*cis* → 11-*trans* isomerization to the twisted all-*trans* photoproduct bathorhodopsin. Bathorhodopsin decays thermally at room temperature through a series of intermediates, finally forming the all-*trans* retinal and opsin. At liquid nitrogen temperatures, the decay of bathorhodopsin to lumirhodopsin is blocked and a photostationary steady-state is formed between 11-*cis* rhodopsin, all-*trans* bathorhodopsin, and 9-*cis* isorhodopsin.

Upon absorption of a photon, the 11-*cis* retinal chromophore isomerizes to a strained, distorted all-*trans* form called bathorhodopsin ([4, 24, 26, 57, 63, 70]; see Fig. 2.2). In the simplest terms, this occurs when half of the molecule torsions $\sim 180°$ around the 11-12 bond. Bathorhodopsin then decays through a series of intermediates called lumirhodopsin, metarhodopsin I, and metarhodopsin II that have been trapped and identified through low-temperature absorption spectroscopy [83]. Metarhodopsin II has been identified as the species that activates transducin and initiates the enzymatic cascade that leads to the perception of vision. After activation of transducin, rhodopsin dissociates into opsin and all-*trans* retinal.

Recent FTIR results show that protein conformational changes occur in the lumirhodopsin, metarhodopsin I, and metarhodopsin II intermediates. In lumirhodopsin, new peaks occur near 1635 cm^{-1} assigned as the amide I band and may reflect a disruption of the α-helical structure [28, 30]. In metarhodopsin I, prominent changes occur in buried carboxyl groups [29, 64]. In metarhodopsin II, both buried and exposed carboxyl groups undergo changes [29, 64]. These changes support the notion that the isomerization of the chromophore prompts a conformational change in the protein that propagates through the opsin protein from the retinal binding pocket inside the membrane to the cytoplasmic loops outside the membrane and activates the transducin protein. For more on the dynamics of the opsin matrix, see the contributions of Hong *et al.* to these proceedings.

Visual pigment photochemistry

A key problem in vision is understanding and elucidating the mechanism and timescale of the rhodopsin→bathorhodopsin transition. Numerous studies of polyenes and retinal derivatives have demonstrated *cis-trans* isomerization can occur on a ~ 10 picosecond timescale. In particular, the excited-state structure and dynamics of *cis-* and *trans*-stilbene have been well characterized. The temperature and viscosity dependence of the isomerization and fluorescence quantum yields of *trans*-stilbene

G. R. Loppnow   33

have shown an activated excited-state isomerization with a barrier of ~
1200 cm^{-1} [13, 31, 66]. These results have been confirmed in the jet-
cooled spectra [85] and picosecond absorption studies [33, 38, 82]. For
cis-stilbene, torsional relaxation times of 7 ps in condensed phase [75]
and 320 fs in vapor [34] have been reported. Analysis of the resonance
Raman intensities of cis-stilbene has shown nuclear motions consistent
with cis-trans isomerization occurring in the excited-state within 20 fs
[52]. More analogous to the visual pigments, time-resolved fluorescence
and absorption of prototypical visual chromophores have seen trans-cis
isomerization of 1,3,5,7-octatetraene and radiationless decay in 225 ns at
4-10 K [1, 32]. Time-resolved fluorescence and absorption has suggested
that excited-state relaxation, presumably to a 90° twisted geometry, in
retinal protonated Schiff bases occurs within 7 ps [36, 37].

Recent work has started to elucidate the excited-state structure and isomerization dynamics of retinal-containing pigments. Analysis of the
resonance Raman intensities of bacteriorhodopsin [51, 54], an analogous
pigment containing an all-trans retinal chromophore, has suggested that
large distortions in low-frequency vibrational modes are responsible for
the rapid trans-cis isomerization. Picosecond resonance Raman spectroscopy of bacteriorhodopsin have shown the appearance of a 13-cis K$_{590}$
species within 7 ps [73]. Femtosecond absorption studies have shown that
a 13-cis precursor to K$_{590}$ (called J$_{620}$) forms in ~ 500 fs and relaxes to K$_{590}$
on a 3-ps time scale [49, 56, 61, 62, 68]. These studies have defined the
kinetics of the isomerization in bacteriorhodopsin but have not produced
a detailed molecular description of the nuclear distortions that give a
femtosecond trans-cis chromophore isomerization in a protein environment. Picosecond absorption [14, 60], Raman [35], and fluorescence [22]
studies of bovine rhodopsin as well as INDO calculations [7] have likewise
indicated a chromophore isomerization on a 6-30 picosecond timescale.
Although a picosecond absorption study in H$_2$O and D$_2$O [60] showed
a deuterium effect on the rate of the initial step, suggesting a proton
translocation mechanism is important during the first few ps, failure to
reproduce the D$_2$O effect in a subsequent experiment [22] has cast doubts
on those results. It is now generally believed that cis-trans isomerization
is the initial photochemical event in rhodopsin. However, the kinetics

of the isomerization have been instrument-limited and no mode-specific technique has been able to resolve the nuclear distortions in time.

The mechanism of the rapid isomerization is also controversial. Given a *cis-trans* isomerization, a one-bond isomerization leads to a large volume being swept out by the chromophore, an unlikely event in the relatively restricted retinal binding pocket on a picosecond timescale. Thus, several models have been proposed to explain the rapid isomerization and minimize the required space for it to occur. Calculations have shown that a bicycle-pedal motion can result in a *cis-trans* isomerization with a very small moment of inertia [78]. In this model, isomerization of the $C_{11}=C_{12}$ and C=N bonds in rhodopsin leads to the all-*trans* photoproduct. This model requires a C=N *syn* configuration in either native 11-*cis* rhodopsin or in bathorhodopsin. Resonance Raman experiments [58, 59] have shown that both rhodopsin and bathorhodopsin are C=N *anti*. Liu *et al.* have alternatively proposed a concerted twist mechanism around a pair of adjoining double and single bonds [43]. For 11-*cis* rhodopsin, this "hula twist" around the C_{10}-C_{11} and $C_{11}=C_{12}$ bonds as proposed would lead to a 10-s-*cis*, 11-*trans* configuration in the photoproduct. Resonance Raman studies of the conformation of bathorhodopsin at low temperatures have shown bathorhodopsin is 10-s-*trans*, indicating the "hula twist" model is not applicable to the photochemistry of visual pigments [59]. Thus far, no isomerization model has been proposed for rhodopsin that adequately describes the available structural and kinetic data

---

Resonance Raman Spectroscopy

To define the structure and dynamics of the retinal chromophore in rhodopsin, a mode-specific probe is needed. Absorption spectroscopy of rhodopsin yields a broad spectrum with no observable vibronic structure, even at temperatures as low as 4 K. Interference from protein vibra-tions make infrared absorption spectroscopy and NMR spectroscopy of rhodopsin difficult, although a significant understanding of the structure of rhodopsin and its intermediates has been achieved from FTIR difference spectroscopy [4, 63, 70]. Resonance Raman spectroscopy provides a sensitive, mode-specific probe of the retinal chromophore structure and dynamics in rhodopsin. Several studies have established the utility of resonance Raman spectroscopy in determining the chromophore structure in rhodopsins [15, 18, 48].

Raman spectroscopy is the inelastic scattering of incident light by a molecule. This chapter will be concerned exclusively with vibrational Raman spectroscopy in which the difference in frequency of the incident and scattered photons is 0 to 3000 cm^{-1}. In Raman spectroscopy, the position of a band along the energy axis determines the frequency of the vibration and yields such information as bond strengths, molecule-environment interactions, vibration-vibration interactions (e.g. stretch-bend interactions), and ground-state structures. The frequencies and normal vibrations of a molecule may be predicted from the molecular structure and force constant approximations for the various internal coordinates (stretch, bend, wag, etc.). However, unless refinement of the force constants and normal vibrations is done, the predictions are usually imprecise. Normal vibration refinement is typically done using the Wilson FG method and isotopically-labeled molecules [80]. In this method, the normal vibration eigenvectors and energies are obtained from the eigenvector equation.

$$(GF - L)A = 0, \qquad\qquad\qquad (2.1)$$

where F is the matrix of force constants, G is the kinetic energy matrix determined by the molecular geometry, L is the diagonal matrix of eigenvectors (i. e. the normal vibrations) and the A matrix relates the internal and normal coordinates. Refinement of the normal vibrations is typically done by cataloguing the frequencies and frequency shifts of isotopically labeled analogues. Because the frequencies are inversely proportional to the square root of the reduced mass, specific analogues can determine directly the normal mode character of a vibration. For large molecules such as retinal, this can be a tedious and time-consuming project. Through the pioneering work of Curry [18], the normal modes of retinal have been completely assigned. Curry's work has been extended to rhodopsin, bathorhodopsin, and isorhodopsin by Eyring [24, 26, 77] and Palings [58]. This invaluable foundation will be used throughout this chapter.

The intensity of a resonance Raman vibration is proportional to the cross-section. As we shall see later, the cross-section and polarizability can be related to a variety of excited-state properties, including the lifetime, inhomogeneous width, and details of the potential energy surface along each normal coordinate. Classically, the Raman scattering process is described as the interaction of the electric field vector with the molecular polarizability [44],

$$P = \alpha \cdot E. \tag{2.2}$$

In the above equation, α is the second-rank molecular polarizability tensor, P is the polarization vector, and E is the electric field vector. Assuming simple harmonic motion and small variations in the polarizability around the ground-state nuclear configuration, the polarizability can be expanded in a Taylor's series and truncated at the linear term to give

$$\alpha_k = \alpha_0 + \alpha'_k Q_k. \tag{2.3}$$

In Eq. (2.3), Q_k is the normal coordinate scalar amplitude, ω_k is the vibrational frequency, α_0 is the equilibrium geometry polarizability, and $\alpha_{k'}$ is the molecular polarizability at the geometry corresponding to a distortion along the kth vibrational coordinate. Assuming a sinusoidal electric field, the polarization becomes

$$P = P_0 \cos(\omega_0 t) + P_\pm \cos[(\omega_0 - \omega_k)t - \delta_k] + P_\pm \cos[(\omega_0 + \omega_k)t + \delta_k], \tag{2.4}$$

where $P_0 = \alpha_0 \cdot E_0$ and $P_\pm = Q_k \alpha_{k'} \cdot E_0/2$ are the magnitudes of the polarization. The first term in Eq. (2.4) is the induced dipole at the frequency of the incident radiation. This is Rayleigh scattering and is typically 10^{-3}-10^{-5} as intense as the incident radiation and 10^{+3}-10^{+5} more intense than the Raman scattering (Fig. 2.3). The two other terms in Eq. (2.4) are the Raman scattering terms, one with scattered radiation at a lower frequency $(\omega_0 - \omega_k)$ called Stokes Raman scattering and one with scattered radiation at a higher frequency $(\omega_0 + \omega_k)$ called anti-Stokes Raman scattering. The difference between the incident and scattered frequency is the frequency of the kth vibrational mode. Thus, the following simple picture of Raman scattering emerges. The incident electric field induces an oscillating dipole in the electron cloud of the molecule. Exchange of energy between the incident field and vibrational states of the molecule lead to oscillations at frequencies lower and higher than the incident frequency by some multiple of the vibrational frequencies of the molecule. The induced dipole re-radiates at an energy consistent with its new, shifted frequency and this is what is detected.

In the quantum mechanical picture of Raman scattering, the incident and final photons interact simultaneously with the molecular ground and excited electronic states, leading to the description of Raman scattering as a "two-photon" process. In the general case, the photons interact with the entire manifold of excited vibronic states. The interaction of the photons with the molecule results in a change in population of the vibrational levels in the ground electronic state of the molecule. This change in energy of the molecule is reflected in the difference in frequency between the incident and scattered photon.

The intensity of a Raman transition from state $|i\rangle$ to state $|f\rangle$, integrated over all directions and polarizations and assuming a randomly oriented sample, is given by [2]

$$P_{\text{scatt},i \to f} = I_{\text{inc}} \sigma_{i \to f}(E_L), \tag{2.5}$$

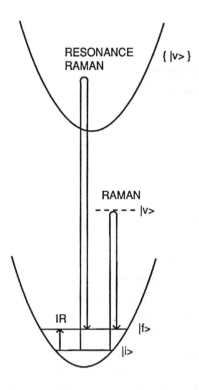

Figure 2.3: Vibrational spectroscopy. In the figure, $|i\rangle$, $|v\rangle$, and $|f\rangle$ de-
note the initial, intermediate or virtual, and final states of the molecule.
Vibrational absorption spectroscopy in the infrared region of the spectrum
involves absorption of light by a molecule and results in a change in the
vibrational quantum number. Raman and resonance Raman are scattering
processes that also involve a change in the vibrational quantum number.
In resonance Raman, the wavelength of the incident light is resonant with
an electronic absorption of the molecule.

where

$$\sigma_{i \to f}(E_L) = \frac{8\pi e^4 E_S^3 E_L}{9\hbar^4 c^4} \sum_{\rho\lambda} |(\alpha_{\rho\lambda})_{if}|^2. \tag{2.6}$$

In the quantum-mechanical picture,

$$(\alpha_{\rho\lambda})_{if} = \sum_{v} \left[\frac{\langle f|m_\rho|v\rangle\langle v|m_\lambda|i\rangle}{E_v - E_i - E_L - i\Gamma} + \frac{\langle f|m_\lambda|v\rangle\langle v|m_\rho|i\rangle}{E_v - E_f - E_L - i\Gamma} \right]. \tag{2.7}$$

In Eq's. (2.5) and (2.6), $P_{scatt,i \to f}$ is the scattered power in photons/sec, I_{inc} is the incident intensity in photons/cm^{2}sec,$\sigma_{i \to f}(E_L)$ is the Raman cross-section in cm^2/molecule, $(\alpha_{\rho\lambda})_{if}$ is the Raman polarizability, and E_S and E_L are the scattered and incident photon energies, respectively. In Eq. (2.7), $|f\rangle$, $|v\rangle$, and $|i\rangle$ are the final, virtual, and initial vibronic states, E_f, E_v, and E_i are their energies, Γ is the homogeneous linewidth, and m_ρ and m_λ are the vector components of the transition dipole length operator. The first term in Eq. (2.7) is the "resonant" term, in which absorption precedes emission. The second term is the "non-resonant" term in which emission precedes absorption and is included since no measurement of the system is made during the intermediate time. Hence, all forms of scattering are added at the amplitude level and squared to yield the experimentally observable cross-section.

Resonance Raman spectroscopy presents several advantages over non-resonant Raman scattering. Enhancement of the Raman scattering on the order of 10^3 from vibrational modes coupled to an electronic transition occurs when the excitation wavelength falls within a strongly allowed electronic absorption band of the molecule. Resonance enhancement allows Raman scattering to be obtained from samples with low concentrations. In the visual pigments, the resonance enhancement effect permits the selective excitation of Raman scattering from the retinal chromophore with no interference from protein scattering, by exciting at a wavelength within a strong chromophore absorption band.

Assuming excitation within a single, strongly allowed absorption band, only the resonant term in Eq. (2.7) still contributes to the resonance Raman polarizability. Equation (2.7) can be further simplified by invoking the adiabatic approximation to separate the electronic and vibrational states and by using the Condon approximation to drop any coordinate dependence in the electronic transition moment and to evaluate the transition moment at the equilibrium geometry, Q_0 [2]. The Raman polarizability and cross-section thus becomes

$$\alpha_{if}(E_L) = M^2 \sum_v \frac{\langle f|v\rangle\langle v|i\rangle}{\epsilon_v - \epsilon_i + E_0 - E_L - i\Gamma}, \tag{2.8}$$

$$\sigma_{i \to f}(E_L) = \frac{8\pi e^4 E_s^3 M^4 E_L}{9\hbar^4 c^4} \left| \sum_v \frac{\langle f|v\rangle\langle v|i\rangle}{\epsilon_v - \epsilon_i + E_0 - E_L - i\Gamma} \right|^2, \tag{2.9}$$

where E_0 is the energy separation between the ground vibrational levels of the excited and ground electronic states (i. e. the "zero-zero energy"), and ϵ_v and ε_i are the energies of the vibrational states $|v\rangle$ and $|i\rangle$. At the same level of approximation, the absorption cross-section is given by

$$\sigma_A(E_L) = \frac{4\pi^2 e^2 M^2 E_L}{3\hbar c n} \sum_v \frac{\Gamma}{\pi} \cdot \frac{|\langle v|i\rangle|^2}{(\epsilon_v - \epsilon_i + E_0 - E_L)^2 + \Gamma^2}, \tag{2.10}$$

where $\sigma_A = 3.824 \times 10^{-5}\epsilon$ and ϵ is the decadic extinction coefficient in $cm^{-1}M^{-1}$. Equations (2.9) and (2.10) are the sum-over-states equations for the Raman and absorption cross-sections. Equation (2.9) is more commonly called the Albrecht A-term epression. Equations (2.9) and (2.10) have been used to elucidate the excited-state structure and dynamics of quite a few molecules including cytochromes [17], nucleotides [8], and carotenoids [69]. The Franck-Condon overlaps $\langle f|v\rangle$ and $\langle v|i\rangle$, in the case of separable, harmonic oscillators, are simply proportional to Δ^2 where Δ is the displacement of the excited-state potential energy minimum from the ground-state potential energy minimum along that normal coordinate.

Thus, the structure of the molecule in the excited-state can be determined by this kind of analysis. The excited-state lifetime can also be determined because

$$\Gamma = \frac{1}{\pi c T_2} = \frac{1}{2\pi c T_1} + \frac{1}{\pi c T_2^*}, \tag{2.11}$$

where T_2 is the total dephasing time (i. e. the excited-state lifetime), T_1 is the population decay time, and T_2^* is the dephasing time. Dephasing is usually thought of as an intermolecular process in which collisions between molecules change the phase of the wavefunction and eventually lead to a loss of phase coherence in the system [69]. Recent femtosecond absorption [67] and holeburning studies [45] of the visual pigments suggest intramolecular dephasing may be important in systems with 10-100 fs excited-state lifetimes ([49, 77]; see below). One disadvantage of these equations is that they are bulky and require extensive cpu time to evaluate the sum over intermediate states [52].

An alternative model for evaluating excited-state properties from absorption spectra and resonance Raman excitation profiles is the time-dependent model. This model has been shown to be more computationally tractable and, in the case of molecules with sub-picosecond dynamics in the excited state, to yield a more intuitive picture of the molecular dynamics. The above equations for the absorption and resonance Raman cross-sections may be written in the time representation by using a simple transformation developed by Lee and Heller [41]. Using the identity

$$\frac{1}{\epsilon_v - \epsilon_i + E_0 - E_l - i\Gamma}$$
$$= \frac{i}{\hbar} \int_0^\infty dt \exp\left[-\frac{it}{\hbar}(\epsilon_v - \epsilon_i + E_0 - E_l - i\Gamma)\right], \tag{2.12}$$

and after a few algebraic steps, the resonance Raman polarizability can be rewritten as

$$\alpha_{if}(E_L) = M^2 \frac{i}{\hbar} \int_0^\infty dt \exp\left[\frac{it}{\hbar}(\epsilon_i + E_L + i\Gamma)\right] \langle f|i(t)\rangle, \tag{2.13}$$

where $|i(t)\rangle = e^{-itH/\hbar}|i\rangle$ is the initial ground-state vibrational wavefunction propagated on the excited-state potential energy surface. Thus, the resonance Raman and absorption cross sections in the Condon approximation can be written in the time-dependent formalism as

$$\sigma_R = \frac{8\pi E_S^3 E_L e^4 M^4}{9\hbar^6 c^4} \sum_I B_I$$

$$\times \left| \int_0^\infty dt \exp\left[i(E_L + \epsilon_i)t/\hbar\right] \langle f|i(t)\rangle G(t) \right|^2, \qquad (2.14)$$

$$\sigma_A = \frac{4\pi e^2 M^2 E_L}{6\hbar^2 cn} \sum_I B_I \int_{-\infty}^\infty dt \exp\left[i(E_L + \epsilon_I)t/\hbar\right] G(t), \qquad (2.15)$$

where $|i\rangle$ and $|f\rangle$ are the initial and final vibrational wavefunctions in the Raman process, ε_i is the energy of the initial vibrational level, E_L and E_S are the energies of the incident and scattered photons, G is a homogeneous linewidth decay function, and B_i is the normalized Boltzmann factor for state i. A one-dimensional example of this wavepacket propagation is presented in Fig. 2.4. The ground state vibrational wavefunction, $|i\rangle$, is carried to the excited- state surface by the incident photon. Because it is not an eigenstate of the excited-state Hamiltonian, the wavepacket propagates on the excited-state surface. The time-dependent overlap of the propagating multimode wavepacket with the initial state, $\langle i|i(t)\rangle$, determines the absorption spectrum, whereas the overlap with the final state, $\langle f|i(t)\rangle$, generates the Raman excitation profiles. The Fourier transforms of these overlaps give the simulated absorption spectrum and Raman excitation profiles.

It should be noted that the absorption $\langle i|i(t)\rangle$ and Raman $\langle f|i(t)\rangle$ overlaps are multimode overlaps. In the special case where the modes are harmonic and separable, the multidimensional absorption overlap becomes simply the product of the one-dimensional overlaps $\langle i|i(t)\rangle$ for each normal mode. The multimode Raman overlap in Eq. (2.11) becomes the product of $\langle f|i(t)\rangle$ in the Raman active mode and $\langle i|i(t)\rangle$ in all other modes. The resonance Raman cross-section of each normal vibration therefore depends on the

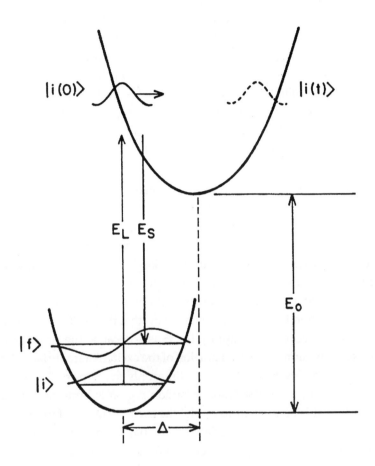

Figure 2.4: The time-dependent picture of resonance Raman scattering for a bound harmonic excited-state surface. The change of the equilibrium geometry upon electronic excitation is denoted by Δ. E_L and E_S are the incident and scattered photon energies, and E_0 is the energy gap between potential minima. The initial and final vibrational wavefunctions in the Raman process are $|i\rangle$ and $|f\rangle$. The excited-state wavepacket $|i(t)\rangle = e^{-iHt/\hbar}|i\rangle$ is the initial ground-state Gaussian vibrational wavefunction propagated on the excited-state potential surface.

excited-state dynamics of *all* modes.

In our treatment, each high-frequency vibrational degree of freedom is described by its frequency and a displacement Δ, in dimensionless normal coordinates, between ground and excited-state harmonic potential surface minima (Fig. 2.4). We are assuming that anharmonicities, Duschinsky rotation [23] of the excited-state normal coordinates, and frequency changes are less important than the potential surface origin displacement. This has been shown to be a reasonable initial assumption for high-frequency modes in the vibronic analysis of polyenes [50].

Vibrations below ~ 500 cm^{-1} can be modeled in two equivalent ways. The first approach is to use a bound harmonic excited-state surface with no excited-state frequency changes. This model is likely to be inaccurate at long propagation times because the excited-state surfaces of these coordinates are expected to have large frequency changes and anharmonicities which would alter $|i(t)\rangle$ at long times. However, as will be shown later, the Raman cross sections indicate that rhodopsin scatters in the "short time limit", which means that the multimode Raman overlap $\langle f|i(t)\rangle$ decays very rapidly (~ 35 fs). Because the propagating wavepacket only samples a small portion of the excited-state surface along low-frequency modes during this short time, only the slope of the excited-state potential surface in the Franck-Condon region is important for the analysis. Thus, it is equivalent to model the excited-state surface for low-frequency modes as a linear dissociative potential. For a harmonic excited-state surface, the relationship between Δ and the excited-state slope in the Franck-Condon region is $\beta/\hbar = \Delta\omega$, where $\beta/\hbar$ is the excited-state slope in cm^{-1} and ω is the excited-state frequency. At short times, the second derivative of the excited-state potential surface is less important than the slope in determining the dynamics. Hence, anharmonicities and frequency changes in the excited state need not be explicitly considered.

Measurement of the Raman excitation profiles permits the differentiation of homogeneous and inhomogeneous contributions to the absorption bandshape [54]. The homogeneous absorption band arises from Franck-

Condon transitions in explicit modes, each of which is broadened by the homogeneous linewidth. As the homogeneous lineshape becomes broader, either through Franck-Condon or lifetime effects, the excitation profiles and the absorption spectrum become diffuse. The resonance Raman cross sections are strongly reduced as the homogeneous width is increased as well. Thus, the resonance Raman cross-sections provide a sensitive probe of the overall homogeneous width. In the case of rhodopsin, the absorption lineshape requires that the homogeneous damping function in Eq's. (2.14) and (2.15) be a Gaussian of the form $\exp(-\Gamma^2 t^2/h^2)$.

In condensed phase we must also include inhomogeneous or 'site' distribution effects. In rhodopsins, it is known that interactions of the retinal chromophore with charge protein residues are responsible for tuning the absorption maxima from 400 to 650 nm [46, 65, 84]. The large effect of protein-chromophore interactions on the electronic state energies suggests that small changes in chromophore-charged residue distances will lead to a modulation of the zero-zero energy (E_0) of the chromophore absorption. Normally at room temperature, these distances will be fluctuating due to protein and chromophore motions. However, on the timescale of the resonance Raman scattering process in rhodopsin (35 fs, see below), these fluctuations for each molecule should be insignificant. Therefore, any differences between molecules in their chromophore-charged residue distances will contribute to the inhomogeneous broadening. In this paper, inhomogeneous effects have been modeled as a Gaussian distribution of zero-zero energies with standard deviation Θ [54]. Inhomogeneous broadening causes both the absorption spectrum and Raman excitation profiles to lose vibronic structure, but it does not change the integrated intensity of either, because inhomogeneous effects enter the calculation as a normalized distribution at the probability level. The equations used in this paper to model the absorption spectrum and resonance Raman excitation profiles of rhodopsin therefore become

$$
\sigma_A = \frac{4\pi e^2 M^2 E_L}{6\hbar^2 cn\sqrt{2\pi\Theta^2}} \sum_i B_i \int dE_0 e^{-(E_0-\langle E_0 \rangle)^2/2\Theta^2}
$$

$$
\times \int_\infty^\infty dt\langle i|i(t)\rangle \exp\left[i(E_L + \epsilon_i)t/\hbar\right] G(t), \tag{2.16}
$$

$$\sigma_R = \frac{8\pi E_S^3 E_L e^4 M^4}{9\hbar^6 c^4 \sqrt{2\pi\Theta^2}} \sum_i B_i \int dE_0 e^{-(E_0 - \langle E_0 \rangle)^2 / 2\Theta^2}$$

$$\times \left| \int_0^\infty dt \langle f | i(t) \rangle \exp\left[i(E_L + \epsilon_i) t / \hbar \right] G(t) \right|^2. \tag{2.17}$$

Thus, measurement of absolute resonance Raman excitation profiles provide a sensitive means for distinguishing between homogeneous and inhomogeneous broadening mechanisms.

Because a number of the Raman-active modes in rhodopsin have frequencies $\leq kT$, we also need to include effects due to molecules whose initial state is vibrationally excited. Modes that are thermally populated will contribute to inhomogeneous ensemble broadening of the system due to the sum over initial states at the probability level in Eq's. (2.16) and (2.17). They may also contribute to homogeneous broadening if the $\langle i | i(t) \rangle$ and $\langle f | i(t) \rangle$ overlaps with $i > 0$ have faster decays. From the Boltzmann equation, the four modes below 300 cm^{-1} in rhodopsin are $\leq 1\%$ populated to i=2 at 277 K. The required Raman and absorption overlaps for these thermally excited states are easily calculated using the linear excited-state model described in Ref. [51]. The derived expressions for the single-mode time-dependent overlaps, $\langle i | i(t) \rangle$ and $\langle f | i(t) \rangle$, are

$$\langle 0 | 0(t) \rangle = \left(\frac{2}{2 + i\omega t} \right)^{1/2} \exp\left[-\frac{\beta^2 t^2}{24\hbar 2}(6 + i\omega t) \right], \tag{2.18}$$

$$\frac{\langle 1 | 0(t) \rangle}{\langle 0 | 0(t) \rangle} = \frac{-i\beta t}{\hbar \sqrt{2}}, \tag{2.19}$$

$$\frac{\langle 1 | 1(t) \rangle}{\langle 0 | 0(t) \rangle} = \left[\frac{2}{2 + i\omega t} - \frac{\beta^2 t^2}{2\hbar^2} \right], \tag{2.20}$$

$$\frac{\langle 2 | 1(t) \rangle}{\langle 0 | 0(t) \rangle} = \frac{i\beta t}{\hbar} \left[\frac{\beta^2 t^2}{4\hbar^2} - \frac{4 + i\omega t}{2(2 + i\omega t)} \right], \tag{2.21}$$

$$\frac{\langle 2 | 2(t) \rangle}{\langle 0 | 0(t) \rangle} = \frac{1}{2} \left[\frac{\beta^4 t^4}{4\hbar^4} - \frac{\beta^2 t^2}{\hbar^2} \left(\frac{4 + i\omega t}{2 + i\omega t} \right) + \frac{8 - \omega^2 t^2}{(2 + i\omega t)^2} \right], \text{ and} \tag{2.22}$$

$$\frac{\langle 3 | 2(t) \rangle}{\langle 0 | 0(t) \rangle} = -\frac{i\beta t}{\hbar \sqrt{6}} \left[\frac{\beta^4 t^4}{8\hbar^4} - \frac{\beta^2 t^2}{\hbar^2} \left(\frac{3 + i\omega t}{2 + i\omega t} \right) + \frac{6}{(2 + i\omega t)^2} + \frac{3}{2} \right]. \tag{2.23}$$

In these expressions β/h is the slope of the excited-state potential surface and ω is the ground-state frequency. The single-mode overlaps which initiate from higher-lying ground-state vibrational levels evolve much faster at short times than the corresponding i=0 overlaps due to the presence of more nodes in the overlapping wavefunctions. When the Raman-active mode is thermally excited, the calculated Raman scattering will be more intense because the single-mode $\langle f|i(t)\rangle$ overlap rises to a maximum much faster, before other homogeneous damping effects become significant. When other modes are thermally excited, the faster fall-off of their $\langle i|i(t)\rangle$ overlaps will damp the multimode $\langle f|i(t)\rangle$ overlap more rapidly, resulting in weaker Raman scattering. The ensemble-averaged cross-sections depend on an interplay of these two effects which is sensitive to temperature, mode frequencies, and the excited-state slopes.

The main purpose of this study is to examine the excited-state structure and isomerization dynamics of the retinal chromophore in rhodopsin. Specifically, we will review the analysis of resonance Raman excitation profiles and holeburning spectroscopy of rhodopsin to develop a model for the multidimensional excited-state potential surface of the retinal prosthetic group, and to set limits on the excited-state isomerization and relaxation times. To do this, the resonance Raman intensities of 25 vibrational modes were measured as a function of excitation wavelength. These excitation profiles were analyzed to determine excited-state geometry changes for each of the observed normal modes and to define the contributions of homogeneous (single molecule) and inhomogeneous (ensemble) broadening to the overall absorption bandshape. An important result of this study is the observation of intense low-frequency skeletal torsional modes whose Franck-Condon progressions contribute significantly to the homogeneous absorption bandshape. The observed $\sim$ 2400-cm^{-1} holes in the holeburning study of rhodopsin were successfully simulated using the parameters obtained from the resonance Raman intensity analysis. The resonance Raman intensity analysis and holeburning study present a complete description of the isomerization of the retinal chromophore in rhodopsin in which the molecule distorts along primarily torsional modes in the first $\sim$ 100-200 fs. The excited-state subsequently crosses to the ground-state on a $\sim$ 500 fs timescale and then thermally relaxes to bathorhodopsin.

Resonance Raman Spectroscopy of Rhodopsin: Techniques

Frozen bovine retinas were purchased from J. A. Lawson (Lincoln, NE), and the rod outer segments were isolated by a sucrose flotation method [65]. The rod outer segments were lysed in water and solubilized in 15% Ammonyx-LO (Onyx Chemical Co., Jersey City, NJ). The resulting rhodopsin solution was purified by hydroxylapatite chromatography [65] and concentrated in Amicon membrane cones to 15-30 μM. Typical yields were 750-1500 nmoles rhodopsin per 100 retinas.

Room temperature, rapid-flow resonance Raman spectra of rhodopsin were obtained with 10-15 ml samples having an absorbance of 0.8-1.5 OD/cm at 500 nm (100 mM phosphate buffer, $< 10mMNH_2OH \cdot HCl$, <1% Ammonyx-LO, pH 6.8-7.0) and containing 0.2-0.6 M potassium nitrate as an internal intensity standard. Addition of potassium nitrate had no noticeable effect on the absorption or resonance Raman spectra of rhodopsin. Raman scattering was excited by spherically focusing the laser beam in the 0.8 mm dia. capillary containing the flowing rhodopsin solution. Laser excitation was obtained with Kr and Ar ion lasers (Spectra-Physics, Mountain View, CA, Models 171 and 2020). Multichannel detection of the Raman scattering was accomplished with a cooled intensified vidicon detector (PAR 1205A/1205D) coupled to a double spectrograph. Single channel photon-counting spectra were obtained with a double monochromator system [58]. Spectral slit widths were 5.0-6.0 cm^{-1}. The laser power (300-600 μ W), flow rate (300-400 cm/s), and beam waist (10 μ) were chosen to minimize the effects of photolysis on the Raman spectra (photoalteration parameter $F \sim 0.1$, Ref. [47]). The actual bulk photolysis rate was determined by measuring absorption spectra of the sample after each Raman scan. The multichannel Raman spectrometer was calibrated using cyclohexene and dicyclopentadiene as external standards. Frequencies are accurate to ±4 cm^{-1}.

Raman data collection was performed with a PDP 11/23 computer. Reported spectra are the sums of 3-9 scans. All spectra were corrected for the wavelength dependence of the spectrometer efficiency using a standard lamp. Fluorescence backgrounds were removed by subtracting a quartic polynomial, and the spectra were digitally smoothed using a three-point sliding average. The integrated intensities of the rhodopsin lines relative to the 1049 cm^{-1} nitrate line were obtained by fitting regions of the spectrum to sums of Lorentzian peaks convoluted with a triangular slit function. The relative intensities of the rhodopsin lines were then corrected for self-absorption using a 0.4 mm path length. The differential self-absorption correction between the nitrate line and any rhodopsin fundamental was no more than 15% Absolute Raman cross sections were found from the relative integrated intensities using:

$$\sigma_{RHO} = \sigma_{NO_3} \frac{\left(\frac{1+2\rho}{1+\rho}\right)_{RHO} [NO_3]I_{RHO}}{\left(\frac{1+2\rho}{1+\rho}\right)_{NO_3} [RHO]I_{NO_3}}, \tag{2.24}$$

where ρ is the depolarization ratio for 90° detection of all polarizations of the scattered light [rho = 0.04 and 0.33 for NO_3 and Rhodopsin, respectively]. Nitrate concentrations were determined by comparing the nitrate scattering of the bleached rhodopsin sample with that of a series of standard nitrate solutions. The average concentration of rhodopsin in each experiment was found by interpolation of the A_{500} values measured before and after each rhodopsin Raman scan.

We initially assumed that the cross sections at 514.5 nm are proportional to Δ^2 and used these Δ's referenced to Δ_{1549} = 1.0 as a starting point in the analysis. The 25 observed modes were used in the time-dependent calculations and the other parameters were selected to best model the experimental absorption spectrum and excitation profiles. The scaling of the Δs is determined by the width of the absorption band. The homogeneous width is determined by the magnitude of the absolute resonance Raman cross sections and the shape of the absorption spectrum on the red edge. The transition moment (M) is dictated by the magnitude of the absorption maximum. Finally, the degree of inhomogeneous broadening Θ and the energy gap between the ground and excited potential surfaces (E_0) are chosen to give the correct absorption bandshape and position. This process is iterated several times until the experimental and calculated excitation profiles and absorption spectra are in agreement. The parameters could not be altered by more than ±5% without significantly reducing the quality of the fit.

Resonance Raman Spectroscopy of Rhodopsin: Results

The resonance Raman spectrum of rhodopsin excited at 514.5 nm is shown in Fig. 2.5A. The symmetric stretch of nitrate, the intensity standard used in these spectra, is seen at 1049 cm^{-1}. Previous studies have suggested that extensive progressions in low-frequency modes may contribute significantly to the overall absorption lineshape [40, 51, 77]. However, the 5 lines at 336, 461, 809, 824, and 859 cm^{-1} have insufficient Raman intensity to contribute substantially to the absorption lineshape (Fig. 2.5B). To reach lower wavenumbers, we used a photon-counting Raman system because of its better stray light rejection. Four intense lines were found at 98, 135, 250, and 260 cm^{-1} (Fig. 2.5C) whose intensities were determined by reference to the intensity of the 970 cm^{-1} line, measured at the same time. These four lines were absent in the bleached spectrum taken immediately after the experiment under identical conditions. Thus, we conclude that these features are chromophore lines and not spurious Rayleigh features or protein lines. These lines are very intense, having cross sections of the same magnitude as the single bond stretches (see Table 2.1). These modes will therefore make a major contribution to the absorption lineshape. INDO and QCFF/PI calculations on the 11-*cis*-retinal chromophore [40, 77] and vibrational assignments of conjugated polyenes [6] indicate that single- and double-bond torsions are found in this region of the Raman spectrum. We therefore ascribe the modes at 98, 135, 250, 260, 336, and 461 cm^{-1} to torsional modes.

Mode	$|\Delta|$
98	80†
135	225†
249	174†
262	231†
336	127†
461	111†
809	0.14
824	0.16
859	0.14
970	0.50
999	0.14
1018	0.32
1189	0.14
1215	0.35
1238	0.40
1268	0.38
1318	0.22
1359	0.20
1389	0.18
1432	0.24
1442	0.25
1549	0.87
1581	0.22
1609	0.29
1659	0.27

Table 2.1: Vibronic parameters of rhodopsin. The Δ's are in units of dimensionless normal coordinates. The dimensionless coordinate, q, is related to the Cartesian coordinate, x, by q=$(\mu\omega/\hbar)^{1/2}$x where μ is the reduced mass. The calculations used E_0=18,500 cm^{-1}, transition length M = 0.2079 nm, temperature T=277K, Gaussian homogeneous linewidth Γ = 170 cm^{-1} half-width, and inhomogeneous linewidth Θ=910 cm^{-1} half-width. Thermally excited initial states to i=2 were considered in modes below 300 cm^{-1}. †For these lines, the calculation was performed using the linear excited-state model. The excited-state slope of $\beta/\hbar$ is given in units of cm^{-1} and the equivalent harmonic surface displacement can be calculated using $\beta/\hbar=\Delta\omega$.

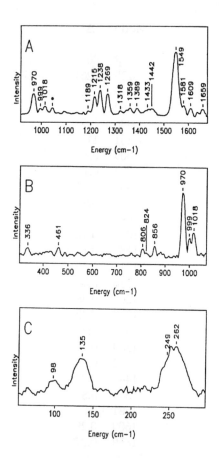

Figure 2.5: Resonance Raman spectra of rhodopsin excited at 514.5 nm in the fingerprint (A), mid- (B), and low-frequency (C) region of the spectrum. Rhodopsin and KNO_3 concentrations in A are 21 μM and 0.18 M, respectively. The 1049 cm^{-1} line is due to NO_3^-. Spectrum B is the sum of four scans of a 25 μM rhodopsin solution exciting at 514.5 nm obtained on the multichannel system and treated as described in the text. The spectrum shown in C is the sum of three scans of a 75 μM rhodopsin solution obtained on the single-channel, double monochromator system exciting at 514.5 nm. The cross-sections of the four lines at 98, 135, 249, and 262 cm^{-1} were determined by referencing to the 970 cm^{-1} line which was measured in the same experiment.

Figure 2.6 demonstrates the good agreement between the experimental Raman excitation profiles and absorption spectra and those calculated using the parameters of Table 2.1. Similarities in the relative intensities at different excitation wavelengths indicate that the enhancement profiles of these modes should be similar and this is borne out in the calculations. The calculated and experimental absorption spectra and excitation profiles are broad and featureless, suggesting rapid dynamics in the excited-state and/or large inhomogeneous effects. Attempts to model these data using a Lorentzian lineshape function whose width was large enough to give the correct Raman cross-sections were unsuccessful because they predicted a large red- edge tail on the absorption that is inconsistent with experiment. The experimentally observed absorption red-edge is consistent with a homogeneous decay lineshape that has a zero slope at t=0. A Gaussian homogeneous lineshape has zero slope at t=0 and this is what was used. A Gaussian lineshape is predicted from strong coupling between electronic states [77] and/or coupling to bath modes [40] which result in loss of overlap due to amplitude decay and dephasing.

Examination of the Raman cross-sections and the Raman overlaps $\langle f|i(t)\rangle$ clearly illustrate the role of torsional deformations and the Gaussian homogeneous decay in the overall molecular dynamics. In Fig. 2.7 it is clear that the area under the multimode $\langle f|i(t)\rangle$, which is proportional to the Raman cross-section, is strongly reduced when the torsions are included because the Raman overlap must be multiplied by a rapidly decaying torsional $\langle i|i(t)\rangle$. However, even with all the torsional degrees of freedom included, the calculated cross-sections are still a factor of 1.5 too high. Thus, the cross-sections require the inclusion of a rapidly decaying homogeneous lineshape function ($t_{1/2}$=45 fs). In summary, *the damped resonance Raman cross-sections require that $\langle f|i(t)\rangle$ must decay rapidly and permanently. The rapid decay at short times is due to large torsional distortion of the chromophore after electronic excitation. The permanence of the decay of $\langle f|i(t)\rangle$ is ensured by the Gaussian homogeneous decay whose origin will be discussed.*

The primary purpose of this study is to examine the dynamics of the

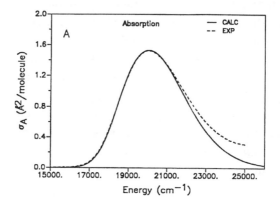

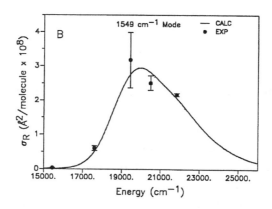

Figure 2.6: Absorption spectra (A) and experimental (points) and calculated 1549 cm^{-1} mode resonance Raman excitation profiles (B) for rhodopsin. The experimental absorption is indicated by a dashed line. The solid line is generated by the final parameters of Table 2.1. Error bars represent the uncertainties in the absolute cross-sections. Deviations of the calculated spectrum from the experimental absorption spectrum at $\sim$ 24000 cm^{-1} are most likely due to the presence of one or more higher-lying electronic states.

excited-state isomerization of the retinal chromophore in rhodopsin. The following temporal picture is suggested by these data. After excitation, the wavepacket leaves the Franck-Condon region rapidly along high-frequency single- and double-bond stretching modes. By the time it returns to the Franck-Condon region along these high-frequency modes ($\sim$ 20 fs), the molecule has distorted significantly about low-frequency torsional modes such that the multimode overlap $\langle i|i(t)\rangle$ is strongly damped. At longer times ($>$ 200 fs), strong radiationless coupling (surface crossing) associated with isomerization must occur that damps out any later recurrences. The strong coupling is represented in this study by dissociative torsions and the Gaussian homogeneous decay. Recently, subpicosecond absorption spectroscopy of bovine rhodopsin has demonstrated that the retinal chromophore isomerizes completely in 200 fs and then gradually relaxes on a $\sim$ 500 fs timescale [67], consistent with the above picture. Also, subpicosecond absorption studies on bacteriorhodopsin have indicated that the torsional distortion of the retinal chromophore occurs on a 100-150 fs timescale and that electronic relaxation occurs with a $t_{1/2} \sim 500$ · fs [49, 62].

The observation of explicit low-frequency torsional modes at 98, 135, 249, 262, 336, and 461 cm^{-1} is an intriguing result of this study. By modeling the excited-state potential of the $C_{11}=C_{12}$ torsion as a linear surface, the intensities of these modes can be used to make a rough estimate of the torsional time. QCFF/PI calculations predice that the $C_{11}=C_{12}$ torsion is primarily localized in a mode at 537 cm^{-1}. Therefore, we will assume that the Raman line at 461 cm^{-1} can be assigned as this torsion. If a wavepacket is propagated on the linear excited-state surface of the 461 cm^{-1} mode, it will reach a dimensionless coordinate of 11, corresponding to a torsional angle of 90°, in only 100 fs [52]. this suggests that the isomerization could occur as fast as 0.2 ps. Because modes below 300 cm^{-1} have very large excited-state slopes, even larger single-bond torsions and skeletal deformations in addition to the $C_{11}=C_{12}$ torsion occur during the isomerization. Normal mode assignments of these lines will provide specific projections of these modes on the isomerization coordinate and introduce further constraints in proposed models of the isomerization mechanism.

The Gaussian homogeneous linewidth used in this study requires com-
ment. The decay of the Gaussian is too fast to be purely electronic
relaxation; other studies have shown 500 fs to be a reasonable electronic
relaxation time for rhodopsins [49, 62]. One possibility is that the Gaussian
decay is due to distortions along torsional modes which are too low in
frequency to be observed in this study. Modes below $\sim 50 \text{ cm}^{-1}$ are very
difficult to observe experimentally even if they are highly displaced. It
seems more likely, however, that the Gaussian decay arises from rapid
dephasing. Strong coupling to bath modes could lead to intermolecular
dephasing on a femtosecond timescale [77]. Intramolecular dephasing may
occur on this timescale. As the polyene torsionally distorts, the excited
electronic-state character will change dramatically, producing concomi-
tant changes in the vibrational states. Because of the Born-Oppenheimer
breakdown, this process should be thought of as "vibronic dephasing."

To more accurately determine the origin of the Gaussian homogeneous
linewidth, holeburning spectroscopy was performed on rhodopsin [45].
Hole-burning spectroscopy provides an alternative technique for examining the short-time isomerization dynamics of these pigments at low
temperature and has been used to study excited-state decay processes of
photosynthetic reaction centers [11, 12, 74], porphyrins [76], and phyco-
erythrin [27]. In a holeburning experiment, a sample is irradiated with
spectrally narrow light and the resulting difference absorption spectrum
recorded. The experiment is performed under conditions in which the
photoproduct is trapped on the experimental timescale; the photoproduct in rhodopsin is trapped by cooling the samples below ~ 130 K. The
resulting difference spectrum yields the partitioning of broadening between homogeneous and inhomogeneous mechanisms in the following
way. If the absorption spectrum is primarily homogeneously broadened
(*i. e.* every molecule exhibits the same or similar absorption spectra), then
the whole absorption spectrum will be bleached upon irradiation with
the narrow-band source. This will occur regardless of the energy of the
narrow-band source, as long as the wavelength of irradiation is within
the absorption band. However, if the absorption spectrum is primarily
inhomogeneously broadened (*i. e.* the absorption spectrum is a composed
of a distribution of single-molecule absorption spectra in energy) then only

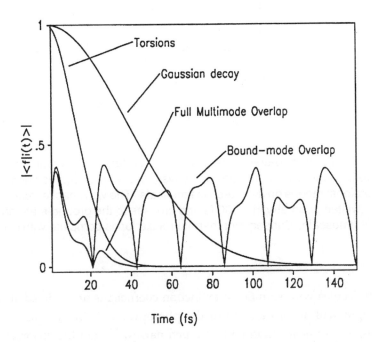

Figure 2.7: Effect of low-frequency torsional modes and the the Gaussian homogeneous decay on the Resonance Raman overlap for the 1549 cm^{-1} mode. The bound mode overlap was calculated after removal of all low-frequency torsional modes (<500 cm^{-1}) and the Gaussian homogeneous decay. The torsion overlap presents $\langle i|i(t)\rangle$ for the six low-frequency torsional modes. The Gaussian decay is for a homogeneous Gaussian linewidth of 170 cm^{-1} HWHM. The full, multimode overlap was calculated using the parameters of Table 2.1 for the state with all vibrational modes populated at i=0.

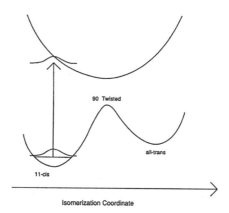

Figure 2.8: The excited-state potential energy surface for the retinal chromophore in rhodopsin. In this figure, the wavepacket propagates primarily along torsional modes on a ~ 40-50 fs timescale. The wavepacket reaches the 90° twisted geometry in ~100 fs and crosses to the ground-state on a ~ 200 fs timescale. Subsequent thermal relaxation is complete within 1-2 ps [67].

those molecules with significant extinction coefficients at the irradiation wavelength will absorb and form the photoproduct. In this case, the resulting loss of absorption will be much narrower than the absorption spectrum and the position of the bleach will depend on the energy of the narrow-band source.

Hole-burning spectra were obtained of rhodopsin at 1.5 K and are shown in Fig. 2.9. Figure 2.9A shows the absorption spectrum at 1.5 K and Fig. 2.9B shows the experimental and simulated hole spectra irradiated 514.5 nm. Note that even at 1.5 K the absorption spectrum is broad and diffuse indicating a large homogeneous linewidth. The observation of ~ 2400 cm^{-1} holewidths with no evidence for a sharp zero-phonon line in the pigment confirms that there is a very rapid vibronic relaxation on a 10-100 fs timescale in this system, even at low temperature. The holes were successfully simulated using the vibronic and lifetime parameters determined from the previous analysis of rhodopsin's resonance Raman intensities. The positions of the hole minima were insensitive to burn wavelength, indicating that minimal site selection occurs at 1.5 K. Indeed, the inhomogeneous linewidth of rhodopsin decreases from 910 cm^{-1} HWHM at room temperature to 350 cm^{-1} HWHM at 1.5 K, suggesting that fluctuations in protein-chromophore interactions responsible for the wavelength regulation in this pigment are significantly reduced at low temperatures. The homogeneous linewidth was relatively unaffected by temperature indicating that the isomerization dynamics are essentially unchanged at low temperature. The similarity of the room and low temperature homogeneous linewidth argues that the phenomenological Gaussian homogeneous broadening is probably not due to unobserved low-frequency modes, as they should lead to a temperature dependent homogeneous linewidth. Thus, the phenomenological Gaussian homogeneous linewidth probably arises from a breakdown of the Born-Oppenheimer approximation due to the rapid excited-state dynamics.

A previous INDO calculation of isomerization trajectories has predicted that rhodopsin distorts along a barrierless $C_{11}=C_{12}$ torsional coordinate on the excited-state potential surface to a 90° twisted geometry from which it crosses to the ground-state potential surface and forms either rhodopsin or bathorhodopsin (Fig. 8, Ref. [7]). Similarly, in our model the wavepacket propagates along low-frequency modes on the excited-state surface until it reaches a geometry where electronic relaxation to the ground state potential surface is optimized. Our predicted time to a 90° twisted geometry ($\sim$ 100 fs) is in good agreement with picosecond fluorescence experiments which suggest 80-170 fs [21] and with recent subpicosecond measurements on rhodopsin [67] and bacteriorhodopsin (bR) [49]. However, INDO calculations [7] predict times of 800-1100 fs to reach the 90° geometry, suggesting that the theoretical excited-state slopes are low.

The two largest geometry changes occur in the 1549 cm^{-1} ethylenic stretch and 970 cm^{-1} 11,12 hydrogen-out-of-plane (HOOP) lines. The large Δ in the 1549 cm^{-1} ethylenic stretch quantifies the change in double-bond

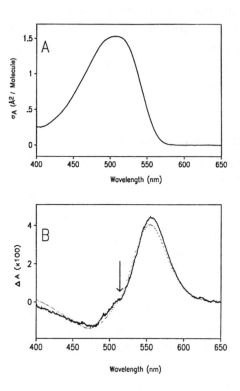

Figure 2.9: 1.5K absorption spectra (A) and hole spectra (B) for rhodopsin. The solid line in B is the rhodopsin experimental hole spectrum and was obtained by irradiating the sample with 10 μW/cm^2 of 514.5 nm (arrow) light for 30 sec. The dashed line is generated by fitting the experimental hole spectrum to the sum of the experimental bathorhodopsin absorption spectrum and calculated rhodopsin hole spectrum. The calculated hole spectrum was obtained from the parameters of Table 2.1 with the following modifications: the Δ's were scaled by 0.92, the homogeneous linewidth was increased from 170 cm^{-1} half-width at half-maximum (HWHM) to 650 cm^{-1} HWHM, the temperature was decreased to 1.5K, and the inhomogeneous broadening was decreased from 910 cm^{-1} HWHM to 350 cm^{-1} HWHM. Deviations of the calculated spectrum from the experimental hole spectrum at $\lambda \leq 435$ nm are most likely due to the presence of one or more higher-lying electronic states.

strength in the excited-state. The large Δ in the 970 cm^{-1} HOOP mode has been interpreted in terms of ground state distortions about the C_{10}-C_{11}, C_{11}=C_{12}, and C_{12}-C_{13} bonds that allow this mode to acquire resonance Raman intensity [24, 26, 79]. A HOOP mode with a Δ of 1.2 has also been seen at 963 cm^{-1} in cis-stilbene, which is distorted around the exocyclic double bond in the ground state [52]. These ground-state distortions have led to the appealing suggestion that the retinal chromophore is twisted around the C_{11}=C_{12} double bond in such a way that the isomerization process is "primed" [58]. This may help to account for the unusually short time [7, 14, 21, 35, 60] required for this isomerization and may contribute to the high quantum yield for isomerization (ϕ=0.67, [19]).

We have previously determined the excited-state structure and dynamics of the bR chromophore from resonance Raman intensities [51, 54]. The most interesting similarities between rhodopsin and bR occur in the calculated excited-state relaxation times for the two proteins. The analysis of rhodopsin is essentially identical to that of bR, except that in bR no low-frequency modes with significant resonance Raman intensity have thus far been detected. Therefore, a Gaussian homogeneous damping function which represented distortion on a highly- displaced torsional surface had to be included to account for the damped Raman cross-sections and broad absorption bandshape. The wavepacket leaves the Franck-Condon region very rapidly in both bR and rhodopsin; $\langle i|i(t)\rangle$ decays to zero in 30 and 35 fs, respectively. Thus, the excited-state relaxation in each is similar. Since both molecules undergo double bond isomerizations in a protein, this suggests that the mechanism for isomerization is analogous. An additional similarity is the relatively large amount of inhomogeneous broadening required in the calculations for bR (Θ=560 cm^{-1}, half-width) and rhodopsin (Θ=910 cm^{-1}, half-width) to account for the spectral diffuseness. This inhomogeneous width could be due to differences in protein-chromophore interactions associated with the opsin shift. This model is further supported by the holeburning data in which the inhomogeneous broadening decreases as a function of temperature in both rhodopsin and bacteriorhodopsin [45].

Conclusions

Resonance Raman spectroscopy has been shown to be a powerful probe of ground-state structure and excited-state structure and dynamics. Resonance Raman intensities have been used to perform a complete Franck-Condon analysis of the absorption bandshape in the visual pigment rhodopsin. The majority of the absorption spectral width is due to Franck-Condon progressions in high-frequency ethylenic stretching modes. Low-frequency torsional modes and the Gaussian homogeneous linewidth contribute to the diffuseness of the absorption spectrum and are responsible for the reduced Raman cross-sections. Holeburning spectra demonstrate that the Gaussian homogeneous linewidth probably arises from breakdown of the Born-Oppenheimer approximation due to the rapid excited-state nuclear distortions in the retinal chromophore of rhodopsin. Rhodopsin is shown to have an optical dephasing time of ~ 35 fs and is predicted to reach a 90° twisted geometry in $\sim 100\text{-}200$ fs. Inhomogeneous site distribution effects also make a large contribution to the absorption bandshape.

We should conclude where we began. We see when the Rhodopsin molecule isomerizes; this molecular "decision" occurs on a femtosecond timescale and is determined by the forces on the retinal chromophore immediately after photon absorptions. The Raman experiments have given us a very detailed picture of these forces which ultimately drive our vision.

Acknowledgements

The author would like to thank Richard A. Mathies for directing this research, Anne B. Myers and W. Thomas Pollard for stimulating discussions, and Thomas R. Middendorf, David Gottfried, and Steven Boxer with whom the holeburning spectroscopy was carried out.

References

[1] J. R. Ackerman, B. E. Kohler, D. Huppert and P. M. Rentzepis (1982). *J. Chem. Phys.* **77**, 3967-3973.

[2] A. C. Albrecht (1961). *J. Chem. Phys.* **34**, 1476-1484.

[3] M. L. Applebury, D. M. Zuckerman, A. A. Lamola and T. M. Jovin (1974). *Biochem.* **13**, 3448-3458.

[4] K. A. Bagley, V. Balogh-Nair, A. A. Croteau, G. Dollinger, T. G. Ebrey, L.Eisenstein, M. K. Hong, K. Nakanishi, and J. Vittitow (1985). *Biochem.* **24**, 6055-6071.

[5] F. W. Billmeyer, Jr. and M. Saltzman (1980). *Color Res. Appl.* **5**, 72.

[6] R. R. Birge, D. F. Bocian and L. M. Hubbard (1982). *J. Am. Chem. Soc.* **104**, 1196-1207.

[7] R. R. Birge and L. M. Hubbard (1980). *J. Am. Chem. Soc.* **102**, 2195-2204.

[8] D. C. Blazej and W. L. Peticolas (1980). *J. Chem. Phys.* **72**, 3134-3142.

[9] J. K. Bowmaker (1984). *Vision Res.* **24**, 1641-1650.

[10] D. Bownds (1967). *Nature* **216**, 1178-1181.

[11] S. G. Boxer, T. R. Middendorf and D. J. Lockhart (1986). *FEBS Lett.* **200**, 237.

[12] S. Boxer, D. Lockhart and T. Middendorf (1986). *Chem.Phys.Lett.* **123**, 476.

[13] L. A. Brey, G. B. Schuster and H. G. Drickamer (1979). *J. Am. Chem. Soc.* **101**, 129.

[14] G. E. Busch, M. L. Applebury, A. A. Lamola and P. M. Rentzepis (1972). *Proc. Nat. Acad. Sci. USA* **69**, 2802-2806.

[15] R. Callender and B. Honig (1977). *Ann. Rev. Biophys. Bioeng.* **6**, 33-55.

[16] M. Chabre (1985). *Ann. Rev. Biophys. Biophys. Chem.* **14**, 331-360.

[17] P. M. Champion and A. C. Albrecht (1981). *J. Chem. Phys.* **75**, 3211-3214.

[18] B. Curry, I. Palings, A. D. Broek, J. A. Pardoen, J. Lugtenburg, and R. Mathies (1985). In *Advances in Infrared and Raman Spectroscopy, vol. 12*, R. J. H. Clark and R. E. Hester, eds., (Wiley, New York).

[19] H. J. Dartnall 1972. In *Handbook of Sensory Physiology. Vol. VII/I*, H. J. Dartnall, ed., pp. 122-145 (Springer-Verlag, New York).

[20] H. Davson, ed. (1977) *Photobiology of Vision, vol. 2B-The Eye* (Academic, San Francisco).

[21] A. G. Doukas, M. R. Junnarkar, R. R. Alfano, R. H. Callender, T. Kakitani and B. Honig (1984). *Proc. Nat. Acad. Sci. USA* **81**, 4790-4794.

[22] A. G. Doukas, M. R. Junnarkar, R. R. Alfano, R. H. Callender and V. Balogh-Nair (1985). *Biophys. J.* **47**, 795-798.

[23] F. Duschinsky (1937). *Acta Physicochim. U. R. S. S.* **7**, 551.

[24] G. Eyring, B. Curry, R. Mathies, R. Fransen, I. Palings and J. Lugtenburg (1980). *Biochem.* **19**, 2410-2418.

[25] G. Eyring, B. Curry, R. Mathies, A. Broek, and J. Lugtenburg (1980). *J. Am. Chem. Soc.* **102**, 5390-5392.

[26] G. Eyring, B. Curry, A. Broek, J. Lugtenburg and R. Mathies (1982). *Biochem.* **21**, 384-393.

[27] J. Friedrich, B. Scheer, B. Zickendraht-Wendelstadt and D. Haarer (1981). *Chem.Phys.* **74**, 2260.

[28] W. J. de Grip, D. Gray, J. Gillespie, P. H. M. Bovee, E. M. M. van den Berb, J. Lugtenburg, and K. J. Rothschild (1988). *Photochem. Photobiol.* **48**, 497-504.

[29] W. J. de Grip, J. Gillespie and K. J. Rothschild (1985). *Biochim. Biophys. Acta* **809**, 97-106.

[30] U. M. Ganter, W. Gartner and F. Siebert (1988). *Biochem.* **27**, 7480-7488.

[31] D. Gegiou, K. A. Muszkat and E. Fischer (1968). *J. Am. Chem. Soc.* **90**, 12.

[32] M. F. Granville, G. R. Holtom and B. E. Kohler (1980). *Proc. Natl. Acad. Sci. USA* **77**, 31-33.

[33] B. I. Greene, R. M. Hochstrasser and R. B. Weisman (1980). *Chem. Phys.* **48**, 289.

[34] B. I. Greene and R. C. Farrow (1983). *J. Chem. Phys.* **78**, 3336.

[35] G. Hayward, W. Carlsen, A. Siegman, and L. Stryer (1981). *Science* **211**, 942-944.

[36] D. Huppert, P. M. Rentzepis and D. S. Kliger (1977). *Photochem. Photobiol.* **25**, 193-197.

[37] D. Huppert and P. M. Rentzepis (1986). *J. Chem. Phys.* **90**, 2813-2816.

[38] R. M. Hochstrasser (1980). *Pure Appl. Chem.* **52**, 2683.

[39] F. Inagaki, M. Tasumi and T. Miyazawa (1974). *J. Mol. Spectrosc.* **50**, 286-303.

[40] R. Kubo (1985). *Pure Appl. Chem.* **57**, 201.

[41] S.-Y. Lee and E. J. Heller (1979). *J. Chem. Phys.* **71**, 4777-4788.

[42] P. A. Liebman and A. M. Granda (1975). *Nature* **253**, 370-372.

[43] R. S. H. Liu and A. E. Asato (1985). *Proc. Natl. Acad. Sci. USA* **82**, 259-263.

[44] D. A. Long (1977). *Raman Spectroscopy* (McGraw-Hill, New York).

[45] G. R. Loppnow, R. A. Mathies, T. R. Middendorf, D. S. Gottfried and S. G. Boxer (1992). *J. Phys. Chem.* in press.

[46] G. R. Loppnow, B. A. Barry and R. A. Mathies (1989). *Proc. Natl. Acad. Sci. USA* **86**, 1515-1518.

[47] R. A. Mathies, A. R. Oseroff and L. Stryer (1976). *Proc. Nat. Acad. Sci. USA* **73**, 1-5.

[48] R. A. Mathies, S. O. Smith and I. Palings (1987). In *Biological Applications of Raman Spectroscopy, Vol. 2-Resonance Raman Spectra of Polyenes and Aromatics*, T. G. Spiro, ed., pp. 59-107 (Wiley, New York).

[49] R. A. Mathies, C. H. Brito Cruz, W. T. Pollard and C. V. Shank (1988). *Science* **240**, 777-779.

[50] A. Myers, R. Mathies, D. J. Tannor and E. J. Heller (1982). *J. Chem. Phys.* **77**, 3857-3866.

[51] A. B. Myers, R. A. Harris, and R. A. Mathies (1983). *J. Chem. Phys.* **79**, 603-613.

[52] A. B. Myers (1984). (Dissertation, University of California at Berkeley).

[53] A. B. Myers and R. A. Mathies (1984). *J. Chem. Phys.* **81**, 1552-1558.

[54] A. B. Myers and R. A. Mathies (1987). In *Biological Applications of Raman Spectroscopy. Vol. 2. Resonance Raman Spectra of Polyenes and Aromatics*, T. G. Spiro, ed., pp. 1-58 (Wiley-Interscience, New York).

[55] M. A. Nardi (1980). *Color Res. Appl.* **5**, 73.

[56] M. C. Nuss, W. Zinth, W. Kaiser, E. Kolling, and D. Oesterhelt (1985). *Chem. Phys. Lett.* **117**, 1-7.

[57] A. R. Oseroff and R. H. Callender (1974). *Biochem.* **13**, 4243-4248.

[58] I. Palings, J. A. Pardoen, E. van den Berg, C. Winkel, J. Lugtenburg and R. A. Mathies (1987). *Biochemistry* **26**, 2544-2556.

[59] I. Palings, E. M. M. van den Berg, J. Lugtenburg and R. A. Mathies (1989). *Biochemistry* **28**, 1498-1507.

[60] K. Peters, M. L. Applebury and P. M. Rentzepis (1977). *Proc. Natl. Acad. Sci. USA* **74**, 3119-3123.

[61] J. W. Petrich, J. Breton, J. L. Martin and A. Antonetti (1987). *Chem. Phys. Lett.* **137**, 369-375.

[62] H. J. Polland, M. A. Franz, W. Zinth, W. Kaiser, E. Kolling and D. Oesterhelt (1986). *Biophys. J.* **49**, 651-662.

[63] K. J. Rothschild, W. A. Cantore, and H. Marrero (1983). *Science* **219**, 1333-1335.

[64] K. J. Rothschild and W. J. DeGrip (1986). *Photobiochem. Photobiophys.* **13**, 245-258.

[65] T. P. Sakmar, R. P. Franke and H. G. Khorana (1989). *Proc. Natl. Acad. Sci. USA* **86**, 8309-8313.

[66] J. Saltiel and J. T. D'Agostino (1972). *J. Am. Chem. Soc.* **94**, 6445.

[67] R. W. Schoenlein, L. A. Peteanu, R. A. Mathies and C. V. Shank (1991). *Science* **254**, 412-415.

[68] A. V. Sharkov, A. V. Pakulev, S. V. Chekalin and Y. A. Matveetz (1985). *Biochim. Biophys. Acta* **808**, 94-102.

[69] A. E. Siegman (1986). *Lasers* (University Science Books, Mill Valley CA).

[70] F. Siebert, W. Mantele and K. Gerwert (1983). *Eur. J. Biochem.* **136**, 119-127.

[71] A. W. Snyder and R. Menzel, eds. (1975). *Photoreceptor Optics* (Springer-Verlag, New York).

[72] W. K. Stell and F. I. Harosi (1976). *Vision Res.* **16**, 647-657.

[73] D. Stern and R. A. Mathies (1985). In *Time-resolved Vibrational Spectroscopy*, M. Stockburger and A. Laubereau, eds., pp. 250-254 (Springer-Verlag, New York).

[74] D. Tang, R. Jankowiak, G. J. Small and D. M. Tiede (1989). *Chem. Phys.* **131**, 99.

[75] O. Teschke, E. P. Ippen and G. R. Holtom (1977). *Chem. Phys. Lett.* **52**, 233.

[76] H. P. Thijssen, R. van den Berg and S. Volker (1983). *Chem. Phys. Lett.* **103**, 23.

[77] M. O. Trulson, G. D. Dollinger and R. A. Mathies (1989). *J. Chem. Phys.* **90**, 4274-4281.

[78] A. Warshel (1976). *Nature* **260**, 679-683.

[79] A. Warshel and N. Barboy (1982). *J. Am. Chem. Soc.* **104**, 1469-1476.

[80] E. B. Wilson, Jr., J. C. Decius and P. C. Cross (1955). *Molecular Vibrations* (Dover, New York).

[81] L. A. Yanuzzi, K. A. Gitter, and H. Schatz, eds. (1978). *The Macula* (Williams and Wilkins, Baltimore MD).

[82] K. Yoshihara, A. Namiki, M. Sumitani and N. Nakashima (1979). *J. Chem. Phys.* **71**, 2892.

[83] T. Yoshizawa (1972). In *Handbook of Sensory Physiology. Vol. VII/I*, H. J. A. Dartnall, ed., pp. 146-179 (Springer-Verlag, New York).

[84] E. A. Zhukovsky and D. D. Oprian (1989). *Science* **246**, 928-930.

[85] T. S. Zwier, E. Carrasquillo and D. H. Levy (1983). *J. Chem. Phys.* **78**, 5493.

Statistical Mechanics and Protein Folding

Hue Sun Chan, Ken A. Dill, and David Shortle

Introduction

The importance of statistical mechanical concepts in our understanding of protein folding has become increasingly apparent. Included here are two representative articles that best summarize the two main topics covered in two lectures given during the first Princeton Lectures on Biophysics in June 1991. The first article, by Chan & Dill and beginning on page 80, provides a survey of a number of statistical mechanical approaches relevant to our understanding of the general physical principles of protein folding. It contains considerable details of the mean-field heteropolymer collapse theory proposed by Dill [33] in 1985, which was discussed in the first lecture. The second lecture considered the general properties of the native structure mapping between a sequence space and its corresponding conformation space, based on exact enumerations using a two-dimensional lattice HP model [14, 82, 83]. This approach is illustrated by the application of this lattice model to the mutational effects on protein stability in the second article by Shortle, Chan & Dill and beginning on page 132.

The following is a very brief overview of the general theoretical framework common to both the mean-field and the lattice approaches. Because concise reviews of the basic biochemical properties of proteins are available in many textbooks, such general background material is omitted.

The HP model

In the simple model protein molecules considered here, monomer units are classified into only two types - H (hydrophobic) and P (polar). These models are based on the understanding that the dominant driving force that causes a protein molecule to fold is the hydrophobic effect, *i. e.*, the H-monomers' aversion of water; and the dominant driving force that causes the molecule to unfold is the entropic effect that favors open conformations, owing simply to the fact that there are more conformations that are open than conformations that are compact [34]. The HP model incorporated only these two driving forces. These models are constructed to capture the basic physics of protein molecules. The kind of coarse-grained modeling employed by HP models is very much similar to that of the Ising models used in the study of magnetism. Obviously, it is not the purpose of these simplified models to provide detailed predictions of microscopic behavior at atomic resolution.

The theoretical consequences of the basic assumptions of HP models are detailed in the Appendix. In particular, it follows that the statistical mechanical description of all equilibrium thermodynamic properties of HP models are completely determined by the distribution function $g(h)$ - the number of chain conformations with h HH contacts (an HH contact is a contact between two H-monomers), and the energy ϵ per HH contact. The contact energy ϵ accounts for the hydrophobic interactions and the density of state $g(h)$ accounts for the conformational freedom of the chain molecules. In all HP models, ϵ is taken as a given quantity that may depend on temperature, solvent condition, etc., whereas the main theoretical effort is directed towards the determination of the distribution $g(h)$. The mean-field and the lattice methods to be discussed below are two different approaches to model $g(h)$.

A mean-field heteropolymer collapse theory

A heteropolymer collapse model has been developed by Dill [33] to account for the balance of forces in protein folding. The model adopts a mean-field approach to account for both the chain conformational freedom and the HH interaction within the basic HP framework. The degrees of freedom in this model are the radius R of the chain conformation and the extent of ordering of the H and P monomers into surface and core regions of the collapsed maximally compact state. In this model, the radius R is given as a function of the monomer density ρ, i. e. $R = R(\rho)$. A discussion of the salient features of this model can be found below. The brief summary of the model below follows closely the original formalism of Dill [33] and Dill, Alonso & Hutchinson [36]. However, additional effort is made to relate the key equations in this mean-field heteropolymer model to the quantity $g(h)$, because the latter is of central importance in any HP formulation.

The mean-field heteropolymer model uses the Flory-Huggins lattice treatment to account for the excluded volume effect and the Flory-Fisk [56] approximation to account for chain connectivity and the resulting elastic entropy. In the model, an n-monomer chain is confined to a spherical volume of radius R, the fraction of the total volume of the R-sphere that is occupied by monomers is the density ρ, hence $0 \leq \rho \leq 1$ [see definition of ρ immediately after Eq. (3.31)]. If the volume of a monomer unit is normalized to unity, it is easy to see that

$$R(\rho) = \left(\frac{3n}{4\pi\rho}\right)^{1/3} . \tag{3.1}$$

The model assumes that the two monomer types - H's and P's - mix randomly when the chain is less than maximally compact (*i. e.* when the monomer density is less than unity, $\rho < 1$). If the maximum possible number of contacts each monomer in the interior of the R-sphere is able to make with other monomers is z, (z is known as the lattice coordination number in Flory-Huggins theories), and the maximum possible number of contacts each monomer on the surface of the R-sphere is able to make with other monomers is σz (the surface correction factor σ is taken to be 2/3 in many applications), then, based on the assumption of random mixing, the number h of HH contacts for a molecule with density ρ is given by

$$h(\rho) = \frac{z\Phi^2\rho}{2}\left\{f_i[R(\rho)] + \sigma f_e[R(\rho)]\right\}, \qquad \rho < 1. \tag{3.2}$$

Here the H-composition Φ ($0 \leq \Phi \leq 1$) is the ratio of the number of H-monomers to the total number of monomers in the chain, and the factor $1/2$ accounts for the fact that the contacts are made between identical monomers. The quantities f_i and f_e are the fractions of volume, respectively, that are in the interior and the surface, of the R-sphere. Clearly, the geometric factors f_i and f_e are functions of R,

$$f_i(R) = \left(\frac{R-1}{R}\right)^3, \qquad f_e(R) = 1 - f_i(R), \tag{3.3}$$

and therefore they are functions of ρ by Eq. (3.1).

The number of conformations g is a function of ρ in the mean-field heteropolymer model. For $\rho < 1$, the logarithm of the number of conformations is given by

$$\ln g[\rho(h)] = -n\left[\left(\frac{1-\rho}{\rho}\right)\ln(1-\rho) + 1\right] - \frac{7}{2}\left(\frac{\rho_0}{\rho}\right)^{2/3} - 2\ln\rho$$
$$+\text{constant}. \tag{3.4}$$

Here the density of state g is a function of monomer density ρ, but ρ is an implicit function of h by Eq. (3.2). Hence the combination of Eq's (3.2) &

(3.4) specifies g as a function of h. In the above expression for g, the first term follows from the Flory-Huggins treatment of the excluded volume effect [Eq. (3.33)], whereas the remaining terms for the elastic entropy follow from the normalized Flory-Fisk [56] distribution

$$P(R_G) = \left(\frac{343}{15}\right)\left(\frac{14}{\pi\langle R_G^2\rangle_0}\right)^{1/2}\left(\frac{R_G^2}{\langle R_G^2\rangle_0}\right)^3 \exp\left(-\frac{7R_G^2}{2\langle R_G^2\rangle_0}\right),$$

$$1 = \int_0^\infty dR_G P(R_G), \tag{3.5}$$

of the radius of gyration R_G of random-flight chains in three dimensions. The simple scaling relation

$$\frac{R_G^2}{\langle R_G^2\rangle_0} = \left(\frac{\rho_0}{\rho}\right)^{2/3}, \tag{3.6}$$

where $\langle R_G^2\rangle_0 = n/6$ is the mean square radius of gyration of random-flight chains, is adopted to relate R_G to the monomer density ρ. It can easily be shown that $\rho_0 \sim O(1/\sqrt{n})$.

As an aisde, we note that if one incorporates the relation between the differential elements dR_G and $d\rho$ that follows from Eq. (3.6), viz. $dR_G = [-\sqrt{\langle R_G^2\rangle_0}/(3\rho_0)](\rho_0/\rho)^{4/3}d\rho$, the term $-2\ln\rho$ in Eq. (3.4) would be modified to $-(10/3)\ln\rho$. Simpler forms for the elastic entropy such as a simple Gaussian distribution for R_G have also been used in other applications of Flory-Huggins theories, as described below. Insofar as the elastic entropy term is capable of confining chain conformations to dimensions dictated by polymer conformational statistics, the prediction of the theory is relatively independent of its exact functional form because the excluded volume term [first term in Eq. (3.4)] plays a much more predominant role in determining the conformational entropy of the chain at intermediate to high monomer densities. Similarly, if one utilizes the fact that R_G is minimum at maximum monomer density, i. e. $\rho = 1$, and this minimum R_G is $\sqrt{3/5}$ that of the minimum R at $\rho = 1$ [Eq. (3.1)], this would imply that ρ_0 in Eq. (3.6) may be taken to be $\rho_0 = 27\sqrt{18}/(10\pi\sqrt{5n})$. In some applications $\rho_0 = \sqrt{19/(27n)}$ has also been used [33, 36]. However, the exact value of ρ_0 is quite inconsequential to the prediction of the theory, because the free energy of folding is nearly independent of ρ_0 over a 30-fold range [33].

In the mean-field heteropolymer model, an extra degree of freedom is available when the chain is maximally compact, *i. e.* when $\rho = 1$. This extra degree of freedom allows the segregation of the two monomer types - H's and P's - into two different regions within the R-sphere: the interior (the core) and the surface. The volume fractions of these two regions are, respectively, $f_i^{(c)} \equiv f_i[R(1)]$ and $f_e^{(c)} \equiv f_e[R(1)]$, which are the $\rho = 1$ values of the geometric factors f_i and f_e defined in Eq. (3.3). The superscript C here and in subsequent discussions labels properties of the maximally compact state.

If x and θ are respectively the fraction of core and surface volume of the $\rho = 1$ R-sphere that is occupied by H-monomers, they must satisfy the constraint

$$x f_i^{(c)} + \theta f_e^{(c)} = \Phi \tag{3.7}$$

that the total number of H-monomers is $n\Phi$. For any specific x and θ, random mixing of monomers in the $\rho = 1$ state results in

$$h^{(c)}(x, \theta) = \frac{z}{2}\left(f_i^{(c)} x^2 + \sigma f_e^{(c)} \theta^2 \right) \tag{3.8}$$

HH contacts. The number $g^{(c)}(x, \theta)$ of $\rho = 1$ chain conformations consistent with specific values of the ordering parameters x and θ is estimated by applying simple binomial combinatorics in the context of the mean-field approximation, leading to the result [33, 36]

$$\ln g^{(c)}(x, \theta) = \ln g(1) \quad - \quad n f_i^{(c)}\left[x \ln \frac{x}{\Phi} + (1-x) \ln \frac{1-x}{1-\Phi} \right]$$
$$- \quad n f_e^{(c)}\left[\theta \ln \frac{\theta}{\Phi} + (1-\theta) \ln \frac{1-\theta}{1-\Phi} \right], \tag{3.9}$$

in which $g(1)$ is the $\rho = 1$ value of g in Eq. (3.4). Hence the relative free energy $G^{(c)}(x, \theta)$ of $\rho = 1$ chain conformations with segregation of H's and P's specified by x and θ is given by

$$G^{(c)}(x, \theta) \equiv \epsilon h^{(c)}(x, \theta) - kT \ln g^{(c)}(x, \theta) , \qquad (3.10)$$

where ϵ is the contact energy per HH contact. [Note that the quantity ϵ used here is equivalent to the quantity $-g$ in Ref. [33] (not to be confused with the number of conformations g used here), and that ϵ is related to the quantity χ used in Ref. [36] by $\epsilon = -2kT\chi/z$.] Thus, the *most probable* value x^* and θ^* of the order parameters x and θ is determined by the minimization of $G^{(c)}$ subject to the constraint Eq. (3.7). For instance, x^* and θ^* may be calculated by first using Eq. (3.7) to express θ as a function of x, then requiring

$$\frac{\partial}{\partial x} G^{(c)}[x, \theta(x)]\bigg|_{\substack{x=x^* \\ \theta=\theta^*}} = 0 . \qquad (3.11)$$

In the mean-field heteropolymer model, the ensemble of $\rho = 1$ conformations with the segregation of H's and P's specified by x^* and θ^* is identified as the native state of the model protein. Therefore, in this model,

$$h_N \equiv h^{(c)}(x^*, \theta^*) \quad \text{and} \quad g(h_N) \equiv g^{(c)}(x^*, \theta^*) \qquad (3.12)$$

are respectively the maximum number h_N of HH contacts and the degeneracy of the $\rho = 1$ native state of the model protein.

When the chain is sufficiently long (n large), the distribution of monomer density ρ for the $\rho < 1$ conformations of the unfolded (denatured) state would have a narrow peak around a maximum ρ^*. This peak value ρ^* is determined by the minimization of the free energy of the $\rho < 1$

conformations,

$$\frac{\partial}{\partial \rho}\left\{\epsilon h(\rho) - kT \ln g[\rho(h)]\right\}\bigg|_{\rho=\rho^*} = 0 \ . \tag{3.13}$$

In that case, properties of the unfolded state that require ensemble averaging over all $\rho < 1$ conformations can be approximated by properties of those chain conformations at $\rho = \rho^*$.

Predictions of protein properties by this mean-field heteropolymer model and the extension of the model to account for electrostatic effects are discussed in below.

An exact two-dimensional lattice model

Another approach to model $g(h)$ of an HP sequence is by exact enumeration of all possible conformations on a lattice. Details of this model can be found elsewhere [14, 82, 83] and the second section below. Despite the two-dimensional lattice model's clear limitations, such as the shortness of the chains and that the number of spatial dimensions in which model protein chains configure is two instead of three, HP sequences in the model exhibit many protein-like properties. We believe that the model is successful in capturing the basic physics of the balance of forces in protein folding.

The lattice approach is valuable as a model by which we can study the exact relationship between a sequence and its density of state $g(h)$. Existing analytical heteropolymer theories do not consider all of the information encoded in a sequence. For example, the mean-field heteropolymer model discussed above considers only the H-composition. In some of the spin-glass models for proteins the distribution of pairwise interaction energies is taken into account. Aside from the few parameters considered, all of these analytical heteropolymer theories assume the sequences to be random in various respects. Therefore it is important to have a means to determine the limitations of these approaches owing to their neglect of part of the sequence information, and to ascertain the conditions for these approximations to be valid.

As an illustration of the differences between approximate analtyical models and exact approaches, Fig. 3.1 shows the the variation of degeneracy of the lowest-energy state of all HP sequences with $n = 18$ monomers as a function of the heterogeneity parameter B used in a number of spin-glass models of proteins [see Eq's (3.67,3.68)]. The lowest-energy (native) state degeneracy of an HP sequence is the number of conformations with the maximum number h_N of HH contacts. The parameter B of a sequence is the standard deviation of the distribution of contact interaction energies between pairs of monomers in that sequence. Therefore, in the HP lattice model, the parameter B in units of the magnitude of the contact energy per HH contact, $|\epsilon|$, is given by

$$ B \equiv \sqrt{\frac{1}{\mathcal{M}(n)} \sum_{\substack{j-i \geq 3 \\ j-i=\text{odd}}} (\Delta_{ij}^{(H)} - B_0)^2 } , \qquad (3.14) $$

where

$$ B_0 \equiv \frac{1}{\mathcal{M}(n)} \sum_{\substack{j-i \geq 3 \\ j-i=\text{odd}}} \Delta_{ij}^{(H)} \qquad (3.15) $$

is the mean interaction energies between monomer pairs in units of ϵ. In the above equations, $\mathcal{M}(n)$ is the total number of possible nearest-neighbor contacts between monomers of a chain of length n configured on a square lattice. It is straightforward to show that $\mathcal{M}(n) = (n-2)^2/4$ for even n, and $\mathcal{M}(n) = (n-1)(n-3)/4$ for odd n. The conditions on the summations over i and j, viz. $j - i \geq 3$ and $j - i = \text{odd}$, restrict the sums to nearest-neighbor contacts that are *realizable* on a square lattice. In other

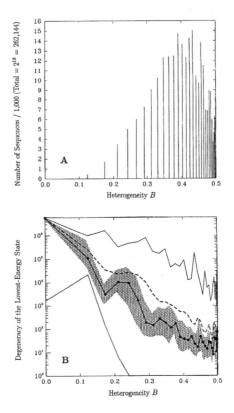

Figure 3.1: Heterogeneity and degeneracy of all possible HP sequences with n=18 monomers configured on the two-dimensional square lattice. (A) Distribution of the heterogeneity parameter B in units of contact energy per HH contact [Eq. (3.14)]. In the lattice HP model B takes discrete values. Each bar of the histogram shows the number of HP sequences with a specific B. (B) Degeneracy of the lowest-energy state as a function of B. For the collection of HP sequences with a specific value of B, the dot connected by solid lines shows the median degeneracy; the shaded area indicates the extent of variation in degeneracy among one half of the sequences with the given B and whose degeneracies are closest to that of the median; the dashed curve shows the average degeneracy; the top solid curve shows the maximum and the bottom curve shows the minimum degeneracy for sequences as functions of B.

words, the summations in Eq's (3.14) & (3.15) are over all monomer pairs (i, j)'s that can form a contact in *some* chain conformations. In the lattice HP model, the quantity $\Delta_{ij}^{(H)}$, which is related to the monomer-monomer interaction energy B_{ij} by $B_{ij} = \epsilon \Delta_{ij}^{(H)}$, is unity when both monomer i and j are H-monomers $(i, j = 1, 2, \ldots, n)$, and zero otherwise.

The comparison in Figure 3.1 underscores both the success and limitation of existing analytical approximations to the protein-folding problem. The figure shows that the parameter B of a sequence has a certain degree of predictive power with regard to the sequence's native degeneracy. The general trend of decrease in median and average degeneracy as B increases is consistent with the predictions of spin-glass models, as described below. However, the considerable variation in degeneracy among sequences with identical B clearly demonstrates that the spin-glass heterogeneity parameter B alone is not sufficient to capture all the subtlety of the information encoded in the HP sequence. Indeed, none of the 6,349 singly-degenerate $n = 18$ HP sequences has the highest value of B, and, among those sequences with the highest B, the lowest degeneracy is equal to 3 but not 1.

Outlook

HP models are a very useful tools for investigating the extent to which protein properties can be accounted for by conformational freedom and hydrophobic interactions alone. In addition to the study of equilibrium protein properties described above, recently we have also used the exact HP lattice approach to study the dynamics of polymer collapse and the folding of model proteins. Owing to the additional variables required in these applications, the density of state $g(h)$ alone is insufficient to determine the chain dynamics. Hence, to account for non-equilibrium phenomena in HP models, a distance or adjacency measure has to be defined on the conformation space [H.S. Chan & K. A. Dill, 1992 in preparation].

Polymer Principles in Protein Structure and Stability

This section is a slightly revised and updated version of a paper by Chan and Dill which appeared in *Ann. Rev. Biophys. Biophys. Chem.* **20**, 447-490 (1991). It is reprinted here with permission of Annual Reviews, Inc..

Proteins are polymers. This review is intended as an introduction to methods and principles of polymer statistical mechanics, old and new, that bear on the large conformational changes and internal organization in protein molecules. We focus on the relative roles played by local vs nonlocal forces. *Local* defines interactions among near neighbors in the chain sequence; these have often been referred to as short-ranged, or secondary forces (see Fig. 3.2). *Nonlocal* describes interactions among monomers distant in the chain sequence, often referred to as long-ranged or tertiary forces.

First we discuss the concept of separability, the idea that interactions in polymers can be separated into two types of force, local and nonlocal. Then we describe local forces. Under conditions in which nonlocal forces are small (the Flory "theta" state), chain conformations are determined by local forces. In these cases, conformational changes depend on an elastic entropy. Effects of monomer sequence are well-modeled using the rotational isomeric state (RIS) model, implemented in one-dimensional Ising model matrix methods. This approach is the basis of helix-coil transition theories, widely applied to polypeptides. Next we describe nonlocal forces. Monomers that are far apart in the sequence interact through their mutual excluded volume and through solvent-mediated interactions. We review the Flory lattice model of nonlocal interactions, then computer lattice simulation results, and path integral methods. We then discuss solvent-driven compactization of homopolymers, heteropolymers, and proteins. Finally, we consider how the compactness of a chain molecule can lead to internal structure.

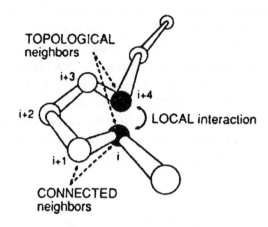

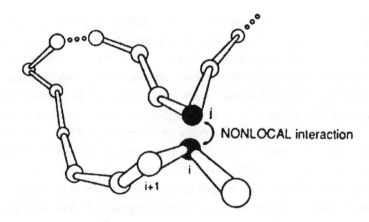

Figure 3.2: Spatially neighboring monomers (i, j) are defined as connected neighbors if they share a backbone bond, $j = i + 1$; otherwise, they are topological neighbors. Interactions are local or nonlocal depending on the separation along the chain of the interacting monomers.

Among the several conclusions summarized here are two recent and unexpected results. First, whereas helices and sheets in globular proteins have been generally assumed to arise from local forces, evidence now indicates that excluded volume instead may be significantly responsible for protein structure. Second, whereas homopolymers in poor solvents will collapse to a relatively large ensemble of compact conformations, heteropolymers are predicted to collapse to an extremely small ensemble: one or two or a few conformations. Thus, the ways in which proteins apparently differ from simpler polymers-(a) that they have unique native structures dictated by the monomer sequence and (b) that native conformations are comprised of helices, sheets, and irregular conformations-appear to arise as natural consequences of heteropolymer collapse processes.

The configurational partition function

In this review, we focus on how the configurations of polymers, particularly proteins, derive from two types of interaction: those among monomers that are near neighbors in the sequence, and those among monomers that are far apart in the sequence. First, we describe the rationale for using the terms "local" and "nonlocal" to describe these two types of interaction. For some years [53], standard usage has been to refer to interactions in chain molecules as "short-ranged" or "long-ranged," depending on whether the interacting monomers are near neighbors in the chain sequence or not. However, this conflicts with terminology, considerably older, wherein the "range" of interaction refers to its spatial distance dependence: ion-ion interactions are long-ranged, and van der Waals interactions are short-ranged, for example. Ambiguity arises because the context may not clarify whether "range" refers to a particular distance dependence of a given type of energy or to the separation of the monomers along the chain sequence. Two monomers of a chain molecule that are far apart in the sequence may interact via either van der Waals or electrostatic interactions; conversely, two monomers that are close in sequence may also interact via either van der Waals or electrostatic interactions. The potential confusion is greatest for proteins, for which the significant interactions can involve both types of distance dependence (van der Waals and electrostatic) and both types of chain separation (together and apart in the sequence). To avoid these ambiguities, we use long-ranged and short-ranged to refer to the distance dependence of the type of energy because that usage is older and more widespread, and we use local and nonlocal to refer to interactions among monomers close or distant in the sequence of a chain molecule. We do not favor "secondary" and "tertiary" in reference to interactions because these terms are vague and refer neither to the range of interaction nor to the positions of interacting monomers in the chain: for example secondary interaction can refer to either helix (local) or sheet (nonlocal) contacts.

The conformational properties of a polymer molecule are described by its partition function. Consider a polymer chain of n bonds ($n + 1$ monomers) in one specific configuration in d-dimensional space, such that monomer 0 is at position r_0, monomer 1 is at position r_1, and monomer n is at position r_n. The conformation of the molecule is completely specified by the set of vectors $\{r_0, r_1, r_2, \ldots, r_n\}$. Alternately, the spatial vectors of the monomers may be labeled by a continuous contour measure τ from one end of the chain, such that $r(\tau_i) \equiv r_i$, where τ_i is the position of monomer i along the contour of the chain; see Fig. 3.3. The latter notation is more general; it is applicable to chains with either discrete or continuous segments (see the path integral section below). The conformation is specified by the values of $\{r(\tau_0), r(\tau_1), r(\tau_2), \ldots, r(\tau_n)\}$.

The partition function, Q, is the total number of conformations accessible to the chain,

$$Q = \mathcal{N} \int dr(\tau_0) \ldots \int dr(\tau_n) \, P[r(\tau_0), r(\tau_1), \ldots, r(\tau_n)] , \qquad (3.16)$$

where $P(\{r(\tau)\}) \equiv P[r(\tau_0), r(\tau_1), \ldots, r(\tau_n)]$ is the relative probability density of a conformation in the $dr(\tau_0)dr(\tau_1) \ldots dr(\tau_n)$ neighborhood specified by $\{r(\tau_0), r(\tau_1), \ldots, r(\tau_n)\}$. The quantity $\mathcal{N}$ is an overall normalization, and $\{r(\tau)\}$ is shorthand for the set of $n + 1$ spatial vector variables $r(\tau_i)$.

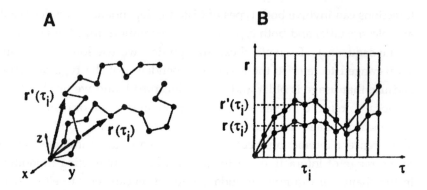

Figure 3.3: Chain conformations (A) as paths in the r-τ space (B). Summation over conformations is equivalent to summation over r-τ paths with appropriate statistical weights.

The quantitative value of the normalization $\mathcal{N}$ is unimportant because it cancels out in most practical applications. The summation over conformations is carried out by integrating P over all possible positions $r(\tau_i)$ for each chain segment $0 \leq i \leq n$. The range of each $\int dr(\tau_i)$ integration is the entire d-dimensional space accessible to the chain.

The probability density $P(\{r(\tau)\})$ is given by the Boltzmann distribution

$$P(\{r(\tau)\}) = \exp\left[-W(\{r(\tau)\})\right] . \tag{3.17}$$

If the chain molecule is immersed in a solvent, $W(\{r(\tau)\})$ is a potential of mean force [60, 70, 74] in units of kT. That is, if the degrees of freedom of the solvent are represented as a vector ξ, then $W(\{r(\tau)\})$ is the reversible work required to put the polymer into configuration $\{r(\tau)\}$ averaged over the ensemble of all the solvent degrees of freedom ξ, $i.\ e.$ the force acting on the ith monomer is

$$-kT\nabla_{r(\tau_i)}W(\{r(\tau)\})$$

$$\equiv \frac{-\int d\xi \, \nabla_{r(\tau_i)} V(\{r(\tau)\},\xi) \, \exp\left[-V(\{r(\tau)\},\xi)/kT\right]}{\int d\xi \, \exp\left[-V(\{r(\tau)\},\xi)/kT\right]}, \qquad (3.18)$$

where $V(\{r(\tau)\},\xi)$ is the potential energy for the chain and the solvent configuration specified by $\{r(\tau)\}$ and ξ.

It is useful to separate the total potential of mean force into a sum of two terms, one owing to local and the other to nonlocal interactions. Local interactions are defined as those that occur among monomers that are separated by up to p units; nonlocal interactions involve monomers separated in the sequence by more than p units,

$$W(\{r(\tau)\}) = W_\ell(\{r(\tau)\}) + W_{n\ell}(\{r(\tau)\})$$

$$W_\ell(\{r(\tau)\}) = \sum_{j=0}^{n-p} w_j[r(\tau_j), r(\tau_{j+1}), \ldots, r(\tau_{j+p})], \qquad (3.19)$$

where w_js account for the local potential of mean force. $W_{n\ell}$ is the nonlocal potential of mean force that accounts for all other interactions not expressible in the functional form of W_ℓ. This separation of terms is useful because two qualitatively different types of behavior are observed in polymers depending on whether local interactions are dominant or not (see below). Since by definition W_ℓ only contains local interactions, p is a small number of order 1. Typically p is chosen to be in the range of 1 to 4. For $p = 1$,

$$W(\{r(\tau)\}) = \sum_{j=0}^{n-1} w_j[r(\tau_j), r(\tau_{j+1})] + W_{n\ell}(\{r(\tau)\}). \qquad (3.20)$$

In this case, the local potential of mean force, W_ℓ, is defined as interactions among connected neighbors, whereas the nonlocal potential of mean force, $W_{n\ell}$, is among topological neighbors [10-12] (see Fig. 3.2); the latter involves excluded volume and solvent interactions (see below).

Local interactions

Under special solvent conditions, referred to as the Flory theta state [53, 54] (see below), the nonlocal interactions are negligible, $W_{n\ell} \simeq 0$, and the configurational states of polymers are governed primarily by local interactions. Local interactions are important for protein structure and stability. They account for stabilities of helices in solution and they contribute an elastic entropy to the folding of globular proteins. In this section, we review the following principles for conformations of chains driven by local interactions: 1. In the most elementary model of local interactions, for which properties of bonds adjacent along the chain are uncorrelated, the chain conformations may be suitably modeled as random flights, the configurational distribution function of the end-to-end distance is Gaussian, which leads to an elastic contribution to the conformational entropy. 2. Even if the bond directions among neighboring links are correlated, provided only that the correlations are local, then the statistics remain those of a random flight. Only an effective bond length is changed; the dependence of end-to-end distance on square root of the chain length remains unaffected, except for a constant. 3. More accurate than the random-flight model is the rotational isomeric state model for capturing the chemical bond details, the effects of monomer sequence, and the finite lengths of short chains.

The simplest classical model for long chains subject only to local interactions assumes that bonds have fixed length and that adjacent bond angles are uncorrelated. This model can be expressed equivalently in terms of a Gaussian bond distribution function [60, 162],

$$
\begin{aligned}
&\exp\left\{-w_j\left[r(\tau_j), r(\tau_{j+1})\right]\right\} \\
&= \left(\frac{d}{2\pi l^2}\right)^{d/2} \exp\left[-\frac{d}{2l^2}|r(\tau_j) - r(\tau_{j+1})|^2\right],
\end{aligned}
\tag{3.21}
$$

where l is the bond length. The distribution of end-to-end lengths $R_n \equiv |r(\tau_n) - r(\tau_0)|$ is obtained by integrating Eq. (3.21) over all spatial coordinates $r(\tau_1), \ldots, r(\tau_{n-1})$ except the ends; the result is a Gaussian distribution,

$$P(R_n) = \left[\frac{d}{2\pi R_0^2}\right]^{d/2} \exp\left[-\frac{d}{2}\left(\frac{R_n}{R_0}\right)^2\right], \qquad (3.22)$$

where $R_0 = \sqrt{\langle R_n^2 \rangle} = ln^{1/2}$ is the root-mean-square, end-to-end length of the random-flight chain.

A random-flight molecule subject to small deformation has a retractive *elastic* entropy, so-called because of its importance in the theory of rubber elasticity [54]. For a uniform isotropic expansion or contraction of the chain by a factor α to a root-mean-square, end-to-end distance $R = \alpha R_0$, the change in (entropic) free energy ΔF_{el} owing to elastic deformation is deduced from simple scaling [54] of the distribution Eq. (3.21):

$$\frac{\Delta F_{el}}{kT} = \frac{d}{2}(\alpha^2 - 1) - d\ln\alpha. \qquad (3.23)$$

To establish from experiments whether local or nonlocal interactions are dominant, it is necessary to measure $\langle R_n^2 \rangle$ vs n. If nonlocal interactions are small, the mean square end-to-end distance $\langle R_n^2 \rangle$ takes the general form

$$\langle R_n^2 \rangle = Cnl^2, \qquad (3.24)$$

where $C = 1$ for chains that obey the random-flight statistics [Eq's (3.21) and (3.22)]. A most remarkable feature of chain conformations is that this dependence, $\langle R_n^2 \rangle \sim n$ for large n, is unaffected even if nearest neighbor bonds are correlated, and also if second neighbor bonds are correlated, *etc.*, provided that correlations extend out to only some finite number of neighbors. In those cases, the effect of local bond correlations is only to change the constant C but not the scaling behavior of $\langle R_n^2 \rangle$ with n [54, 60, 162]. Thus the distinction between local and nonlocal is not sharply defined in terms of some particular number of neighbors distant in the sequence over which interactions occur (*i. e.* the exact choice of p in Eq. (3.19) is not critical as long as $p \ll n$). Rather, the critical evidence that indicates when chain behavior is dominated by local interactions is experimental observation that the mean square end-to-end length scales with the first power of the chain length. The scaling behavior is determined by the nature of the solvent and the temperature (see below).

A more accurate molecular treatment of polymer conformations that are subject only to local interactions W_ℓ is given by the rotational isomeric state theory [6, 55, 153]. In this model, the continuum of conformational states of sequential bonds is approximated as a set of discrete rotational isomers. Now a pair of sequential bonds need no longer be represented as uncorrelated; local details of bond chemistry can be more suitably represented. The main advantages of this approach are: (*a*) the constant C and other moments of a chain configurational distribution function can be calculated from molecular details of the chemical bonds and are found to be generally in good agreement with experiments, (*b*) short chains of any length can be treated, and (*c*) the effects of the monomer sequence can be treated explicitly.

The degrees of freedom considered by rotational isomeric state models are the bond rotation angles ϕ_i between pairs of planes defined by sequential pairs of bonds, $\phi_i \equiv \cos^{-1} [b_{i-2} \times b_{i-1}) \cdot (b_{i-1} \times b_i)]/[|b_{i-2} \times b_{i-1}| \, |b_{i-1} \times b_i|]$, $2 \leq i \leq n$, where $0 \leq \phi_i \leq \pi$ when $b_{i-2} \cdot (b_{i-1} \times b_i) \geq 0$, $\pi \leq \phi_i \leq 2\pi$ when $b_{i-2} \cdot (b_{i-1} \times b_i) \leq 0$; and $b_i \equiv r(\tau_{i+1}) - r(\tau_i)$ is the bond vector from monomer i to monomer $i + 1$. Bond lengths $|b_i|$ and bond angles $\theta_i \equiv \cos^{-1} (b_{i-1} \cdot b_i/|b_{i-1}||b_i|)$ are fixed. The model makes the simplifying assumption that ϕ_i may adopt any one of a finite set of discrete values, $\phi_i = \phi_i^{(k_i)}$, $k_i = 1, 2, \ldots, s$, hence the integrations over $r(\tau_i)$s in the expression [Eq. (3.16)] for the partition function Q can be approximated by summations over discrete values of ϕ_is. Positions of four sequential monomers are required to determine a single bond rotation angle; therefore when pairs of sequential bond rotation angles ϕ_i and ϕ_{i+1} are correlated, the local interaction terms w_j in Eq. (3.19) have $p = 4$,

$$
\begin{aligned}
W_\ell(\{r(\tau)\}) &= \sum_{j=0}^{n-4} w_j[r(\tau_j), r(\tau_{j+1}), \ldots, r(\tau_{j+4})] \\
&= w_2^{(1)}(\phi_2) + \sum_{j=2}^{n-1} w_j^{(2)}(\phi_j, \phi_{j+1}) .
\end{aligned}
\tag{3.25}
$$

Here $w_2^{(1)}(\phi_2)$ is the potential of mean force for a single rotation angle ϕ_2, and $w_j^{(2)}(\phi_j, \phi_{j+1})$ is the potential of mean force for a pair of sequential rotation angles. The term $w_2^{(1)}$ for ϕ_2 is necessary in addition to the $w_j^{(2)}$ terms because ϕ_2 has no predecessor; the rotation angle of the first internal bond, ϕ_1, is undefined. In this representation, the partition function is given by the summation of the local interaction Boltzmann factor $\exp(-W_\ell)$ over all possible combinations of bond rotation angles,

$$
\begin{aligned}
Q = \sum_{k_2=1}^{s} \cdots \sum_{k_n=1}^{s} &\left\{ \exp[-w_2^{(1)}(\phi_2^{(k_2)})] \right. \\
\times \prod_{j=2}^{n-1} &\left. \exp[-w_j^{(2)}(\phi_j^{(k_j)}, \phi_{j+1}^{(k_{j+1})})] \right\} .
\end{aligned}
\tag{3.26}
$$

Using one-dimensional Ising model [77] matrix methods [55, 81], Eq. (3.26) can be expressed more conveniently as

$$
\begin{aligned}
Q &= \begin{pmatrix} 1 & 0 & 0 & \cdots & 0 \end{pmatrix} \left[\prod_{j=1}^{n-1} U_j \right] \begin{pmatrix} 1 \\ 1 \\ 1 \\ \vdots \\ 1 \end{pmatrix} \\
&= \begin{pmatrix} 1 & 0 & 0 & \cdots & 0 \end{pmatrix} U^{n-2} \begin{pmatrix} 1 \\ 1 \\ 1 \\ \vdots \\ 1 \end{pmatrix} .
\end{aligned}
\tag{3.27}
$$

In this expression, both the row and column matrices consist of s elements, and U_j is an $s \times s$ matrix of statistical weights, representing relative populations of all possible rotational isomers for monomer j in the sequence. The matrix elements of U_j are given by $(U_2)_{1,k} = \exp[-w_2^{(1)}(\phi_2^{(k)})]$,

$(U_2)_{k,k'} = 0$ for $k > 1$, and $(U_j)_{k,k'} = \exp[-w_j^{(2)}(\phi_j^{(k)}, \phi_{j+1}^{(k')})]$ for $j > 2$. The last equality in Eq. (3.27) holds if all monomer units are identical (*i. e.*, in a homopolymer) and the numbering of the rotational isomeric states is such that $w_2^{(1)}(\phi_2^{(k)}) = w_j^{(2)}(\phi_j^{(1)}, \phi_{j+1}^{(k)})$. For a symmetric chain with three rotational isomers, such as *n*-alkanes, U takes the form

$$
U = \begin{array}{c} \\ t \\ g^+ \\ g^- \end{array}
\begin{array}{c} \begin{array}{ccc} t & g^+ & g^- \end{array} \\
\left(\begin{array}{ccc} 1 & \sigma & \sigma \\ 1 & \sigma\psi & \sigma\omega \\ 1 & \sigma\omega & \sigma\psi \end{array} \right) \end{array},
\tag{3.28}
$$

where t denotes the *trans* state and $g^\pm$ denotes the two *gauche* states; σ, $\sigma\psi$ and $\sigma\omega$ are statistical weights of combinations of these rotational states, the details of which can be found in [55].

The methods described above provide the basis for the helix-coil transition theories widely applied to polypeptides in solution [55, 87, 117, 119, 164]. For example, in the Zimm-Bragg model [164], the two possible conformations of a bond pair are taken to be helix (h) or coil (c). Then U becomes

$$
U = \begin{array}{c} \\ c \\ h \end{array}
\begin{array}{c} \begin{array}{cc} c & h \end{array} \\
\left(\begin{array}{cc} 1 & \sigma s \\ 1 & s \end{array} \right) \end{array},
\tag{3.29}
$$

where σ is a helix initiation equilibrium constant, representing the bond-pair conformation . . .ch. . ., and s is a helix-coil equilibrium constant for an h unit preceded by either h or c. Thus, the helix-coil partition function is given by substituting the matrix in Eq. (3.29) into Eq. (3.27).

From the partition function, various configurational averages and their temperature dependences are obtained using standard methods [55, 164]. If the helical configuration of a bond pair is energetically favored (for example by hydrogen bonding between monomers i and $i + 3$, as in the α-helix), and if the coil configuration of the bond pair is entropically favored by having more accessible states than the helix, then the helix will be stable at low temperatures and will undergo a transition to the coil state at high temperatures. The transition between helix and coil will be sharp if the chain is long. This type of theory has been highly successful in predicting solvent and temperature dependences of helix stabilities of polymers and peptides [119].

Thus the helix-coil behavior of a chain is governed by factors local in the sequence because the chain partition function is a product of the elementary partition functions of bonded neighbors taken pairwise, and because the use of this product of bond-pair partition functions is based on the assumption that the configuration of any bond pair is independent of any other.

Nonlocal interactions

Flory [53, 54] was the first to appreciate the importance of nonlocal interactions in governing polymer conformations. He made a simple argument to show that a homopolymer molecule should be more expanded in an inert solvent (e.g. a solvent of chain monomers) than would be predicted by the random-flight model. Consider a random-flight conformation. Any two particular monomers i and j distant in the sequence are not likely to be close in space. But a long chain contains many monomer pairs. On balance, a random flight is likely to have at least one steric violation, whereby some pair of monomers would occupy the same volume element in space. Because this is physically impossible for a real chain, the random-flight model errs and requires correction. Such physical violations will be more common in compact conformations. Therefore steric exclusions will eliminate more of the compact conformations than open ones, leading to a net expansion of the average molecular radius relative to the random flight prediction. The distribution function, P in Eq. (3.22) above, accounts only for local interactions of connected neighboring monomers along the chain and neglects these consequences of nonlocal interactions among monomers distant in the sequence.

The first and simplest calculation for the magnitude of this effect was given by Flory based on the Flory-Huggins lattice model [52-54, 75, 76]. If the local and nonlocal interactions are independent, *i. e.*, have additive free energies, then the effects of excluded volume can be described by a factor, ω_{ex}, that multiplies the random-flight distribution to give the number of accessible conformations of the chain. The required excluded-volume factor is defined as the ratio of the number of conformations, Ω, that are accessible when one considers excluded volume, divided by the number of conformations, Ω_0, accessible when excluded volume is ignored, *i. e.*, $\omega_{ex} = \Omega/\Omega_0$. Assuming that the monomers can be distributed uniformly throughout the available (d-dimensional) volume R^d, this ratio is calculated by parceling the volume occupied by the chain into $m = R^d/v$ sites, each having a volume v equal to that of one of the n chain segments. The excluded volume contribution to the distribution function can be calculated through use of a fictitious process whereby each segment is considered to be inserted, one at a time, into a lattice of sites. The first chain segment can be inserted into any one of m accessible sites. The second chain segment can only be inserted into any of the remaining $m - 1$ sites. The third segment can be inserted into any of the remaining $m - 2$ sites, and so on, until the final insertion of chain segment n occurs into any of the remaining $m - n + 1$ sites. Thus the total number of configurations resulting from this insertion process is

$$\Omega = m(m-1)(m-2)\cdots(m-n+1) = \frac{m!}{(m-n)!} \, . \qquad (3.30)$$

This quantity must be divided by the number of configurations accessible to the chain if excluded volume were ignored. In that case, the first segment could occupy any of the m sites. If excluded volume is neglected, then the second and all subsequent segments will also have access to any of the m sites, hence $\Omega_0 = m^n$. Thus the ratio required to correct for the excluded volume contribution to the number of accessible conformation is

$$\omega_{ex} = \frac{\Omega}{\Omega_0} = \frac{m!}{m^n(m-n)!} = \frac{(n/\rho)!}{(n/\rho)^n(n/\rho - n)!} , \tag{3.31}$$

where the latter equality follows from the the definition $\rho \equiv n/m$. If we use Stirling's approximation $x! = \sqrt{2\pi x}(x/e^x)$ and the Boltzmann-Planck equation $S = k \ln \Omega$, then this contribution to the conformational entropy is

$$\frac{\Delta S_{ex}(\rho)}{nk} = -\left[\left(\frac{1-\rho}{\rho}\right)\ln(1-\rho) + 1\right] , \tag{3.32}$$

where terms of order n^{-1} are neglected. This formula estimates the effect of nonlocal steric repulsions on chain conformational entropy.

There is a second nonlocal contribution if the solvent is not inert. This contribution is described in terms of the relative interactions of monomer-monomer, monomer-solvent, and solvent-solvent, expressed by the Flory χ parameter [54]. The quantity χ is the free energy (divided by $2kT$) for the following thermodynamic exchange process: one monomer from a pure medium of monomers is transferred into a pure solvent and one solvent molecule is similarly transferred from the pure solvent into a medium of pure monomers. Combined with the nonlocal excluded volume, the total nonlocal free energy in the Flory model is,

$$\frac{\Delta F_{nonlocal}}{nkT} = \left(\frac{1-\rho}{\rho}\right)\ln(1-\rho) + 1 - \chi\rho . \tag{3.33}$$

Expansion of the nonlocal free energy in a Taylor series around the low-density $\rho = 0$ conformations yields,

$$\frac{\Delta F_{nonlocal}}{kT} = n\rho\left[\frac{1}{2} - \chi\right] + \sum_{s=2}^{\infty} \frac{n\rho^s}{s(s+1)} . \tag{3.34}$$

If only the term linear in ρ is kept ($\rho = nv/R^d$ corresponds to the probability of forming a monomer-monomer contact), the sum of local [Eq. (3.23)] and nonlocal contributions [Eq. (3.34)] gives the total free energy

$$\frac{\Delta F_{\text{total}}}{kT} = \frac{\Delta F_{\text{el}}}{kT} + \frac{\Delta F_{\text{nonlocal}}}{kT} = \frac{d}{2}(\alpha^2-1) - d\ln\alpha + \frac{n^2 v}{\alpha^d R_0^d}\left(\frac{1}{2}-\chi\right), (3.35)$$

in terms of $R = \alpha R_0$. By setting the derivative $\partial\Delta F_{\text{total}}/\partial\alpha = 0$, we obtain the equilibrium expansion, α, of the molecule. The Flory result generalized to d dimensions is

$$\alpha^{d+2} - \alpha^d = \frac{n^2 v}{R_0^d}\left(\frac{1}{2} - \chi\right) \propto n^{2-d/2}\left(\frac{1}{2} - \chi\right), \qquad (3.36)$$

where the second step follows from the random-flight relation $R_0 \sim n^{1/2}$.

The important conclusion from this result is that it defines three types of solvent, depending on their effects on the chain expansion. First, the so-called Flory theta solvent is defined as that solvent for which $\chi = 1/2$, or alternatively the Flory theta temperature is that temperature, for a given solvent, for which $\chi = 1/2$. For the theta solvent, Eq. (3.36) shows that $\alpha = 1$, and the chain adopts its ideal unperturbed distribution. This solvent is somewhat unfavorable, such that chain contraction resulting from monomer-monomer attractions (relative to monomer-solvent attractions) cancels the expansion resulting from nonlocal steric repulsions; thus in the theta solvent, only the local interactions remain.[1] The second is referred to as a "good" solvent, $\chi < 1/2$ ($\chi = 0$ for an inert solvent), because then, for sufficiently long chains and $d < 4$, the $(d+2)$th power term clearly dominates the left side of Eq. (3.36), $\alpha^{d+2} \sim n^{2-d/2}$, resulting in an

---

[1] The conclusion that solvent interactions and excluded volume cancel exactly is only valid to the second order, i. e., to the level of the two-body interactions. The second virial coefficient is zero, but the third virial term is not [58, 124, 152]. Hence, exact cancellation does not occur, so the distribution will not be exactly Gaussian in any solvent condition.

expanded radius relative to the ideal chain,

$$R = \alpha R_0 \sim \left[n^{(2-d/2)/(d+2)} \right] \left[n^{1/2} \right] = n^{3/(d+2)} . \tag{3.37}$$

This expression [49] reduces to $R \sim n$, in $d = 1$ dimension, $R \sim n^{3/4}$ in $d = 2$ dimensions, and the well-known result that $R \sim n^{3/5}$ in $d = 3$ dimensions. Equation (3.36) shows that for dimensions $d \geq 4$, effects of excluded volume are negligible; for all $d > 4$, $\alpha \to 1$ with increasing n, and for $d = 4$, α is independent of n, indicating that the chain is ideal in four dimensions, $i.\,e.$, $R = \alpha R_0 \sim n^{1/2}$. Thus, in a good solvent (for $d < 4$), the balance of nonlocal interactions is such that the excluded volume expansion exceeds the solvent-driven contraction. The third case is referred to as a "poor" solvent, $\chi > 1/2$, because the monomer-monomer attractions are favored so strongly relative to monomer-solvent attractions that $\alpha < 1$ and the chain contracts. Thus in a poor solvent, the solvent-driven contraction exceeds the excluded volume expansion, and, as $n \to \infty$, the radius becomes zero. Clearly, the model is unphysical in this latter case because the chain radius cannot diminish beyond the steric size dictated by the sum of the monomer sizes. Moreover, as the polymer collapses to its minimum size, ρ approaches 1, which violates the small-ρ requirement of Eq. (3.35). This situation is repaired by including the three-body repulsive interaction, which is the basis for the simplest homopolymer collapse theories (see below).

Exhaustive lattice simulations

The Flory approximation, Eq. (3.31), estimates the effect of excluded volume by supposing that the chain monomers are distributed uniformly in the available volume; in this way, it neglects the severity of steric constraints among monomers close together in the sequence that is greater than among those far apart in the sequence. Despite its simplicity, it has led to remarkably accurate predictions of scaling behavior for chain expansion; for example, the exponent 0.6 for $d = 3$ in Eq. (3.37) is now known in better approximation to be 0.588 [84]. One of the most important tools for refinement of the theory of polymer excluded volume has been the use of exhaustive simulations of self-avoiding chains on lattices. Such methods permit the prediction of long-wavelength properties of chains that depend little on microscopic details such as bond angles and local steric effects.

In 1947, W. J. C. Orr [109] first exhaustively enumerated all the conformations of short chains on simple lattices subject to excluded volume. He considered chains of lengths $n \leq 9$ on two-dimensional square lattices and $n \leq 7$ on three-dimensional simple cubic lattices. With advances in computer technology in the 1950s, Domb and his colleagues [39] extended the exhaustive simulations; when combined with sophisticated counting theorems of Sykes [147], this work led to improved enumeration methods for longer chains. The main emphasis of these systematic efforts has been the determination of scaling exponents, connective constants, and the distributions of various chain properties in the presence of excluded volume. The most thoroughly studied of these properties include the exponent ν in the asymptotic scaling of the mean square end-to-end distance $\langle R_n^2 \rangle \sim n^{2\nu}$ [38, 50], the exponent γ and the connective constant μ in the asymptotic form of the total number of conformations $Q_0(n) \sim n^\gamma \mu^n$ [50, 148], the exponent g in the total number of end-to-end cycles or rings $Q(n; 1, n) \sim n^g \mu^n$ [which is equivalent to $Q(N; 1, N+1)$ [12], where $N = n - 1$ is the number of bonds while n is the number of monomers] [73, 89, 128], and the distribution of end-to-end vectors R_n [41].

The study of lattice self-avoiding walks [4] has also been extended to incorporate solvent interactions [50] and nearest-neighbor attractive energies [92] to account for monomer-monomer association. One noteworthy finding is that even at the theta point when $\langle R_n^2 \rangle$ scales as n, the distribution of R_n is not Gaussian [41, 48, 50, 93, 154, 155]. Thus no attraction between monomers can fully compensate for the excluded volume effects. These exhaustive simulation studies contributed significantly to the development of modern polymer theory, including scaling [28] and renormalization group methods [24, 25, 27, 30, 31, 60, 61, 103-107]. Recently, exhaustive simulations have also been applied to chains with several constraints as a function of compactness [10-13], as well as to compact chains confined to specific shapes [11, 13, 23]. Exhaustive simulations are of special importance to the study of compact chains because the efficiency and ergodicity of Monte Carlo samplings of compact conformations are problematic [64, 112, 113]. Compact conformations are of interest because of their relevance to globular native states of proteins.

The path-integral method

More satisfactory than the Flory approximation for the nonlocal interactions is the self-consistent field path-integral method, introduced by S.F. Edwards [43]. In addition to self-consistent field considerations, the path integral is often analysed using perturbation theory, which is applicable to chains with few intrachain contacts [59, 60]. This approach is important for proteins for at least two reasons. First, near the theta point, it currently provides the most accurate description of excluded volume effects and may provide the best approach for refining polymer collapse theories, or at least the initial stages of collapse from the open conformations. Second, it provides an off-lattice alternative for the study of internal architectures in polymers and proteins (see below).

In general, the nonlocal potential of mean force $W_{n\ell}$ in Eq. (3.20) is expressed in terms of two-monomer, three-monomer, and higher many-monomer interaction terms as follows:

$$
\begin{aligned}
W_{n\ell}(\{r(\tau)\}) &= \sum_{i<j} u^{(2)}[r(\tau_i) - r(\tau_j)] \\
&+ \sum_{i<j<k} u^{(3)}[r(\tau_i), r(\tau_j), r(\tau_k)] + \dots .
\end{aligned} \tag{3.38}
$$

In the path integral approach, interactions among connected neighbors along the chain sequence are usually assumed to obey random-flight statistics [Eq. (3.21)]. In many applications, the series for $W_{n\ell}$ in Eq. (3.38)

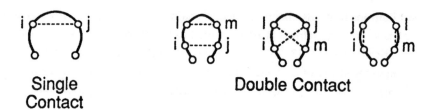

Single   **Double Contact**
Contact

Figure 3.4: Cluster diagrams for the single- and double-contact terms. Note that there are three possibilities in the order of arrangment of the four indices i, j, l, and m under the conditions $i < j$, $l < m$, and $i < l$.

is truncated to retain only the binary term,

$$W_{n\ell}(\{r(\tau)\}) = \sum_{i<j} u[r(\tau_i) - r(\tau_j)] . \qquad (3.39)$$

This restriction is equivalent to the superposition approximation in the theory of simple fluids [74].

When the monomer-monomer interactions $u(r)$ are spatially short-ranged, then u is sharply peaked at $r = 0$. Thus the pair interaction is usually approximated by the Dirac δ-function, which may be expressed either in terms of the Mayer function f_{ij},

$$e^{-u[r(\tau_i)-r(\tau_j)]} = 1 - \beta_0 \delta(r(\tau_i) - r(\tau_j)) = 1 + f_{ij} , \qquad (3.40)$$

as in the Ursell-Mayer theory of nonideal gas [91, 116], first introduced by Zimm [163], or in the form

$$u[r(\tau_i) - r(\tau_j)] = \beta_0 \delta[r(\tau_i) - r(\tau_j)] . \qquad (3.41)$$

In both cases, β_0 is a constant determined by the strength of interaction of the chain and its environment. Provided β_0 is small and the δ-functions are handled appropriately to avoid unphysical singularities, Eq's (3.40) and (3.41) are equivalent in perturbation theory, as noted by Fixman [51].

Fixman studied chains with excluded volume ($\beta_0 > 0$) [51] by using a method resembling the Mayer cluster expansion technique [91, 116], which consists of expanding the interaction term as

$$
\begin{aligned}
\exp\left[-W_{n\ell}(\{r(\tau)\})\right] &= \prod_{i<j}(1 + f_{ij}) \\
&= 1 + \sum_{i<j} f_{ij} + \underbrace{\sum_{i<j}\sum_{l<m} f_{ij}f_{lm}}_{i<l} + \cdots \\
&\quad + \underbrace{\sum_{i<j}\sum_{l<m}\cdots\sum_{r<s} f_{ij}f_{lm}\cdots f_{rs}}_{i<l<\cdots<r} + \cdots . \quad (3.42)
\end{aligned}
$$

Because each Mayer function f_{ij} is associated with a two-segment excluded volume interaction, a term with k f_{ij} factors is referred to as a k-contact term. In this way, the partition function Q in Eq. (3.16) is given by the series

$$
\begin{aligned}
Q &= 1 - \beta_0 \sum_{i<j} P_0(0_{ij}) + \beta_0^2 \underbrace{\sum_{i<j}\sum_{l<m} P_0(0_{ij})P_0(0_{lm})}_{i<l} + \cdots \\
&\quad + (-\beta_0)^k \underbrace{\sum_{i<j}\sum_{l<m}\cdots\sum_{r<s} P_0(\underbrace{0_{ij}, 0_{lm}, \ldots 0_{rs}}_{k\ 0s})}_{i<l<\cdots<r} + \cdots , \quad (3.43)
\end{aligned}
$$

obtained by substitution of Eq's (3.42) and (21) into Eq. (3.20) then into Eq. (3.17) for $P(\{r(\tau)\})$. Because each Mayer function f_{ij} contains a δ-function of the spatial separation between $r(\tau_i)$ and $r(\tau_j)$, straightforward integrations of the f_{ij}s in Eq. (3.42) with the Gaussian bond probability [Eq. (3.21)] give rise to factors of P_0s in Eq. (3.43), which are random-flight (unperturbed, $\beta_0 = 0$) probabilities for the chain to adopt conformations with specific self-contacts. For example, $P_0(0_{ij})$ is the unperturbed probability that segments i and j are in contact, given by the

Jacobson-Stockmayer [78] factor

$$P_0(0_{ij}) = \left[\frac{d}{2\pi l^2 |i - j|}\right]^{d/2} . \tag{3.44}$$

Similarly $P_0(0_{ij}, 0_{lm}, \ldots, 0_{rs})$ is the unperturbed probability for the chain to have the contact pairs (i, j), (l, m), $\ldots,(r, s)$, which may readily be computed by the Wang-Uhlenbeck theorem [156]. The terms in expansions such as Eq. (3.43) are often represented by cluster diagrams (Fig. 3.4), a convenient device for combinatorics bookkeeping. In Fig. 3.4, a solid line represents the chain and each dotted line indicates the pair of segments that interact with repulsion β_0. Theories of this type are called two-parameter models because they depend only on the parameters nl^2 and $n^2\beta_0$, the unperturbed mean square end-to-end distance and the total excluded volume among chain segments, respectively [162]. The main result is that the partition function in Eq. (3.43) now contains not only the local interactions, but also contains, in series expansion, the two-, three-, four-self-contact contributions, etc, from the nonlocal interactions.

The transition from the discrete bonds to the continuum path integral formulation is straightforward. The procedure consists of letting the segment lengths shrink to zero ($\Delta\tau_j \equiv \tau_j - \tau_{j-1} \to 0$) and the number of segments increase to infinity while keeping the total contour length $N \equiv \sum_{j=1}^n \Delta\tau_j$ and the step length $l \equiv \langle R_n^2 \rangle / N$ finite. Despite the fact that real chains are comprised of discrete segments, adoption of a continuum model has advantages. Even in the discrete two-parameter model, final summations in expressions such as Eq. (3.43) are often approximated by integrals [162]. Also, the continuum path integral has been applied extensively in quantum mechanics and quantum field theory [45, 46, 63, 127], so a whole arsenal of techniques is available, including renormalization group methods [60, 161].

Using the bond probability [Eq. (3.21)] and excluded volume interaction

[Eq. (3.41)], the partition function is given by

$$Q = \mathcal{N} \int dr(\tau_0) \quad \cdots \quad \int dr(\tau_n) \exp\left\{ -\frac{d}{2l^2} \sum_{j=1}^{n} |r(\tau_j) - r(\tau_{j-1})|^2 \right.$$

$$\left. - \frac{\beta_0}{2} \underbrace{\sum_{i=0}^{n} \sum_{j=0}^{n}}_{i \neq j} \delta[r(\tau_i) - r(\tau_j)] \right\}, \tag{3.45}$$

where the factors of $(d/2\pi l^2)^{d/2}$ are absorbed into the normalization $\mathcal{N}$, and the summation $\sum_{i<j}$ in Eq. (3.39) is replaced by the symmetric summation $1/2 \sum_i \sum_j$. Because $l = \Delta\tau_j$ for discrete chains (with $N = nl$), and because, by the definition of l, the mean-square length of a generalized segment is $l\Delta\tau_j$, the continuum limit of the bond probability term in the exponent is

$$\lim_{\Delta\tau_j \to 0} \quad -\frac{d}{2l^2} \sum_{j=1}^{n} |r(\tau_j) - r(\tau_{j-1})|^2$$

$$= \lim_{\Delta\tau_j \to 0} -\frac{d}{2l} \sum_{j=1}^{n} (\Delta\tau_j) \left| \frac{r(\tau_j) - r(\tau_{j-1})}{\Delta\tau_j} \right|^2$$

$$= -\frac{d}{2l} \int_0^N d\tau \left| \frac{dr(\tau)}{d\tau} \right|^2. \tag{3.46}$$

In the excluded volume interaction term in Eq. (3.45), β_0 is the repulsive strength per chain segment. Therefore, to generalize to the continuum case, β_0/l^2 is identified as the repulsive strength per square unit of continuous contour length τ. Hence, by substituting the equality $l = \Delta\tau_j$ for discrete chains, the interaction term may be rewritten as

$$-\frac{\beta_0}{2l^2} \underbrace{\sum_{i=0}^{n} \Delta\tau_i \sum_{j=0}^{n} \Delta\tau_j}_{i \neq j} \delta[r(\tau_i) - r(\tau_j)]. \tag{3.47}$$

The continuum limit is obtained by replacing the double summation by integrals,

$$-\frac{\beta_0}{2l^2} \underbrace{\int_0^N d\tau \int_0^N d\tau' \, \delta[r(\tau) - r(\tau')]}_{|\tau-\tau'| \geq a} \,, \tag{3.48}$$

where the phenomenological constant a ($\simeq l$) is introduced to prevent the unphysical situation of a chain segment interacting with itself. The *path integral* or *functional integral* $\int \mathcal{D}[r(\tau)]$ is defined as the $|\tau_j - \tau_{j-1}| \to 0$ limit of the multiple integrals over $dr(\tau_j)$s,

$$\int \mathcal{D}[r(\tau)] \equiv \lim_{\substack{n \to \infty \\ \Delta\tau_j \to 0}} \prod_{j=0}^{n} \int dr(\tau_j) \,. \tag{3.49}$$

This is sometimes schematically represented as the product $\prod_\tau \int dr(\tau)$ over continuous values of τ. Collecting continuum limits from Eq's (3.45), (3.48) and (3.49), we arrive at the path-integral partition function

$$Q = \mathcal{N} \int \mathcal{D}[c(\tau)] \, e^{-H[c(\tau)]} \,,$$

$$H[c(\tau)] = \frac{1}{2} \int_0^N d\tau \left| \frac{dc(\tau)}{d\tau} \right|^2 + \frac{v_0}{2} \underbrace{\int_0^N d\tau \int_0^N d\tau' \, \delta[c(\tau) - c(\tau')]}_{|\tau-\tau'| \geq a} \,.$$

$$\tag{3.50}$$

Here $v_0 \equiv \beta_0 d^{d/2}/l^{2+d/2}$ is the repulsive excluded volume strength. The variable $c(\tau) \equiv (d/l)^{1/2} r(\tau)$ is adopted instead of $r(\tau)$ because then the path integral does not contain the spatial dimension d explicitly, which is often useful.

When v_0 is small-for chains near the theta condition-conformational properties may be deduced from the continuum path integral [Eq. (3.50)] by systematic perturbation expansion in powers of v_0. For instance, Muthukumar and Nickel [98] determined the mean-square end-to-end distance $\langle R_n^2 \rangle$

through the sixth order in v_0 for polymers with excluded volume in three dimensions. A detailed discussion of the Feynman-diagrammatic [45, 46] techniques of the perturbation procedure can be found elsewhere [10, 60].

Thus the two main avenues to refined modeling of nonlocal interactions in polymer chains beyond the Flory approximation are lattice simulations and continuum path integral methods. Each method has its advantages. Lattices offer more realistic treatment of local excluded volume and effects of finite monomer size. Finite-volume monomer units and finite-ranged infinite repulsive potentials are difficult to implement in analytic models: the path integral model treats chains as mathematical curves with vanishing volume, and excluded volume is modeled by a two-body δ-function repulsion of vanishing range. On the other hand, for properties that are not sensitive to local details, the path integral approach is not restricted to short chains nor is it constrained as a lattice is to fixed bond angles and periodic spatial arrangements of monomers. A review by McKenzie provides an insightful discussion of the similarities and differences between lattices and analytic continuum models [94].

Homopolymer collapse theories

As noted earlier, the Flory model predicts that homopolymers will collapse to compact states in poor solvents [143]. However, until the work of Ptitsyn et al. [3, 123, 124] and Lifshitz [85] in the late 1960s, polymer collapse was apparently studied less than polymer expansion, perhaps partly because collapse was assumed to be difficult to observe in competition with precipitation from solution in poor solvents. Ptitsyn et al. [3, 123, 124] first proposed homopolymer collapse as a model for protein folding. Several homopolymer collapse models [3, 25, 120, 123, 124, 160] are based on addition of a three-body monomer repulsion term to Eq. (3.35),

$$\frac{\Delta F_{\text{nonlocal}}^{(3)}}{kT} = \frac{n\rho^2}{6} = \frac{n^3 v^2}{6R^{2d}} \, , \tag{3.51}$$

which is the $s = 2$ term in the summation of Eq. (3.34), and $n\rho^2$ is the probability that three monomers are in the same neighborhood. With the addition of this term, Eq. (3.36) now becomes

$$\alpha^{2d+2} - \alpha^{2d} + \frac{n^2 v \alpha^d}{R_0^d}\left(\chi - \frac{1}{2}\right) = \frac{n^3 v^2}{3R_0^{2d}} . \tag{3.52}$$

In the poor solvent limit $\chi > 1/2$, now the three-body lowest order term dominates, yielding

$$\alpha^d = \frac{2nv}{3(2\chi - 1)R_0^d} , \tag{3.53}$$

implying that $R \sim n^{1/d}$, the molecule approaches maximum physical collapse in a poor solvent in the limit of infinite chain length.[2]

Equation (3.52) is the basis for several models of homopolymer collapse [3, 25, 120, 123, 124, 160]. They predict that flexible chains will have a second-order phase transition (*i.e.*, a single broad minimum in free energy) but that stiff chains of finite length will have a first-order transition (*i.e.*, two stable states with a free energy barrier between them). They also predict that the transition for all chains, flexible or stiff, will approach second order in the limit of infinite chain length. These models adopt the Flory approximation for the nonlocal interactions. They do not attempt to address highly compact states such as those of globular proteins, because they approximate the Flory expression by retaining only the two-body and three-body terms in the perturbation expansion, as noted above [see Eq's (3.35) and (3.52)]. More refined path integral treatments of homopolymer collapse [42, 80, 97] do not rely on the Flory approximation, but they also retain only two- and three-body terms and focus on perturbation

---

[2] In addition to the two-monomer χ parameter, Orofino and Flory [108] have also introduced other χ parameters in series expansions such as Eq. (3.52), in which many-monomer interactions are taken into consideration. In that formulation, a parameter χ_i is associated with the i-monomer term that is proportional to $n^i v^{i-1}$.

around the theta state. For example, in the three-parameter model of Kholodenko & Freed [80], the extra three-monomer interaction term

$$\frac{\omega_0}{3!} \int_0^N d\tau \int_0^N d\tau' \int_0^N d\tau'' \delta[c(\tau) - c(\tau')]\delta[c(\tau') - c(\tau'')] , \qquad (3.54)$$

where ω_0 is the three-monomer excluded volume interaction parameter, is added to the two-parameter path-integral Boltzmann factor Eq. (3.50).

For processes of chain collapse to highly compact states, as in protein folding, it is necessary to further include the higher many-body terms. The homopolymer collapse model of Sanchez [129] retains the full Flory expression [Eq. (3.33)] rather than approximating it with the two- and three-body terms. The Sanchez model also differs from those discussed above in the use of the Flory-Fisk empirical distribution [56] for the radius of gyration, instead of a Gaussian function for end-to-end distance distribution, for chain elasticity. Interestingly, the combined use of the Flory-Fisk distribution for the local interactions with the Flory approximation for excluded volume [Eq. (3.32)] reproduces remarkably well the exact self-avoiding walk distribution function for the radius of gyration; we compare that combination of factors with our exhaustive simulations in Fig. 3.5.

An important issue is whether the collapse transition is first order or more gradual [40, 42, 80, 86, 97, 129, 160]. Experiments have addressed this issue, most often with polystyrene in cyclohexane [18-22, 115, 144, 146, 152]. Experimental observation of the collapsed state of homopolymers has been extremely challenging because collapse competes with phase separation and therefore one must observe extremely low polymer concentrations. The order of the transition is not yet fully resolved. Sun *et al.* [144] have found good agreement of their experimental data with the model of Sanchez (Fig. 3.6), provided that χ is taken to be an empirically chosen free energy; they consider the required added entropy contribution to account for the restriction of the benzene pendant groups of the polystyrene when intrachain contacts are formed.

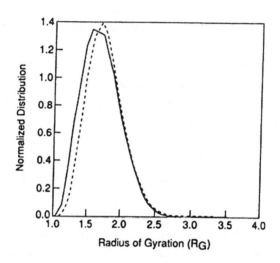

Figure 3.5: Distribution of radius of gyration R_G calculated from exhaustive simulation of chains configured on three-dimensional simple cubic lattices (solid curve), and the Flory-Fisk approximation [56] in conjunction with Flory's treatment of excluded volume (dotted curve). The chains consist of $n = 14$ segments. R_G is given in units of the minimum possible radius $(R_G)_{\min}$ of 14-segment chain on the lattice. The parameter ρ in the Flory equation [Eq. (3.32)] is given by $\rho = [(R_G)_{\min}/R_G]^3$, so the maximum density $\rho = 1$ coincides with the minimum radius.

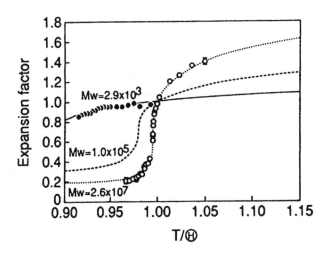

Figure 3.6: Collapse transition of polystyrene in cyclohexane vs temperature. The data of Sun *et al.* [144] is compared with the modified model of Sanchez [129] (from [144]).

Protein folding is generally observed to involve a first-order (two-state) collapse transition even for relatively short chains (as short as 102 monomers) [121]. To model the first-order transition of protein folding, Shakhnovich & Finkelstein [47, 135] added a term to a simple homopolymer collapse model to account for side-chain freezing upon collapse. They devised a potential in which a small volume increase from the compact state permits side chains to surmount a free-energy barrier. If this free energy function is sufficiently abrupt, the Shakhnovich-Finkelstein model predicts a first order transition. It also predicts that the activation enthalpy of unfolding should be independent of solvent; this independence has been observed experimentally by Segawa & Sugihara [133]. However, if side-chain freezing is the principal origin of a first-order transition, then their model would predict first-order transitions in other homopolymers because most polymer molecules also have pendant groups similar to side chains on amino acids.

Heteropolymer collapse

Proteins are heteropolymers. Unlike homopolymers, they are comprised of different monomer types. Heteropolymers and homopolymers differ in important respects, including the collapse process. The 20 types of amino acids may be usefully divided into two classes-nonpolar and other-based on their strong or weak aversion for aqueous solvents. This division is natural because an organizing principle in native proteins is the segregation of nonpolar monomers into an interior core and of polar monomers into an exterior shell [69]. This form of organization represents a degree of freedom available to heteropolymers that is not available to homopolymers, and is thus one important respect in which heteropolymer collapse differs significantly from homopolymer collapse.

A heteropolymer collapse model has been developed to predict the free energies of protein folding [33, 36]. In this model, the degrees of freedom are the radius of the chain and the extent of interfacial ordering of the two monomer types into surface and core regions of the collapsed state. The free energy depends on chain length, χ parameters for the different monomer types, and on the composition, $i.\,e.$, the numbers of each of the two monomer types, based on the assumption that the monomer sequence is random. In the homopolymer limit of pure composition, this model is identical to that of Sanchez [129], with the elastic entropy given by the Flory-Fisk [56] distribution function and the excluded volume entropy given by the Flory approximation. As noted in Fig. 3.5, the combination of these two functions predicts quite well the exact distribution of radii. For the heteropolymer, the configurational free energy also depends upon segregation of some monomers into a core and others to the surface [33, 35]; this segregation is treated as a mixing process of the two monomer types into two "solutions," one in the core and one at the surface. The thermodynamic properties of the compact and denatured states are calculated using a fictitious thermodynamic pathway of collapse then interfacial ordering. Other possible pathways are not treated in the model, so it does not address questions of kinetics, or of barriers, or therefore of whether the phase transition is first order or gradual. A barrier is found

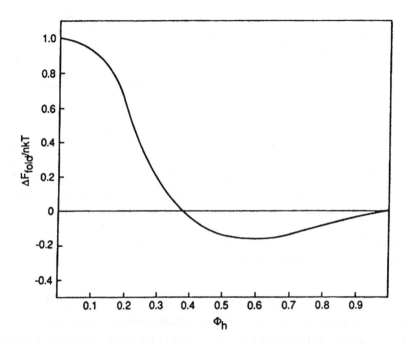

Figure 3.7: Free energy of folding vs fraction, Φ_h, of monomers that are solvent-averse (large positive χ), for n = 100 from the heteropolymer collapse model [33].

along the chosen pathway, but no further exploration has been attempted of the order of the transition.

This model makes several predictions for the thermodynamics of protein collapse [1, 33, 35, 36]. Figures 3.7 and 3.8 show that the collapsed state with nonpolar core is favored only if the number of solvophobic monomers is sufficient and the chain is sufficiently long. If χ depends on temperature or solvent character, then the free energy of collapse will depend on those quantities. Based on amino acid oil-water transfer experiments [66, 100, 101, 149], the temperature dependence of χ has been estimated to be

$$\chi(T) = \frac{1.4}{kT}\left\{\Delta H_0 + \Delta C_P(T-T_0) - T\left[\Delta S_0 + \Delta C_P \ln(T/T_0)\right]\right\} , (3.55)$$

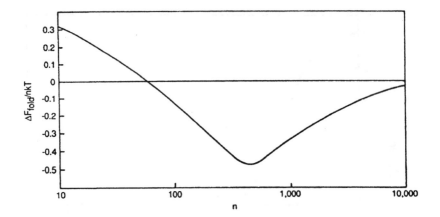

Figure 3.8: Free energy of folding vs chain length for $\Phi_h = 0.45$ from the heteropolymer collapse model [33].

where ΔH_0 and ΔS_0 are the enthalpy and entropy, respectively, of the exchange process at $T = T_0$. The temperature dependence Equation (3.40) is obtained by straightforward integration of the thermodynamic relation $C_P \equiv T(\partial S/\partial T)_{N,P} = (\partial H/\partial T)_{N,P}$, assuming constant ΔC_P. On this basis, Fig. 3.9 shows the predicted temperature dependence of protein stability in comparison with differential scanning calorimetry experiments of Privalov and coworkers [121].

In a similar way, the effects of nonspecific denaturants (such as urea and guanidinium hydrochloride GuHCl) have been treated based on the assumption that they reduce χ, i.e., that solutions of denaturant + water are better solvents for the nonpolar amino acids than water alone. For example, χ is estimated to have the following dependence on denaturant concentration, c at $T = 25°C$,

$$\frac{kT}{1.4}\chi(c) = g_{tr} = \begin{cases} 2.0 \times 10^3 - 111c + 5.76c^2, & \text{for GuHCl} \\ 2.0 \times 10^3 - 36.8c & \text{for urea,} \end{cases} \qquad (3.56)$$

where c is in units of moles/liter and g_{tr} is in calories per mole of hydrophobic side chains. On this basis, Fig. 3.10 shows the prediction for

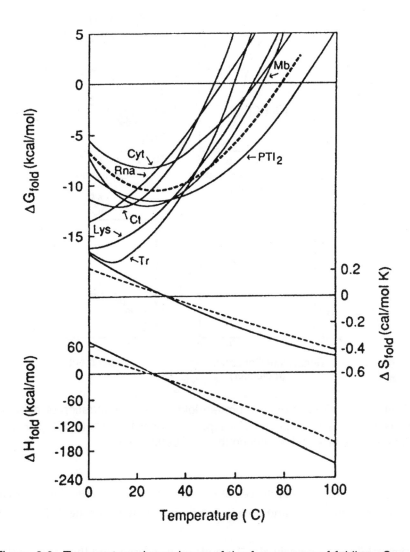

Figure 3.9: Temperature dependence of the free energy of folding. Continuous lines show the scanning calorimetry experiments of Privalov *et al.* [121]; dashed lines show the prediction of the heteropolymer collapse theory [36] with $\chi(T)$ given by Eq. (3.55). Denaturation at high temperatures results largely from the gain of conformational entropy; "cold" denaturation results from the temperature dependence of nonpolar association [36].

Protein Folding

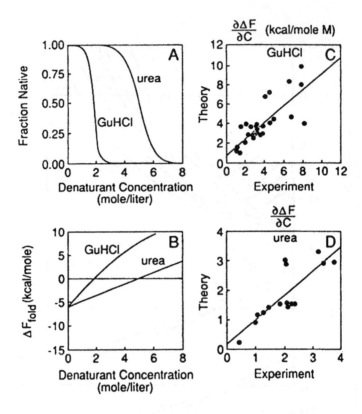

Figure 3.10: Free energy of protein folding vs concentration of denaturants, from the heteropolymer collapse theory, where Eq. (3.56) gives the dependence of χ on the concentration of denaturant.

the concentration dependences of the stabilities, $\partial \Delta F_{\text{fold}}/\partial c$, for globular proteins in guanidine and urea in comparison with experiments [1].

The stability of a protein also depends on its charge, and therefore on the pH and ionic strength of a solution. Proteins in acidic or basic solutions will have a net charge. Net charge tends to destabilize the folded state relative to the denatured state because the process of unfolding leads to reduction of the charge density, and thus to a lowering of the electrostatic free energy. The effects of charge have been modeled [141, 142] using a variant of the Linderstrom-Lang charged sphere model [88] to characterize

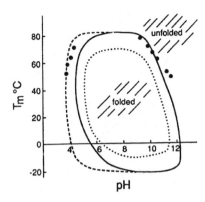

Figure 3.11: Prediction for the denaturation temperature of myoglobin vs pH, given by the heteropolymer collapse theory with Poisson-Boltzmann treatment of electrostatics contributions [142], compared with scanning calorimetry experiments (*filled circles*) of Privalov et al [122] on sperm whale metmyoglobin in aqueous solutions of ionic strength 0.01M. *Continuous curve*: theory with 50% of the residues (monomers) assumed nonpolar, six histidines buried, and six exposed at the surface. Dotted curve: same as above except hydrophobic fraction reduced by 10%. Dashed curve: theory with all 12 histidines assumed exposed at the surface.

the native state, and a variant of the Hermans-Overbeek polyelectrolyte porous sphere model [72] to characterize the denatured state. The electrostatic field and the charge distribution are found by self-consistent solution of the Poisson-Boltzmann equation [142]. This free energy resulting from electrostatics is added to the other free energies in the heteropolymer collapse model. Figures 3.11 and 3.12 show predictions for the dependence of myoglobin stability on pH and ionic strength compared with experiments [67, 68]. One unexpected result is the predicted coexistence of two different denatured states under some conditions: one ensemble of open conformations, stabilized by a balance of long-ranged charge repulsions and elastic entropy, and one ensemble of compact denatured conformations, stabilized by the nonlocal interactions [142]. This prediction is compared to the experimental pH-ionic strength phase diagram for myoglobin [67, 68] in Fig. 3.12.

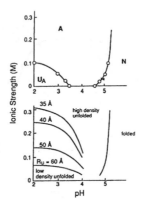

Figure 3.12: Prediction for the phase diagram of myoglobin as a function of pH and ionic strength; same model as in Fig. 3.11, compared with the data of Goto & Fink [67,68].

The collapsed conformations of heteropolymers are predicted to differ from those of homopolymers in one other important respect. The ensemble of compact conformations of a chain increases approximately as κ^n with chain length n. For example, $\kappa \simeq 1.41$ and $\kappa \simeq 1.74$ for maximally compact chains on two-dimensional square lattices and three-dimensional cubic lattices, respectively [11, 12]. Therefore, even though a homopolymer driven to compactness in a poor solvent will have a considerably restricted configurational space relative to the full ensemble, the number of accessible compact conformations is still exponentially large. For a heteropolymer, the situation is qualitatively and quantitatively different. Exhaustive lattice simulations [14, 82, 83] and an analytical random energy heteropolymer model of Shakhnovich & Gutin [137, 138] (see below) show that the ensemble of lowest-energy compact conformations of heteropolymers is small (of the order of 1 or a few conformations), and does not grow rapidly with chain length. This observation appears to account for one of the most remarkable features of the native states of biological globular proteins, namely that there is only a single compact conformation, and that it is determined uniquely by the monomer sequence.

Spin-glass models

Some recent models are based on the analogy that aspects of protein folding may resemble the spin-glass transition from a disordered to a glassy state. These models aim to predict the nature of the transition from disorder to order, and the distribution of energy states, for an ensemble of different sequences of chain molecules with random intermonomer interactions.

Edwards & Anderson [44] first proposed a theory of spin glasses to account for the magnetic properties of dilute solutions (alloys) of Mn in Cu and other nonmagnetic metals. The interaction Hamiltonian for the spins (*i. e.*, among the magnetic moment vectors S_i), is taken to be the well-known Heisenberg [71] spin-spin interaction

$$H = \sum_i \sum_j J_{ij} S_i \cdot S_j , \qquad (3.57)$$

in which the interaction strength J_{ij} is assumed to be a random variable [44] instead of the constant nearest-neighbor interaction commonly used in Ising models [77, 116]. The randomization of J_{ij} corresponds to the uncertainty in the positions of the Mn atoms in a dilute solution in Cu, which are not symmetrically spaced as in a crystal lattice. Although values of J_{ij} are fixed for any one particular sample by the particular conditions of preparation, to ascertain the behavior of a typical sample, one must average over macroscopic samples, which corresponds to averaging over samples with different values of the J_{ij}s.

Following the standard canonical ensemble formulation of statistical mechanics [116], the Helmholtz free energy A of one particular spin glass sample is

$$A = -kT \ln Q = -kT \ln \left[\sum_{\text{spin states}} \exp(-\frac{H}{kT}) \right]$$

$$= -kT \ln\left[\int [\mathcal{D}S_i] \exp(-\frac{H}{kT})\right], \tag{3.58}$$

where

$$Q = \sum_{\text{spin states}} \exp(-\frac{H}{kT}) = \int [\mathcal{D}S_i] \exp(-\frac{H}{kT}) \tag{3.59}$$

is the partition function. In these equations, the notation $\int [\mathcal{D}S_i] \equiv \prod_i \int dS_i$ denotes the summation over spin states. The average free energy over samples, F, is obtained by performing an additional averaging over J_{ij},

$$\begin{aligned} F &= \int [\mathcal{D}J_{ij}] P(J_{ij}) A \\ &= -\int [\mathcal{D}J_{ij}] P(J_{ij}) \ln Q \\ &= -\int [\mathcal{D}J_{ij}] P(J_{ij}) \left\{ \ln \int [\mathcal{D}S_i] \exp(-\frac{H}{kT}) \right\}, \end{aligned} \tag{3.60}$$

where $P(J_{ij})$ is the probability distribution of J_{ij}. However, the logarithm in the above equation poses serious difficulties for the integration over J_{ij}. To circumvent this, Edwards & Anderson used the identity

$$\ln Q = \lim_{m \to 0} \frac{Q^m - 1}{m} \tag{3.61}$$

and assumed that the average free energy F can be obtained by first computing

$$\int [\mathcal{D}J_{ij}] P(J_{ij}) Q^m \tag{3.62}$$

for any integer value of m, and then they analytically continued the result to $m \to 0$. Note that the mth power of Q is given by

$$Q^m = \left[\int [\mathcal{D}S_i] \exp(-\frac{1}{kT} \sum_i \sum_j J_{ij} S_i \cdot S_j) \right]^m$$

$$= \int [\mathcal{D}S_i^\alpha] \exp(-\frac{1}{kT} \sum_{\alpha=1}^{m} \sum_i \sum_j J_{ij} S_i^\alpha \cdot S_j^\alpha) , \tag{3.63}$$

so that the integration over J_{ij} in Eq. (3.62) can be performed if simple functional forms for $P(J_{ij})$ are assumed. For example, F becomes

$$F = \lim_{m \to 0} \frac{1}{m} \left\{ 1 - \int [\mathcal{D}S_i^\alpha] \exp \left[\frac{J^2}{2(kT)^2} \left(\sum_{\alpha=1}^{m} \sum_i \sum_j S_i^\alpha \cdot S_j^\alpha \right)^2 \right] \right\} \tag{3.64}$$

for the Gaussian distribution

$$P(J_{ij}) = \frac{1}{\sqrt{2\pi J^2}} \exp(-J_{ij}^2/2J^2) . \tag{3.65}$$

To arrive at Eq. (3.64), the order of integration of $[\mathcal{D}J_{ij}]$ and $[\mathcal{D}S_i^\alpha]$ is interchanged. The spin-glass problem thus reduces to the statistical mechanics problem specified by Eq. (3.64). The m systems (summation over α) are known as *replicas*, named by the mathematical trick adopted to perform integrations over logarithms.

Applications of spin-glass methods to proteins begin with the premise that the pairwise interactions of amino acid monomers can be described in terms of a distribution function of different energies. For example, using a path-integral approach, Garel & Orland [62] employ an interaction term of the form $v_{ij}\delta(r_i - r_j)$ where v_{ij} is a random variable instead of the constant positive v_0 used in the usual excluded volume treatment. If the distribution of random interactions has a particular form (often a Gaussian is assumed), then averaging over all possible interactions can lead to equivalence with the spin glass.

Bryngelson & Wolynes [7, 8] mapped the problem of heteropolymer collapse onto the random-energy model of Derrida [29]. The energy of the protein is expressed as

$$E = -\sum_i \epsilon_i(\alpha_i) - \sum_i J_{i,i+1}(\alpha_i, \alpha_{i+1}) - \sum_{i,j} K_{ij}(\alpha_i, \alpha_j, r_i, r_j) \,, (3.66)$$

where i labels the amino acid monomer along the sequence, α_i refers to the state of the ith amino acid, and r_i is its spatial position. Eq. (3.66) has three separate contributions to the total energy: (a) $\epsilon_i(\alpha_i)$ is the primary structure energy, depending on whether monomer i lies in a favorable section of the Ramachandran map [126]; (b) $J_{i,i+1}(\alpha_i, \alpha_{i+1})$ is the connected-neighbor interaction between a pair of monomers i and $i+1$ adjacent in the sequence; and (c) $K_{ij}(\alpha_i, \alpha_j, r_i, r_j)$ accounts for the nonlocal interaction. Each type of energy ϵ, J, and K is assumed to be a random variable with a given mean value and standard deviation. Chain connectivity is neglected in this model [7, 8]. The main predictions are: (a) a first-order transition from native to random-coil states and (b) the possible existence of a frozen phase in which different molecules would have different stable misfolded structures. The existence of this glassy phase in the model is a direct consequence of the random energy distribution; the transition temperature to this phase is nonzero only when the standard deviations of some of the energy functions in Eq. (3.66) are nonvanishing. This glassy phase is different from both the unfolded phase and the native phase. Bryngelson & Wolynes [7] suggested that this glassy state may account for the apparently irreversible denaturation of elastase not attributable to aggregation [65].

Shakhnovich & Gutin [136-138] used a more realistic chain model with a two-body random nonlocal interaction, a three-body excluded volume term, and with the local interactions given by random-flight chain statistics. In their model, the interaction terms of Eq. (3.38) are given by

$$W_{n\ell} = \frac{1}{2} \sum_{i \neq j} B_{ij} \Delta(r_i - r_j) + \frac{1}{6} C \sum_{i \neq j,\, j \neq k} \Delta(r_i - r_j) \Delta(r_k - r_j) \,, (3.67)$$

where r_i is the position vector of the ith monomer and Δ is the Kronecker delta; the latter equals unity when its argument is zero and zero otherwise [see also Eq's (3.50) and (3.54)]. The interaction heterogeneity owing to the different monomer types in a chain is characterized by the random two-monomer interaction term B_{ij}, which is assumed to have a Gaussian distribution

$$P(B_{ij}) = \frac{1}{\sqrt{2\pi B^2}} \exp\left[-\frac{(B_{ij} - B_0)^2}{2B^2}\right] , \qquad (3.68)$$

with mean value B_0 and standard deviation B. The correspondence with spin glass methods is apparent when B_{ij} is compared with J_{ij} in Equation (3.50). The positive constant C in the three-monomer term in Equation (3.52) accounts for excluded volume; as in the homopolymer models described above, it prevents the chain from collapsing to zero radius when the average two-monomer interaction is attractive ($B_0 < 0$).

Using the replica method [95, 114], Shakhnovich & Gutin [136-138] obtained a phase diagram from this model for three states: coil, random globule and frozen globule. A most interesting result follows from this model. If the chain is homogeneous (B is small or zero), then the chain will collapse from the coil state to the random globule as the average attraction between monomer increases, *i. e.*, as B_0 becomes more negative. However many accessible conformations remain in the random globule state; the chain does not have a unique native structure. This state is expected for a compact homopolymer. Only when the chain is a heteropolymer with a large value of B, *i. e.*, with a large standard deviation in the distribution of intermonomer interactions, will the chain undergo the transition to the frozen globule state. In this case, it is dominated by only a few conformations (of order 1). The prediction of few stable states for heteropolymers in a solvent that is poor for one of the monomer types is consistent with exhaustive simulations of maximally compact conformations for various sequences of 27-monomer chains on a three-dimensional simple cubic lattice [137], and for all possible sequences of several chain lengths on two-dimensional square lattices [14, 82, 83]. Both sets of lattice simulations also show that a significant fraction of all possible sequences have unique native structures. Moreover, the simulations show that when there are multiple native structures, those native conformations generally have little structural similarity with each other [14, 82, 83], and that native structures have little structural similarity with conformations of the next higher energy level [137, 136].

The analytical models that have been applied to protein collapse-homopolymer, heteropolymer, and spin-glass models-do not consider the sequence of monomers and are therefore not applicable to the prediction of a protein structure from its amino acid sequence. Current heteropolymer and spin-glass models represent a protein as a random ensemble of sequences. They are presently more suited to predicting folding stabilities and kinetics than specific native structures. Computer-simulation methods are needed to treat the specific effects of monomer sequences and the specific compact conformations in polymer molecules. Some results of these simulations are described below.

Internal structure in compact polymers

It follows from the preceding discussion that local forces are mainly responsible for helix formation in polymers in solution whereas nonlocal forces are mainly responsible for polymer collapse to compact states. Because globular proteins are both compact and often comprised of considerable amounts of helix, the question arises: is it the local or nonlocal forces that determine the structures of globular proteins? Until recently, the predominant view was that the local forces determine the internal structures in proteins while the nonlocal forces stabilize the compact state [2, 131, 132]. For example, according to Anfinsen and Scheraga [2]:

> Evidence is now accumulating to suggest that nearest-neighbor, short-range interactions play the dominant role in determining the conformational preferences of the backbones of the various

amino acids, but that next-nearest neighbor (medium-range) interactions and, to a lesser extent, long-range interactions involving the rest of the protein chain are required to provide the incremental free energy to stabilize the backbone of the native structure. The basis for this view comes from helix-probability profiles of denatured proteins, from the ability to predict with some success the location of the helix, extended structures, and β-turns in native proteins, from conformational energy calculations on short oligopeptides, and from experiments on the folding of protein fragments and sets of complementary protein fragments.

In contrast, recent work described below leads to the opposite view that nonlocal forces may be important, perhaps dominant, determinants of structure in globular proteins. This view has several bases. First, some globular proteins have little or no helix content [79]. Parallel and antiparallel sheets in proteins involve only nonlocal interactions, except at turns. Second, evidence summarized elsewhere suggests that local interactions are weaker than nonlocal interactions for folding globular proteins [34].

Third, results reviewed in the following section show that protein-like internal architectures will arise naturally from the severe steric restrictions that occur in compact chain molecules. To discuss internal structure in chain molecules, we must first show how to characterize it. A convenient way to represent forms of organization in polymer molecules is through use of a *contact map*, a list of all the spatial neighbor pairs in a particular configuration (see Fig. 3.13). For chains with n monomers, the contact map is an $n \times n$ matrix, in which an 1 at matrix position (i, j) (shown as a dot) indicates that monomers i and j are spatial neighbors in a given configuration, and a 0 indicates that those monomers are not spatial neighbors (shown as a blank). The four different types of pattern that appear on this map represent all forms of secondary structures in proteins (see Fig. 3.14) (for three-dimensional lattice representations of secondary structures, see Fig. 9 of Ref. [12]). Note that only helices and turns

Chain Geometry   Chain Topology

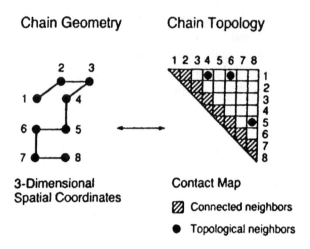

3-Dimensional   Contact Map
Spatial Coordinates
                                 ▨ Connected neighbors
                                 ● Topological neighbors

Figure 3.13: Example of a contact map. Dots indicate pairs of monomers which are spatially adjacent for the configuration shown on the left. For this conformation the topological contacts are [(1,4)(1,6)(5,8)].

among these four types of structure involve local interactions; parallel and antiparallel sheets involve nonlocal contacts (*i. e.*, dots that are not adjacent to the main diagonal). Because each dot on the contact map represents an intrachain contact, the number of these dots defines the "compactness" of the chain configuration. Compactness is correlated with, but not identical to, the density. We describe below path integral and exhaustive simulation studies of the conformational spaces for chains with one self-contact, then for chains with two self-contacts, then for chains of increasing compactness, up to the maximum possible.

The full conformational space of a chain subject to a given single self-contact was first described by Jacobson & Stockmayer in 1950 [78] for Gaussian random-flight chains. The principal findings of that study are as follows. (*a*) The probability of a contact between monomers i and j diminishes as $k^{-3/2}$ in three dimensions with their sequence separation $k \equiv |i-j|$ [see Eq. (3.44)]. (*b*) This probability is independent of the position of the monomers within the chain sequence, *i. e.*, it depends only on $|i - j|$ and not on i or j individually. These predictions are in good agreement

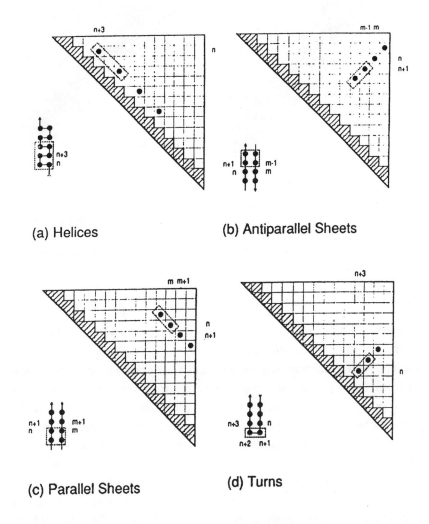

Figure 3.14: Secondary structures as contact patterns. Square lattice models of secondary structures [11]. The dotted boxes enclose minimal units of secondary structures. (a) Contacts for helices are along the third diagonal. (b & d) Contacts for antiparallel sheets and turns form strings that are perpendicular to the main diagonal. (c) Contacts for parallel sheets form strings that are parallel to the main diagonal.

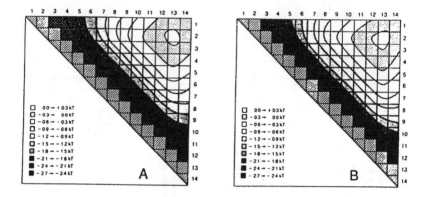

Figure 3.15: Entropic free energy surfaces for chains with 14 monomers. Contours show relative free energy of formation of all possible contact pairs. (A) Chains configured on a simple cubic lattice subject to excluded volume. (B) First-order path integral prediction with the excluded volume interaction strength v_0 =0.41 in three dimensions. Contours are given in 0.3 kT steps.

with experiments on polymers in theta solvents [9, 57, 58, 134, 145]. The results are somewhat different, however, for chains subject to excluded volume. The formation of a single contact between a pair of monomers far from the two ends of a chain with excluded volume and separated by k bonds reduces the number of accessible conformations by a factor proportional to $(k+1)^{-\nu_3}$, where $\nu_3 \simeq 2.4$ for chains in three dimensions [12]. This value of ν_3 estimated from short-chain enumerations and is quite close to the asymptotic ($n \to \infty$) exponent predicted by des Cloizeaux [32] using renormalization group methods, and agrees with the light-scattering experiments of Tsunashima & Kurata [151]. The exponent ν_3 enters into calculations [90] of stability gain of the folded states when disulfide or other crosslinks are formed in globular proteins [5, 90, 102, 110, 118, 150, 158, 159]. The value $\nu_3 \simeq 2.4$ gives a better fit [10] for the data from five crosslinks studied by Pace *et al.* [110] than the random-flight exponent $\nu_3 = 3/2$.

Figure 3.15A shows a convenient way to display the results of calculations of chain conformations subject to a constraint resulting from the adjacency of neighbor pair (i, j). The value

$$-\ln Q(i,j) + \text{constant} , \tag{3.69}$$

where $Q(i,j)$ is the number of conformations with monomer i and j in contact, is inserted in the matrix position (i, j). The quantity in Eq. (3.69) represents the (entropic) free energy, in units of kT, of forming the contact (i, j). We then construct a contour free-energy diagram directly on the contact map by connecting all the points of equal free energy by a contour line.[3] For example, Fig. 3.15A shows contours of free energies describing the relative favorabilities of forming all the possible monomer-monomer contacts. Figure 3.15B shows that the path integral treatment leads to essentially the same results as exhaustive lattice simulations (the path-integral results are given by Eq's 4.3 and 4.4 in Ref. [12]). The principal conclusions are: (a) smaller loops are more probable than larger ones, i. e. the most favored contacts are the most local; (b) excluded volume leads to a larger exponent than for random-flight chains; and (c) loop formation is more favored near chain ends than near the middle of the chain [10, 12].

Next, consider the full ensemble of a chain molecule that has two self-contacts. For a chain with a given contact that forms a small loop, the most probable second contact, taken over all possible conformations, is that which is closest in the chain sequence to the first contact. Among all possible two-contact conformations, these have the greatest entropy. These are in helix and antiparallel sheet configurations relative to the first contact (see Fig. 3.16A and Figure 2 in Ref. [13]). The conclusion that helices and sheets are entropically most favorable among two-contact chains is confirmed by path integral calculations (Fig. 3.16B); hence this phenomenon is not an artifact of the lattice. The contour plot in Fig. 3.16B is deduced from a first-order perturbation calculation of the path integral partition function with two presumed contacts (results given by Eq's 7.1-7.6 in Ref. [12]). The path integral results for $v_0 \simeq 0.25$ are in good agreement with

---

[3] For details of free energy contour plots on contact maps, see Ref. [12]. A discussion of the differences between contour plots of path-integral and lattice results is given at the end of Section VI of Ref. [12].

t	$n = 17$	$n = 18$	$n = 19$	$n = 20$
0	2,201,636	5,175,268	12,195,660	28,632,804
1	3,916,960	9,769,072	24,321,552	60,199,464
2	4,072,664	10,725,424	28,035,128	72,831,272
3	3,096,160	8,599,096	23,578,096	64,185,456
4	2,003,704	5,838,576	16,673,016	47,317,032
5	1,100,928	3,367,560	10,150,912	30,120,520
6	538,440	1,778,304	5,451,800	16,961,984
7	212,160	758,600	2,634,304	8,682,840
8	87,560	318,528	1,071,432	3,875,112
9	15,120	122,864	420,848	1,549,512
10		13,384	121,632	619,840
11			4,352	136,760
12				4,024
Total	17,245,332	46,466,676	124,658,732	335,116,620

Table 3.1: Number of conformations $Q^{(t)}(n)$ on square lattices as a function of chain length n and number of nearest-neighbor intrachain contacts t. The related quantity $\Omega^{(t)}(N)$, where $N \equiv n - 1$ is given in Ref. [11] for $n = 4\text{-}16$. Chains related by rigid rotations and reflections are counted as distinguishable by $Q^{(t)}$ but are considered to be identical by $\Omega^{(t)}$. Hence $Q^{(0)}(n) = 4[2\Omega^{(0)}(N = n-1) - 1]$ and $Q^{(t)}(n) = 8\Omega^{(t)}(N = n-1)$ for $t > 0$.

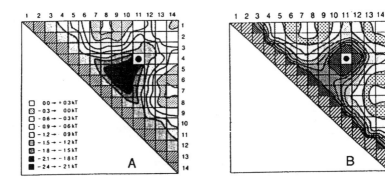

Figure 3.16: Contour plots showing relative entropic free energy of formation for all possible second contacts given the first contact at (4,11). Panel A is obtained from exhaustive simulation of n=14 chains on three-dimensional simple cubic lattices, and panel B is from first-order path integral calculations in three dimensions with the excluded volume interaction strength v_0 =0.25. Contours are given in 0.3 kT steps.

the lattice results; they predict the $(i, j) = (5, 10)$ antiparallel sheet contact is the most favorable second contact given the first contact at $(4, 11)$. The path integral also predicts the least favored contacts around $(1, 8)$ and $(7, 14)$, in agreement with the lattice simulations.

Now consider ensembles of chains of increasing compactness. Exhaustive lattice simulations show that the number of configurations accessible to a chain molecule diminishes rapidly with increasing numbers of intrachain contacts; see Table 3.1. While the total number of all conformations of n-segment chains increases approximately as $n^\gamma \mu^n$, where $\gamma \simeq 1/3$ in two dimensions and $\gamma \simeq 1/6$ in three dimensions, and $\mu \simeq 2.64$ for square lattices and $\mu \simeq 4.68$ for simple cubic lattices [50, 148], the number of maximally compact conformations varies approximately as κ^n, with $\kappa \simeq 1.41$ for square lattices [11] and $\kappa \simeq 1.74$ for simple cubic lattices [12], as noted above. In the Flory-Huggins [52-54, 75, 76] notation $\kappa = (z-1)/a$, where z is the lattice coordination number, the results indicate that $a \simeq 2.88$ in three dimensions. The restriction on conformational freedom by compactness is somewhat more severe than is estimated by the Flory approximation, which predicts $a = e = 2.71828\ldots$.

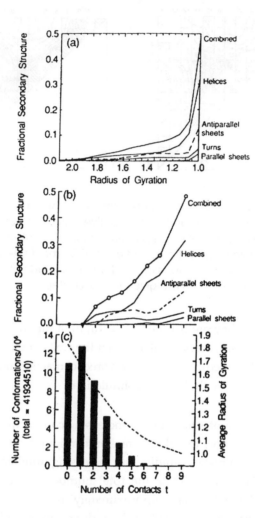

Figure 3.17: Amount of secondary structure over the full ensemble of all conformations of 12-segment chains: (a) as a function of the radius of gyration of the chains, measured in units of the minimum possible radius, and (b) as a function of chain compactness. (c) The histogram shows the number of accessible conformations as a function of the compactness of the molecule. The dotted curve shows the average radius of gyration for all chain compactnesses. Increasing compactness produces fewer accessible conformations, decreasing average radius, and increased amounts of secondary structure [13].

A most interesting result is that complete ensembles of chain conformations, as a function of increasing chain compactness, have increasing and
relatively large amounts of secondary structures in two or three dimensions [11, 13] (see Fig. 3.17). The distribution of the lengths of these
structures in the full ensemble of maximally compact chains closely resembles those in globular proteins [79] (see Fig. 3.18). Therefore three main
conclusions from these studies are: (a) compactness of a chain molecule
leads to internal structure; (b) the distribution of structures closely resembles that observed in globular proteins; and (c) some helix and turn
structure is even favored at the level of only two intrachain contacts.

Hence, if the nonlocal hydrophobic interactions drive proteins to collapse,
then the collapsed state is likely to have much internal structure comprised
of helices and sheets. In this regard, the role of density in causing internal
organization in compact polymers resembles the role of density in small
molecule systems: increasing density causes hard sphere liquids to become
crystalline solids, or hard rod liquids to align into liquid crystalline phases.
This implies that protein-like internal organization should arise in compact
polymers even in the absence of local forces, and thus that nonlocal forces
may be dominant determinants of structure in globular proteins. It does
not imply, however, that local forces are unimportant determinants of
protein structure. They undoubtedly account for the partial successes of
helix predictions in globular proteins [2, 17, 125]. Whereas nonlocal forces
should favor general helices, i. e., repetitions of $(i, i + 3)$ and related near-
neighbor contacts, we believe that the local forces dictate that polyamino
acids will specifically form α-helices, whereas other types of monomers
would lead to helices of other geometric structures.

Simulations also show that polymers near impenetrable surfaces (a) have
diminished conformational freedom and (b) prefer conformations with
increased amounts of secondary structure. Exhaustive simulations show
that the number of accessible conformations diminishes sharply as the
chain center of mass approaches a wall [16, 157] and references therein]
(see Fig. 3.19A). At the same time, proximity to the surface increases the
propensity for formation of helices in two-dimensional chains near a one-

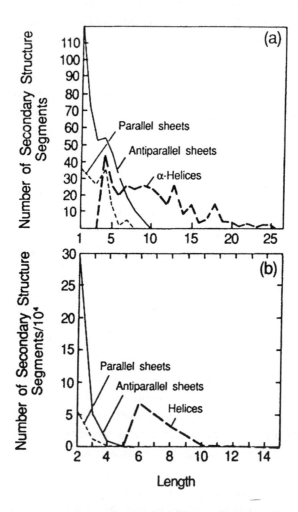

Figure 3.18: Length distributions of secondary structures. (a) Database observations of Kabsch & Sander [79] on 62 proteins of different chain lengths. (b) Exhaustive simulations of maximally compact chains of 26 residues on a two-dimensional square lattice. "Length" refers to the number of residues in a secondary structural element [13].

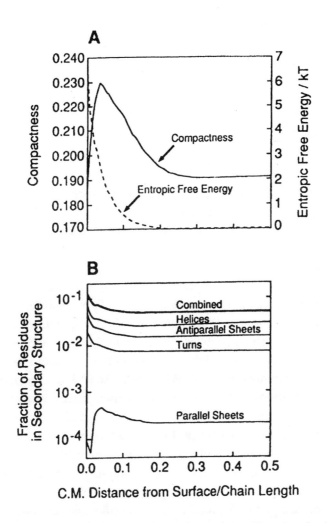

Figure 3.19: Surface effects on chain conformations. Results of exhaustive simulations of $n = 14$ chains on simple cubic lattices. (A) Compactness [11, 12] and entropic free energy relative to the bulk (negative logarithm of the number of accessible conformations divided by the number of accessible conformations in the bulk) as functions of the distance of the chain center of mass (C.M.) from the impenetrable planar surface. (B) Enhancement of secondary structure near the surface.

dimensional wall [157], and for helices and sheets in three-dimensional chains near a two-dimensional wall [16] (see Fig. 3.19B).

Conclusions

Our main purpose here has been to review methods and results of polymer theory that illuminate the roles of local and nonlocal interactions in determining the conformations of polymers, particularly proteins. The main conclusions are as follows. First, the interactions that cause polymers and polypeptides to form helices in solution are primarily local forces: e.g. hydrogen bonding between monomers $(i, i + 3)$ in α-helices. Second, the interactions that cause polymers to expand or collapse in different solvents (relative to the random coil) are nonlocal. Third, the collapse process in heteropolymers such as proteins differs from that in homopolymers in at least two respects. (a) Collapsed heteropolymers should have a core of solvent-averse monomers. (b) Whereas compact homopolymers have many conformations of low free energy (exponentially increasing with chain length), compact heteropolymers are predicted to have very few conformations of low free energy (1 or 2 or a few, relatively independent of chain length). Few heteropolymer configurations achieve the maximum possible number of contacts among solvophobic monomers; those configurations invariably involve a compact core of solvophobic monomers. Fourth, because globular proteins have considerable amounts of helix, local forces have generally been assumed to be dominant determinants of protein structure. However, recent exhaustive lattice simulations show that nonlocal forces alone are sufficient to account for much of the internal architecture, as well as the compactness and hydrophobic cores, of globular proteins. An individual chain molecule, when driven to compactness by nonlocal forces, will be comprised of considerable amounts of helix and sheet, simply due to steric constraints. In that regard, the unique structures of globular proteins, and their characteristic forms of internal organization, appear to be not so much a special consequence of amino acids as monomers in their abilities to pack or interact in special ways, but instead appear to arise simply as a general property of any heteropolymer containing strongly solvent-averse monomers.

Modeling the Effects of Mutations on the Denatured States of Proteins

This section is a slightly revised and updated version of a paper by Shortle, Chan and Dill which appeared in *Protein Science* 1, 201-215 (1992). It is reprinted here with permission of Cambridge University Press.

The stability of a protein is defined as the difference in free energy between its native and denatured states. Mutational changes in a protein affect its stability only through this free energy difference; there is no direct experiment to determine whether a mutation has more effect on one or the other of the native or denatured states. However, one way to learn about the relative importance of native and denatured states in mutational processes is through the use of theoretical models. The purpose of this paper is to describe an elementary model of protein stability that can be used to assess how mutations can affect the native and denatured state contributions to stability and the denaturation processes. The model shows that mutations can have large effects on the denatured states, and it provides a conceptual basis for understanding two puzzling experimental observations: (1) that replacing a polar residue by a hydrophobic residue on the surface of a protein can *destabilize* the molecule, and (2) that a single site mutation can affect the "denaturation slope" (*i. e.*, the m value; see below) by as much as 30%; previous theory has not been able to account for this [1, 37].

The Lattice Model

The philosophy underlying our choice of a model for protein denaturation is as follows: it should be based on the dominant physical driving forces - the hydrophobic interactions, conformational freedom of the chain, and the steric restrictions imposed by the excluded volume of the chain. It should provide results from which we can draw rigorous conclusions, without adjustable parameters or arbitrary assumptions, about native and denatured states for different sequences of monomers. These requirements preclude the use of atomic resolution models or the use of mean-field or long-chain lattice models. These requirements are satisfied, however, by an elementary model of short chains configured on two-dimensional square lattices that are sequences of H (hydrophobic) and P (other) monomers [14, 82, 83]. In this model, a contact between two H monomers is favorable by a free energy, ϵ ($\epsilon < 0$), envisioned to arise mainly from hydrophobic interactions [15, 34]. The contact interaction free energy is a potential of mean force [15], and implicitly accounts for the desolvation of the H monomers that must occur prior to formation of HH contacts. Because the chains are sufficiently short ($n = 16$ monomers in the present study), exhaustive computer enumeration of all the possible self-avoiding conformations permits unequivocal identification of the native conformations, *i. e.*, those at the global minimum of free energy and therefore also of all other conformations that by definition constitute the denatured state. In this model, native conformations are those that have the greatest number of HH contacts. Because the chains are sufficiently short, we can then identify the native structure(s) for every possible sequence of H and P monomers, *i. e.*, for the full sequence space.

Despite the obvious simplifications - that the chains are short, two-dimensional, and low-resolution - the model displays the following general properties of real proteins. (1) For a substantial fraction of sequences, small changes in the solvent character or temperature near a point of marginal stability lead to a relatively sharp transition from a large ensemble of denatured conformations to a relatively small ensemble of highly compact native conformations with a hydrophobic core; for a fraction of these folding sequences the ensemble of native structures is very small - of the order of one or a few configurations [14, 82, 83] (2) The two-dimensional compact configurations have the same distribution of helix and parallel and antiparallel sheet topologies as in the known proteins [11, 13]. (3) The mutational properties of the model are consistent with those of real proteins: (a) most mutations among sequences with unique native structures are neutral in that they do not change the native states (see below), (b) the surface residues are more mutable than core sites, (c) there are second-site revertants, and (d) there is much sequence convergence; *i. e.*, a given native structure will be encoded often in many different monomer sequences [14, 83]. (4) The kinetics of folding to the native state resembles that of proteins in the following respects: (a) there are favored folding pathways that are sequence-dependent [96], and (b) the kinetic intermediates are highly compact states (H. Chan & K. A. Dill 1992 in preparation). We believe the shortness and the two-dimensionality of the chains are not as severe a limitation as it might seem, at least for questions of principle, because a most important quantity relevant to the driving forces is the ratio of surface-to-interior sites in the compact structure. For real three-dimensional proteins of chain length equal to 100 amino acids, the ratio of the number of surface/interior sites is about 2.3 - 4.0; for two-dimensional model chains, the same value of this ratio is obtained for chain lengths of $n = 16$, the chain length we explore here.

The purpose of the present paper is to apply this model to problems of the reversible denaturation of proteins and of the effects of mutations on stability. Our study here of H/P two-dimensional square lattice chains of length $n = 16$ monomers does not include the full sequence space; we focus only on the subset of sequences that fold to unique native structures. Whereas there are $2^{16} = 65,536$ sequences in the $n = 16$ sequence space, only 1,539 sequences fold uniquely to a single native conformation. We believe that this subset of sequences is most representative of real biological proteins. In the present analysis, the number of sequences counts both a sequence and its distinguishable reverse sequence, *i. e.*,

$$PHPHHPHHHPHHHHHH \quad \text{and} \quad HHHHHHPHHHPHHPHP \quad (3.70)$$

are treated as distinct. For any $n = 16$ unique sequence there are $\Omega_0(16) = 802,075$ conformations on the square lattice [11], one of which is the unique native (N) state and the ensemble of all the rest of the conformations are defined to be the denatured (D) state. Each one of the $\Omega_0(16) - 1 = 802,074$ conformations in the D state will be referred to as a D conformation (see Fig. 3.20).

For any reversible process between two states, the equilibrium constant K_{eq} is the ratio of the probabilities that the system will be in one or the other state. Hence for the reversible folding-denaturation process, $D \rightleftharpoons N$,

$$K_{eq} = \frac{P_N}{P_D} , \qquad (3.71)$$

where P_N and P_D are the equilibrium probabilities of the N and D states, respectively. The free energy of folding ΔG, is given by

$$\Delta G = -kT \ln K_{eq} , \qquad (3.72)$$

where k is the Boltzmann constant and T is the absolute temperature. If h_N is the number of HH contacts in the N state, standard statistical mechanical considerations imply that

$$
\begin{aligned}
P_N &= Q^{-1} \exp\left[-\frac{h_N \epsilon}{kT}\right] , \\
P_D &= Q^{-1} \sum_{h=0}^{h_N-1} g(h) \exp\left[-\frac{h\epsilon}{kT}\right] ,
\end{aligned}
\qquad (3.73)
$$

where $g(h)$ is the number of conformations with h HH contacts and Q is the partition function defined in the Appendix. For $n = 16$ square-lattice chains, h ranges from 0 to the maximum [11] of 9, depending on the monomer sequence. It follows from Eq's (3.71-3.73) that three factors are responsible for determining the equilibrium between the N and D states: (1) ϵ, the strength of the HH potential of mean force in the solvent, (2) h_N, the maximum number of HH contacts that are achievable in the unique

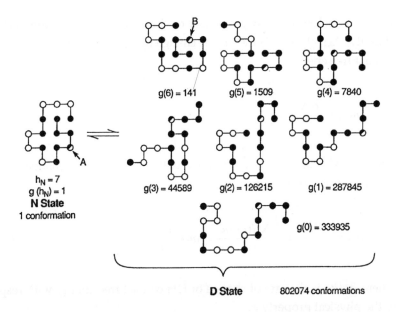

Figure 3.20: The model N and D conformations for a particular sequence. Open, filled, and half-filled circles represent P and H monomers and the mutation site, respectively. The number of conformations with h HH contacts is $g(h)$. One example D conformation is shown for each h. Site A is a position that has no H interaction in the N state (a corner site); site B shows that the same monomer does have one H interaction (an edge site) in some of the dominant D conformations (the $g(6) = 141$ conformations are dominant under strong folding conditions). Thus H at that position *destabilizes* the N state. Note that the H/P sequence determines the number and the structure of these $h = 6$ HH-contact D conformations.

N state of a given sequence, and (3) $g(h)$, the sequence-dependent D state distribution of the number of conformations as a function of the number of HH contacts. The statistical mechanics given in Eq. (3.73) and in the Appendix is general for H/P copolymers, and is not restricted to lattice models or to two-dimensional models.

We focus below on two properties of the D $\rightleftharpoons$ N equilibrium: (1) the stability, $\Delta G(\epsilon)$, and (2) the rate of change of stability with respect to HH contact free energy,

$$m(\epsilon) = \frac{\partial \Delta G(\epsilon)}{\partial \epsilon} . \tag{3.74}$$

The rate of change, m_x, of stability with respect to the change in any physical property x, is given by

$$m_x = \frac{\partial \Delta G}{\partial x} = \frac{\partial \Delta G(\epsilon)}{\partial \epsilon} \frac{\partial \epsilon}{\partial x} = m(\epsilon) \frac{\partial \epsilon}{\partial x} , \tag{3.75}$$

where $\partial \epsilon / \partial x$ is the rate of change of HH contact free energy with respect to the physical property x.

The study of these quantities in the model provides a conceptual framework for understanding the protein-folding equilibrium, especially with regard to native state stability and the experimentally determined "solvent denaturation slope,"

$$m_{\text{exp}} = \frac{\partial \Delta G}{\partial c} , \tag{3.76}$$

where c is the concentration of some denaturing agent in the solution. Commonly used in experiments are the solutes urea or guanidine hydrochloride (GuHCl) [37, 139, 140, 149]. In particular, we consider here only the simplest denaturing agents, for which the hydrophobic interaction decreases essentially linearly with increasing denaturant concentration. Data from experiments on free energies of transfer of hydrophobic residues indicates that this approximation of linear functional dependence of ϵ on c is reasonably good for both urea and GuHCl, whereas the deviation from linearity is slightly larger for GuHCl than for urea [99-101]. Hence for our purposes, the rate of change $\partial\epsilon/\partial c$ of ϵ with respect to concentration c may be taken to be constant. Therefore because $m_{\text{exp}} = m(\epsilon)\partial\epsilon/\partial c$, with $\partial\epsilon/\partial c$ a constant, it is most convenient to focus here simply on the behavior of $m(\epsilon)$.

Solvent Effects on Denaturation

How does the denaturation equilibrium depend on the strength of the HH attraction? It is clear from Eq. (3.73) that if the HH attraction is strengthened ($|\epsilon|$ is increased, ϵ more negative), by changing the solvent conditions or other physical parameters, the N state will be favored relative to every D conformation. This is because the number h_N of HH contacts in the N state is by definition greater than the number h of HH contacts in any D conformation. The free energy of folding is shown as a function of ϵ in the inset of Fig. 3.21. It is shown in the Appendix that the slope of this curve is

$$m(\epsilon) = h_N - \langle h \rangle_D(\epsilon) \tag{3.77}$$

for any H/P sequence, where $\langle h \rangle_D(\epsilon)$ is the ensemble-averaged number of HH contacts in the D state. In strong denaturing conditions ($|\epsilon| \approx 0$), there are on average relatively few HH contacts in the D state (*i. e.*, $\langle h \rangle_D$ is small) and the slope m is mainly determined by h_N, the number of HH contacts, *i. e.*, the amount of hydrophobic burial, in the N state. For some H/P sequences with $n = 16$, h_N is as small as 4 (relatively less stable N states); for other sequences, h_N is as large as 9 (relatively more stable N states). Under strong denaturing conditions ($|\epsilon| \approx 0$), the more hydrophobic contacts that a native protein has, the more strongly it is driven to fold up as the solvent is made to be increasingly poor for the

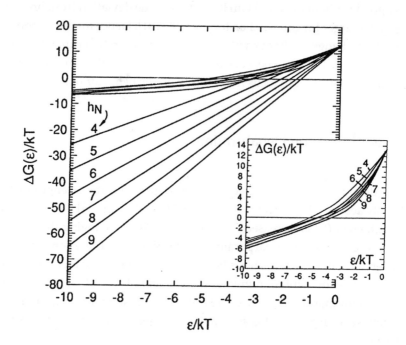

Figure 3.21: Stability, $\Delta G(\epsilon) = G_{\mathrm{N}}(\epsilon) - G_{\mathrm{D}}(\epsilon)$, as functions of HH contact free energy ϵ for H/P unique sequences ($n = 16$). Both $\Delta G(\epsilon)$ and ϵ are given in units of kT. More negative values of $\Delta G(\epsilon)$ imply that the N states are more stable. Each curve that is concave upwards shows $\Delta G(\epsilon)$ averaged over sequences with a specific number h_{N} of N state HH contacts, and is labelled by h_{N} in the inset. These curves are computed from the physically correct model that accounts for HH interactions in both the N and the D states. The straight lines in the main figure (labeled by h_{N}, which is equal to the slope of these straight lines) follows from the physically incorrect model discussed in the text, which arbitrarily assumes that there is no HH interaction in the D state. The inset shows more clearly the ϵ-dependence of the correct $\Delta G(\epsilon)$ on an expanded vertical scale.

hydrophobic residues, and thus $m(\epsilon)$ increases.

A more interesting result from the inset of Fig. 3.21 is the behavior of chains in solvent conditions for which there is large HH attraction (large $|\epsilon|$). As $|\epsilon|$ increases, the average number h of HH contacts in the D state, $\langle h \rangle_{\rm D}(\epsilon)$, increases, limiting at the highest h possible for the D conformations as $\epsilon \rightarrow -\infty$. For all the $n = 16$ unique sequences, the highest possible h is $h_{\rm N} - 1$, one fewer than the number of HH contacts in the N state. It follows that in the strong HH attraction limit, the very small class of D conformations that have $h_{\rm N} - 1$ HH contacts constitute the dominant population of the D state, hence $\langle h \rangle_{\rm D}$ is approximately given by $h_{\rm N} - 1$. Thus, by Eq. (3.77) the slopes $m = h_{\rm N} - \langle h \rangle_{\rm D} \approx 1$ for all sequences. Independent of their stabilities, the m value becomes identical for all sequences in this limit. In the same large $|\epsilon|$ limit the stability of the N state is approximately given by [see Eq. (3.94) in the Appendix]

$$\Delta G(\epsilon) \approx \epsilon + kT \ln g(h_{\rm N} - 1) , \tag{3.78}$$

thus the stability of the N state is principally determined by the number $g(h_{\rm N} - 1)$ of D conformations that have one fewer HH contact than the N state. In this limit, the denaturation midpoint ($\Delta G = 0$) is the point at which the extra contact gained by the N state is just counterbalanced by the higher degeneracy (conformational freedom) of the compact D conformations with $h = h_{\rm N} - 1$.

It has been common to view denatured states of proteins as random flight chains that are highly expanded and that are highly exposed to the solvent; such conformations are observed for many different proteins in 6M GuHCl [149]. In contrast, the results above indicate that the ensemble of denatured conformations depends on: (1) ϵ, the solvent-dependent strength of the hydrophobic interaction, and (2) $g(h)$, the property of the monomer sequence which describes how the chain can configure to form all the possible nonnative clusters of H monomers. In solvents that strongly favor folding, the stability of the native state is principally determined by the number of denatured conformations that have the highest number of hydrophobic contacts [corresponding to $g(h_N - 1)$ in the model, see also Eq. (3.92) in the Appendix]. This is a relatively small ensemble of denatured conformations; they have contact energies close to that of the native structure. Although we prefer the term "compact denatured states," similar conformations have also been referred to as the "molten globule" state.

To illustrate how different from this is the view of a fully exposed denatured state, we construct for comparison a hypothetical but physically incorrect model in which all denatured conformations are arbitrarily assumed to have zero contact interaction energy, as if all H residues were fully exposed to the solvent in the D state. According to this incorrect model, HH contact interactions are present only in the N state. The probability P_N for the N state is therefore identical to that of Eq. (3.73),

$$P_N = Q^{-1} \exp\left[-\frac{h_N \epsilon}{kT}\right].$$ (3.79)

However, because the D state is assumed to have no HH interaction, the $h\epsilon$ term in the second equation of Eq. (3.73) is zero, thus

$$P_D = Q^{-1} \sum_{h=0}^{h_N-1} g(h) \exp[0] = Q^{-1}[\Omega_0 - 1],$$ (3.80)

resulting in a sequence-independent D state that depends only on the total number of D conformations. This hypothetical model of D $\rightleftharpoons$ N equilibrium is incorrect because it assumes that the HH interaction can be turned on or off depending on whether the chain is in the N state or in one of the D conformations. This is clearly physically unrealistic. The consequence of this "wrong" model for the D states is shown in Fig. 3.21, and is compared with the physically correct Boltzmann-weighted D states model of Eq. (3.73). In this "wrong" model, the D state carries no sequence information; only the sequence-dependent number h_N of HH contacts in the N state is important for stability. According to the "wrong" model, the slopes m are always equal to h_N and independent of ϵ. Consequently the stabilities predicted by assuming that conformations of the D state are always fully exposed to the solvent are always larger than the true stabilities (see Fig. 3.21).

Single Site Mutations

How are the stability and the denaturant slope of a protein affected by changes in its sequence of residues? To address these questions, we have taken as wild type all 1,539 16-mer sequences that encode unique native structures, and we have mutated every possible position, one-at-a-time, from H to P or P to H, whichever is appropriate. Each mutant sequence was analyzed to determine if it encoded the same unique N state as the wild-type sequence by exhaustive conformational search. Of the $1,539 \times 16 = 24,624$ mutant sequences thus generated, 21,192 (86.06%) did not encode a unique N conformation and 704 (2.86%) encoded a unique N state different than that of the wild type, leaving 2,728 (11.08%) mutants that uniquely fold to the wild-type N structure; it is these mutations we selected for characterization. Thus most $(2,728/3,432 = 79.49\%)$ mutations which generate sequences with unique N states are neutral in that they do not change the wild-type N states. However, in contrast to earlier studies of the maximally compact conformations [83], a large fraction of all possible mutations either do not retain a single N state or alter the native states.

The effects of mutations depend on ϵ. Most of the results below are presented for $\epsilon = -4kT$. For this value a majority of sequences have a stable N state; $i.\,e.$, $\Delta G(\epsilon) < 0$. This value of ϵ, however, has no particular significance: qualitatively similar results are obtained for values of ϵ ranging from -1 to $-5kT$.

By symmetry, each P to H substitution that converts a wild-type sequence S to mutant sequence S′ will also be found when H to P substitution in S′ converts it to S. We consider only one direction here. In the present model, substitution mutations are classified into three types:

1. H0 → P0: In this class of substitution, an H residue that contacts no other residue in the N state is replaced by a P. These correspond to the corner positions in the square-lattice conformations. There are 824 such substitutions (and 824 P0 → H0 substitutions in the reverse direction). Two of the 824 substitutions do not have any effect on stability because the mutated monomer cannot form a contact with any H residue along the sequence and therefore do not change the D state distribution $g(h)$. Because this is an artifact of the even-odd effect of the square lattice [10, 12], we only consider below the other 822 H0 → P0 substitutions that lead to changes in $g(h)$.

2. HP → PP: In this class of substitution, an H residue that forms a topological contact with a P residue in the N state is replaced by a P. Because these residues are involved in contacts, they cannot be at lattice corners; rather they occur along outside edges of the lattice configuration of the N state. There are 104 such substitutions.

3. PH → HH: In this class of substitution, a P residue that forms a topological contact with an H in the N state is replaced by an H. These also correspond to edge positions in the square-lattice conformations. There are 434 such substitutions.

Only two of the $2,728/2 = 1,364$ wild-type/mutant pairs involve a position that forms two contacts in the N state (in both cases with two P monomers). Presumably interior mutations of monomers that have two or more N state contacts with other monomers almost invariably lead to multiply-degenerate N states or an N state with a different structure.

Effects of Mutations on Stability

H0 → P0 Mutations

Without exception (822/822), the replacement of a non-contacting surface H monomer with a P monomer *increases* the stability of the N state ($\Delta\Delta G <$ 0). The stability increase is found to be quite substantial on average (top panel of Fig. 3.22A). The average gain in stability is approximately $0.25|\epsilon|$, and the maximum is approximately $0.6|\epsilon|$. This is a significant fraction of $|\epsilon|$, the maximum change possible for forming a single HH contact. These mutants by definition have the same N structure as the wild type and the mutated monomer is at a position that makes no intrachain contacts in the N conformation. Therefore the free energy of the N state must be unchanged by the mutation, so any change in the D $\rightleftharpoons$ N equilibrium *must be* due to changes in the D states. This is illustrated in Fig. 3.20. Although the mutated monomer forms no contacts with other monomers in the N state, the same monomer is involved in intrachain contacts in the D state. It is clear that the number h of HH contacts in some of the D conformations in Fig. 3.20 depends on whether the mutated monomer is an H or a P. It follows that the mutation must alter $g(h)$, the distribution of conformations in the D state. The reason that this type of mutation always stabilizes the N state is because replacing an H monomer by a P necessarily causes fewer HH contacts in the ensemble of D conformations, increasing the free energy of D. Figure 3.23A(i) shows the distribution $g(h)$ and the N structure of the sequence that undergoes the largest increase in stability after a single mutation of this type. The D conformations shift towards those with fewer HH contacts. The numbers of D conformations with one fewer ($h = h_N - 1$) and two fewer ($h = h_N - 2$) HH contacts than the N state are also shown in the same figure for the wild-type/mutant pair of sequences. These numbers, $g(h_N - 1)$ and $g(h_N - 2)$, are major determinants of stability under strong folding conditions [see Eq's (3.78) and (3.94)].

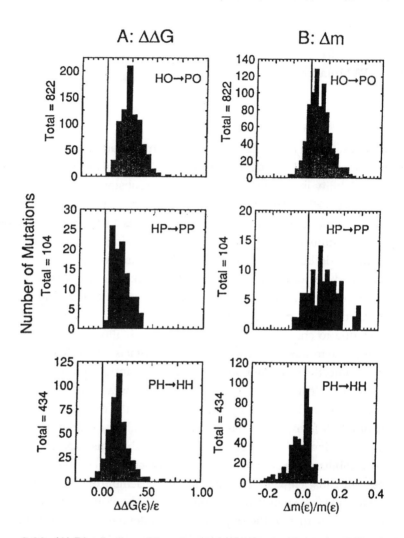

Figure 3.22: (A) Distribution of the change in the free energy of stabilization
$\Delta\Delta G(\epsilon)$ due to mutations (given in units of the HH contact free energy ϵ,
for $\epsilon = -4kT$). Because $\epsilon < 0$, positive changes in these plots indicate
increases in N state stability; negative changes indicate decreases in N
state stability. (B) Fractional change $\Delta m(\epsilon)/m(\epsilon)$ ($\epsilon = -4kT$).

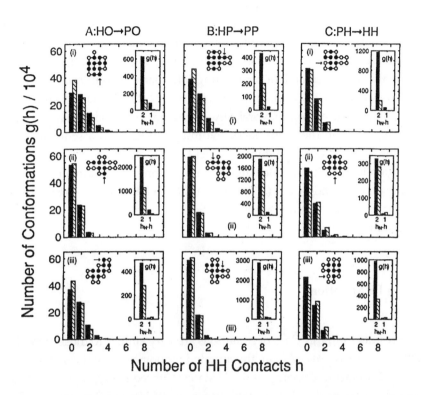

Figure 3.23: Changes in the distributions $g(h)$ of the number h of HH contacts in the D state for sequences before (solid bars) and after (striped bars) mutations. Each wild-type sequence is shown in its N structure, mutation sites are denoted by arrows. The insets show $g(h)$ for $h = h_{\mathrm{N}} - 1$ and $h = h_{\mathrm{N}} - 2$, i. e., the number of D conformations with one-fewer and two-fewer HH contacts than the N states, before and after mutations. Note that the number of N-state HH contacts h_{N} is unchanged by mutations of classes A and B whereas class-C mutations always increase h_{N} by one.

How does the mutation in Fig. 3.23A(i) increase the stability so much? The answer is most clearly seen by considering strongly folding conditions, for which the D $\rightleftharpoons$ N equilibrium is determined by the balance between the unique N conformation ($g(h_N) = 1$, zero conformational entropy) with h_N HH contacts (low contact energy), and the larger number $g(h_N - 1)$ of D conformations (higher conformational entropy) that have one fewer HH contacts than the N state (higher contact energy). Figure 3.23A(i) shows that the wild type is capable of configuring into many D conformations with $h = h_N - 1$ ($g(h_N - 1) = 94$). Thus the wild-type N state is relatively unstable due to this large entropic favorability of the compact D conformations with $h = h_N - 1$. On the other hand, the mutation leads to a sequence that has far fewer accessible highly compact D conformations ($g(h_N - 1) = 7$). The gain in stability, as calculated by substituting these numbers into the approximate relation Eq. (3.78), is accurate to 8% for $\epsilon = -4kT$. Thus, the main effect of the mutation is on the entropy of the highly compact D conformations with $h = h_N - 1$.

HP $\rightarrow$ PP Mutations

The results from this class of mutations are very similar to those from the H0 $\rightarrow$ P0 class above. Again, the mutated monomers do not contribute to the HH contact free energy of the N states. Consequently, replacement of H by P increases the N state stability for all 104 HP $\rightarrow$ PP mutations. The distribution of stability gain is given in the middle panel of Fig. 3.22A. The effect of this type of mutation on $g(h)$ is shown in Fig. 3.23B(i), for a mutant that causes a maximal gain in stability. The number of D conformations with $h = h_N - 1$ is 31 before and 4 after the H0 $\rightarrow$ P0 mutation. This large decrease in $g(h_N - 1)$ accounts for the large gain in stability.

The main difference between the H0 $\rightarrow$ P0 above and the HP $\rightarrow$ PP mutations is that the average $\Delta\Delta G$ is significantly smaller for the latter. For instance the average $\Delta\Delta G(\epsilon)$ over all HP $\rightarrow$ PP mutations shown in the middle panel of Fig. 3.22A for $\epsilon = -4kT$ is 0.174ϵ, only about 70% of that for the H0 $\rightarrow$ P0 mutations. Similar shifts in $\Delta\Delta G$ are observed for other ϵ. For ϵ ranging from -1 to $-5kT$, the average $\Delta\Delta G$ over the HP $\rightarrow$ PP mutations is 44%-75% of the average $\Delta\Delta G$ from the H0 $\rightarrow$ P0 mutations. Because these mutated monomers can contribute to HH contact energy only in the D states, this observation implies that on average an H monomer before an HP $\rightarrow$ PP mutation is less efficient in forming HH contacts in the D state than an H monomer before an H0 $\rightarrow$ P0 mutation. This phenomenon can be explained in the context of the model. For any given jth monomer along the sequence, only contacts with monomers numbered $j-3$, $j-5$, ..., and $j+3$, $j+5$, ... are possible. Each of these potential contacting monomers can either be an H or a P, depending on the sequence. If an H monomer to be mutated is already in contact with a P monomer in the N state (for HP $\rightarrow$ PP mutations), on average the number of its potential contacting partners that happen to be H's will be reduced relative to an H monomer to be mutated that does not have any intrachain contacts in the N state (in H0 $\rightarrow$ P0 mutations). This is confirmed by the observation that the average number of H monomers that are able to contact the mutated monomer in the D state for the H0 $\rightarrow$ P0 and HP $\rightarrow$ PP mutations are 2.52 (standard deviation 0.79) and 1.77 (standard deviation 0.72), respectively.

PH $\rightarrow$ HH Mutations

This class of mutations differs from the other two in that it alters the energetics of the N state - an additional HH contact is formed in the N state. However, even these mutations do not always lead to a net stabilization of the N state. In 14 out of 434 cases, the outcome is a net loss of stability when $\epsilon = -4kT$. Also noteworthy is the fact that the average increase in stability (bottom panel of Fig. 3.22A) is only approximately 0.17ϵ for $\epsilon = -4kT$, less than the stability gain for the corresponding H0 $\rightarrow$ P0 class. Surprisingly, on average, N is stabilized more by removing a noncontacting H at the surface than by adding an additional HH contact.

Obviously, the stabilization that could be added by introducing a new HH contact in the N state is at most ϵ. However, any mutation of P to H will contribute not only to new HH contacts in the N state but also to new HH contacts in the D state, so in general the increased stability will be less than ϵ. How much smaller depends both on the value of ϵ and on the details of the sequence, through the distribution function $g(h)$. Because this type of mutation affects both the N and the D states, the change in stability is more complex than the two types of mutations discussed above. Although the other two types of mutations always stabilize, it is possible, as noted above, for PH $\rightarrow$ HH mutations to destabilize the N states.

Fig. 3.23C(i) shows a pair of histograms of $g(h)$ for a wild type and its mutant, for the mutation that is maximally stabilizing. At $\epsilon = -4kT$, stability is mainly dependent upon $g(h_N-1)$, the number of highly compact D conformations with $h = h_N-1$. The PH $\rightarrow$ HH mutation changes $g(h_N-1)$ from 72 to 4, which leads to the large gain in stability. The main effect of this mutation on stability is therefore caused by the decreased entropy of the highly compact D state rather than on the additional HH contact in the N state. Similar histograms for the most destabilizing mutation are shown in Fig. 3.23C(ii). As anticipated from the arguments above, this mutation leads to an increase in $g(h_N - 1)$, in this case from 7 to 15.

Effects of Mutations on the Slope $m(\epsilon)$

The denaturant slope m_{exp} defined in Eq. (3.76) and the stability ΔG reflect different aspects of the distribution of protein conformations, and can be affected differently by mutations. A principal conclusion of this section is that whereas under strong folding conditions $g(h_N - 1)$ is a major determinant of stability, it is the ratio $g(h_N - 2)/g(h_N - 1)$ that is a major determinant of the denaturant slope. It is shown in the Appendix that if a wild-type sequence S is mutated to the mutant sequence S', the change in $m(\epsilon)$ is

$$\Delta m(\epsilon) = m(\epsilon)' - m(\epsilon) = \Delta h_N + \left[\langle h \rangle_D(\epsilon)\right] - \left[\langle h \rangle_D(\epsilon)\right]' . \qquad (3.81)$$

Here $\Delta h_{\mathrm{N}} = h_{\mathrm{N}}' - h_{\mathrm{N}}$ is the change in the number h_{N} of HH contacts in the N state upon mutation, and $\langle h \rangle_{\mathrm{D}}$ is the ensemble-averaged number of HH contacts in the D state.

Under strong folding conditions (large $|\epsilon|$), the mutational change $\Delta m(\epsilon)$ is approximately [see Eq. (3.100) in the Appendix]

$$\Delta m(\epsilon) \approx \left\{ \left[\frac{g(h_{\mathrm{N}} - 2)}{g(h_{\mathrm{N}} - 1)} \right]' - \left[\frac{g(h_{\mathrm{N}} - 2)}{g(h_{\mathrm{N}} - 1)} \right] \right\} e^{\epsilon/kT} . \tag{3.82}$$

This shows that the behavior of $m(\epsilon)$ is more subtle than $\Delta G(\epsilon)$ at large $|\epsilon|$. Whereas $\Delta\Delta G$ is determined mainly by the change in the number $g(h_{\mathrm{N}} - 1)$ of highly compact D conformations (see above), Δm is determined mainly by the change in $g(h_{\mathrm{N}} - 2)/g(h_{\mathrm{N}} - 1)$, the ratio of the number of D conformations with $h = h_{\mathrm{N}} - 2$ relative to that with $h = h_{\mathrm{N}} - 1$. In other words, the large $|\epsilon|$ limit of the m value of a sequence depends on the shape of the distribution $g(h)$ near $h = h_{\mathrm{N}}$.

H0 → P0 Mutations

As noted earlier, all of the mutations in this class increase the stability of the N state. However their effects on m are more subtle and diverse. The top panel of Fig. 3.22B shows that for the 822 mutants in this class, mutation at $\epsilon = -4kT$ leads to a wide distribution of changes in m, some positive and some negative. Under strong folding conditions (large $|\epsilon|$), the change in m mainly depends on $g(h_{\mathrm{N}} - 2)/g(h_{\mathrm{N}} - 1)$. A majority of mutations leads to an increase in m ($\Delta m > 0$), which according to Eq. (3.82) implies that $g(h_{\mathrm{N}} - 2)/g(h_{\mathrm{N}} - 1)$ increases. On the other hand a minority of mutations leads to $\Delta m < 0$, i. e., for these mutations $g(h_{\mathrm{N}} - 2)/g(h_{\mathrm{N}} - 1)$ decreases instead. Although all these mutations shift the D state ensemble to a smaller numbers of HH contacts, the relative partitioning of these changes among the sub-populations of compact conformations varies widely depending on the monomer sequence. The top panel of Fig. 3.22B shows that a single mutation can change m by as much as 31% relative to the wild type. Because the HH interactions in the N state are not affected by this class of mutations, this remarkably large change in m is solely attributable to the way the mutation affects the distribution of conformations in the the D state. Figures 3.23A(ii) 3.23A(iii) show examples of mutations that lead to a large increase and decrease in m respectively. The ratio $g(h_N - 2)/g(h_N - 1)$ changes from $2{,}436/201 = 12.12$ to $1{,}116/34 = 32.82$ for the mutations with the largest increase ($\Delta m/m = 31\% > 0$) and $g(h_N - 2)/g(h_N - 1)$ changes from $477/15 = 29.8$ to $291/15 = 19.4$ for the largest decrease ($\Delta m/m = -10.4\% < 0$).

HP $\rightarrow$ PP Mutations

This class of mutations shows the same patterns as the H0 $\rightarrow$ P0 mutations. A majority of mutations increase the value of m, but a significant fraction (16/104) decrease m (see the middle panel of Fig. 3.22B). Examples of the maximum increase and decrease in m at $\epsilon = -4kT$ are given in Fig. 3.23B (ii, iii respectively). The $g(h_N - 2)/g(h_N - 1)$ changes in the two cases are $1{,}904/106 = 17.96$ to $1{,}480/25 = 59.2$ ($\Delta m/m = 30\% > 0$), and $2{,}871/146 = 19.66$ to $1{,}183/104 = 11.38$ ($\Delta m/m = -8\% < 0$), respectively.

PH $\rightarrow$ HH Mutations

Although most of the mutations of this class increase the stability of the N state (see above), the effects on m are about evenly divided - about half the mutants have higher m than wild type, and about half have lower m (see the bottom panel of Fig. 3.22B). Because the mutant chain has an extra HH contact in the N state, Δh_N in Eq. (3.81) is equal to $+1$, in contrast with the other two classes of mutations, which have $\Delta h_N = 0$. However, the average number of HH contacts $\langle h \rangle_{\mathrm{D}}$ always increases because of the additional H monomer, resulting in a negative contribution to $\Delta m(\epsilon)$. In many cases the latter negative contribution overcompensates the positive $\Delta h_{\mathrm{N}} = +1$ contribution. Consequently, the distribution of Δm for this class of mutations is shifted towards more negative values relative to the distribution in the previous two classes of mutations. An example of this class of mutations which leads to the largest decrease in m at $\epsilon = -4kT$ is given in Fig. 3.23C(iii), the change in $g(h_{\mathrm{N}} - 2)/g(h_{\mathrm{N}} - 1)$ is from $972/14 = 69.43$ (wild type) to $334/18 = 18.56$ (mutant), with $\Delta m/m = -24\%$.

Summary

We have developed a simple model for the protein-folding equilibrium. It is based on short self-avoiding chains of H/P (hydrophobic/polar) copolymers on a two-dimensional square lattice. We compute the partition function for the D $\rightleftharpoons$ N (denatured/native) equilibrium exactly by exhaustive computer enumeration, and we explore all the possible single-site mutations for the sequences that encode unique native conformations. The main results of this model are as follows.

1. Mutations can have large effects on the entropy of the denatured states. By definition, the stability of a protein under native conditions is the free energy difference, $\Delta G = -kT \ln[\mathrm{N}]/[\mathrm{D}_0]$, where $[\mathrm{N}]$ is the concentration of the folded (native) species and $[\mathrm{D}_0]$ is the concentration of the denatured species under the same solution conditions [37] The present model study indicates that the denatured population D_0 is dominated by a relatively small ensemble of compact conformations with many hydrophobic contacts. Under strong folding conditions, the dominant denatured species will be highly compact and will have the second highest possible number of hydrophobic contacts h. In the two-dimensional lattice model analyzed here, the second highest h is always one fewer than the native number of HH contacts ($h_{\mathrm{N}} - 1$), and there are $g(h_{\mathrm{N}} - 1)$ such conformations. These highly compact denatured conformations are determined by the amino acid sequence, just as the native conformation

is. A major conclusion of the present work is that the amino acid sequence also determines *how many* highly compact denatured conformations there are. That is, the entropy of the denatured state of a protein under strong folding conditions is sequence-dependent. A single mutation can have a large effect on the stability of the protein by controlling how many highly compact denatured conformations are available, (*i. e.*, $g(h_N - 1)$).

2. The model shows that there are upper limits to stabilization that can be expected from addition of hydrophobic monomers, in agreement with an earlier mean-field model [33, 37]. As more H monomers are added to the sequence, more HH contacts become possible in the N state, but more HH contacts also occur in the conformations of the D state.

3. Two puzzling experimental results can be rationalized by the principles that emerge from the model. First, the model shows that a protein can be *destabilized* by replacing a surface residue by a more hydrophobic amino acid. This situation arises if the mutation site is at a position in the structure that makes no hydrophobic contacts in the native state, because the added hydrophobic residue can potentially make hydrophobic contacts in the denatured state, resulting in destabilization. Such a result has been observed experimentally [111], where it has been referred to as the "reverse hydrophobic effect." In our view, it is important not to misconstrue the term "reverse hydrophobic effect" to imply some special solvation process, because, according to the present model, the effect can arise simply from greater burial of nonpolar surface in the denatured than in the native state. Likewise, replacing a buried polar residue by a hydrophobic residue may not stabilize a protein, because the free energy of the denatured state may be lowered by a greater amount than the free energy of the native state. Second, there is now much experimental evidence for mutational changes in denaturation behavior that cannot be readily accounted for by changes in the native state alone [139, 140]. One example is the very broad distribution of $m_{exp} = \partial \Delta G / \partial c$ values for the mutational changes of stabilities of staphylococcal nuclease (see Fig. 3.24B). Simple thermodynamic models and mean-field statistical mechanical models [1, 37] would predict that if one monomer is altered in a chain of n monomers, then the change in

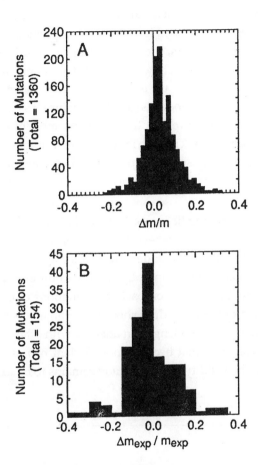

Figure 3.24: A: Distribution of fractional change $\Delta m(\epsilon)/m(\epsilon)$ ($\epsilon = -4kT$) for all mutations. B: Distribution of fractional change $\Delta m_{\mathrm{exp}}/m_{\mathrm{exp}}$ in denaturant slope for 154 single mutations on staphylococcal nuclease (D. Shortle et al., unpublished results). This includes substitutions of phenylalanine, isoleucine, leucine, methioine, asparagine, proline, glutamine, serine, threonine, valine and tyrosine residues to both alanine and glycine, as well as substitutions of alanine to glycine and glycine to alanine.

m value should be approximately of magnitude $1/n$, i.e, a few percent at most. However, the present model, which takes into account the effects of amino acid sequence on the denatured states, shows that the effect can be considerably larger than this; Fig. 3.24A shows predicted changes of up to 31%. Consistent with this, the experiments by Shortle and coworkers on staphylococcal nuclease show similar distributions, and with single mutational change on m by up to 30 - 40%. The reason for such large changes, according to the model, is that the denaturant slope is determined by both the native structure and the small ensemble of highly compact denatured conformations. The distribution of hydrophobic contacts in this small ensemble is highly sensitive to changes in the amino acid sequence and therefore can be changed dramatically by single mutations.

Although the present prediction of large mutational effects is based on results on short chains, we believe the same principles should apply to longer chains. This is because under strong folding conditions the $D \rightleftharpoons N$ equilibrium is mainly determined by the distribution $g(h)$ for h values slightly less than h_N. In the compact conformations of copolymers, the clustering of hydrophobic residues should be spatially localized and relatively independent of chain length, in analogy with other conformational properties of polymers that become independent of n at high densities [28]. However, the validity of this opinion remains to be tested on longer chains.

Acknowledgements

We thank Eugene Shakhnovich for helpful discussions. We thank the NIH, the URI Program of DARPA, and the Pew Scholars Program in the Biomedical Sciences for financial support.

Appendix: The Statistical Mechanics of the Model

The following is the general theory of conformational change in H/P copolymers and proteins. The theory is not restricted to lattice models and applies to any spatial dimensionality.

Stability and the Partition Function

The probability of a single conformation with h HH contacts is proportional to

$$\exp[-\frac{h\epsilon}{kT}],$$

(3.83)

where $\epsilon < 0$ is the contact free energy per HH contact, k is the Boltzmann constant and T is the absolute temperature. Hence the probability that the chain adopts conformation(s) with h HH contacts is

$$P(h) = g(h)e^{-h\epsilon/kT}/Q,$$

(3.84)

where $g(h) \geq 0$ is the number of conformations that have h HH contacts, and the summation

$$Q \equiv \sum_{h=0}^{h_N} g(h)e^{-h\epsilon/kT}$$

(3.85)

is the partition function; hence $\sum_{h=0}^{h_N} P(h) = 1$. Here h_N is the maximum number of HH contacts for the specific sequence. Because we are primarily interested in sequences with unique native structures, i. e., $g(h_N) = 1$, h_N is the number of HH contacts in the unique native structure of a particular sequence. (The quantity $g(h_N)$ is identical to the degeneracy g in Chan & Dill [14] As the denatured (D) state is the ensemble of all conformations except the unique native (N) state, the probability P_N and P_D of the N and the D state are, respectively,

$$P_N(\epsilon) = \frac{e^{-h_N \epsilon / kT}}{Q} \tag{3.86}$$

and

$$P_D(\epsilon) = \frac{\sum_{h=0}^{h_N - 1} g(h) e^{-h\epsilon / kT}}{Q} . \tag{3.87}$$

The uniqueness of the N state implies that its free energy $G_N(\epsilon)$ is essentially determined by the sum of HH contact free energies,

$$G_N(\epsilon) = -kT \ln P_N(\epsilon) = h_N \epsilon + kT \ln Q , \tag{3.88}$$

whereas the free energy $G_D(\epsilon)$ of the D state consists of contributions from the ensemble of all other conformations except the N state,

$$G_D(\epsilon) = -kT \ln P_D(\epsilon) = -kT \ln \left[\sum_{h=0}^{h_N - 1} g(h) e^{-h\epsilon / kT} \right] + kT \ln Q . \tag{3.89}$$

Thus the stability of the N state relative to the D state is given by the difference in free energy,

$$\Delta G(\epsilon) \equiv G_{\mathrm{N}}(\epsilon) - G_{\mathrm{D}}(\epsilon) = h_{\mathrm{N}}\epsilon + kT \ln\left[\sum_{h=0}^{h_{\mathrm{N}}-1} g(h)e^{-h\epsilon/kT}\right]. \qquad (3.90)$$

Hence the stability of the N conformation is a balance: the N state is stabilized by its h_{N} HH contacts, but is destabilized by the multiplicity $g(h)$ of D conformations. The N state becomes highly populated as the HH attraction becomes strong (large $|\epsilon|/kT$), but under this condition the dominant species of D in the summation of Eq. (3.90) are the few conformations with the largest h (*i. e.*, the most compact D states). Under these conditions of strong HH attraction, q. (3.90) may be expressed as

$$
\begin{aligned}
\Delta G(\epsilon) &= h_{\mathrm{N}}\epsilon + kT \ln\left[e^{-(h_{\mathrm{N}}-1)\epsilon/kT}g(h_{\mathrm{N}}-1)\right. \\
&\quad \times\left.\left\{1 + e^{\epsilon/kT}\frac{g(h_{\mathrm{N}}-2)}{g(h_{\mathrm{N}}-1)} + e^{2\epsilon/kT}\frac{g(h_{\mathrm{N}}-3)}{g(h_{\mathrm{N}}-1)} + \cdots\right\}\right] \\
&= \epsilon + kT \ln g(h_{\mathrm{N}}-1) \\
&\quad + kT \ln\left[1 + e^{\epsilon/kT}\frac{g(h_{\mathrm{N}}-2)}{g(h_{\mathrm{N}}-1)} + e^{2\epsilon/kT}\frac{g(h_{\mathrm{N}}-3)}{g(h_{\mathrm{N}}-1)} + \cdots\right],
\end{aligned}
$$
$$(3.91)$$

for sequences with $g(h_{\mathrm{N}}-1) \neq 0$. The generalization of this result is straightforward; if $h_{\mathrm{N}} - j$ is the highest number of HH contacts possible among the D state conformations, the above expression is instead

$$
\begin{aligned}
\Delta G(\epsilon) &= j\epsilon + kT \ln g(h_{\mathrm{N}}-j) \\
&\quad + kT \ln\left[1 + e^{\epsilon/kT}\frac{g(h_{\mathrm{N}}-j-1)}{g(h_{\mathrm{N}}-j)} + e^{2\epsilon/kT}\frac{g(h_{\mathrm{N}}-j-2)}{g(h_{\mathrm{N}}-j)} + \cdots\right].
\end{aligned}
$$
$$(3.92)$$

Because $j = 1$ for all $n = 16$ sequences considered here, subsequent analysis will only be presented for the case of Eq. (3.91). It is straightforward to obtain results for the general case by using Eq. (3.92) instead of Eq. (3.91) in subsequent derivations.

The last logarithmic term in Eq. (3.91) can be further expanded using the Taylor expansion,

$$\ln(1 + x) = -\sum_{s=1}^{\infty} \frac{(-x)^s}{s} = x - \frac{x^2}{2} + \frac{x^3}{3} - \dots . \qquad (3.93)$$

The stability ΔG can then be expressed as a power series in $e^{\epsilon/kT}$,

$$\begin{aligned}
\Delta G(\epsilon) &= \epsilon + kT \ln g(h_{\rm N} - 1) \\
&+ kT \left[e^{\epsilon/kT} \frac{g(h_{\rm N} - 2)}{g(h_{\rm N} - 1)} + e^{2\epsilon/kT} \left\{ \frac{g(h_{\rm N} - 3)}{g(h_{\rm N} - 1)} \right. \right. \\
&- \left. \frac{1}{2} \left[\frac{g(h_{\rm N} - 2)}{g(h_{\rm N} - 1)} \right]^2 \right\} \\
&+ \left. O(e^{3\epsilon/kT}) \right] , \qquad (3.94)
\end{aligned}$$

where $O(e^{3\epsilon/kT})$ stands for all terms with $e^{\epsilon/kT}$ raised to the third and higher powers. When the HH attraction is very strong, $\epsilon/kT \to -\infty$, all powers of $e^{\epsilon/kT}$ tend to zero. Consequently, the stability ΔG of the N state is primarily determined by the relative magnitude of ϵ and the entropic free energy contributed by the conformational freedom of the $h_{\rm N} - 1$ ("first excited" state) conformations. Only when the HH attraction is not too strong can conformations that have fewer than $h_{\rm N} - 1$ HH contacts have more significant effects on stability.

Rate of Change of Stability with respect to HH Contact Energy

The rate of change of ΔG with respect to the contact free energy ϵ is given by the derivative of Eq. (3.90),

$$m(\epsilon) \equiv \frac{\partial \Delta G(\epsilon)}{\partial \epsilon} = h_{\rm N} - \frac{\sum_{h=0}^{h_{\rm N}-1} h \, g(h) e^{-h\epsilon/kT}}{\sum_{h=0}^{h_{\rm N}-1} g(h) e^{-h\epsilon/kT}} = h_{\rm N} - \langle h \rangle_{\rm D}(\epsilon), (3.95)$$

where $\langle h \rangle_\text{D}(\epsilon)$ is the average number of HH contacts in the denatured (D) state. Thus $m(\epsilon)$ is the difference between the number of HH contacts in the N state and the average number of HH contacts in the D state. If the contacting area per HH contact is given by C_A, and the exposed area of an H monomer is defined as a quantity proportional to the number of contacts it makes with P monomers or solvent, $i.\,e.$, all contacts except those with other H monomers, then the exposed H area A_N in the N state is

$$A_\text{N} = A_0 - 2C_A h_\text{N} \,, \tag{3.96}$$

where A_0 is the exposed H area when all H monomer are fully exposed. This is because two units of exposed area are buried per contact pair. Similarly, the average exposed H area in the D state is

$$\langle A \rangle_\text{D}(\epsilon) = A_0 - 2C_A \langle h \rangle_\text{D}(\epsilon) \,. \tag{3.97}$$

Hence it follows from Eq. (3.95) that

$$m(\epsilon) = \frac{1}{2C_A}[\langle A \rangle_\text{D}(\epsilon) - A_\text{N}] \,, \tag{3.98}$$

$i.\,e.$, $m(\epsilon)$ is proportional to the difference between the average exposed area of H monomers in the D state and the exposed area of H monomers in the N state, as proposed by Schellman [130].

The second derivative of ΔG yields

$$
\begin{aligned}
\frac{\partial m(\epsilon)}{\partial \epsilon} &\equiv \frac{\partial^2 \Delta G(\epsilon)}{\partial \epsilon^2} \\
&= \frac{1}{kT}\left\{ \frac{\sum_{h=0}^{h_\text{N}-1} h^2 g(h) e^{-h\epsilon/kT}}{\sum_{h=0}^{h_\text{N}-1} g(h) e^{-h\epsilon/kT}} - \left[\frac{\sum_{h=0}^{h_\text{N}-1} h\, g(h) e^{-h\epsilon/kT}}{\sum_{h=0}^{h_\text{N}-1} g(h) e^{-h\epsilon/kT}} \right]^2 \right\} \\
&= \frac{1}{kT}\left\{ \langle h^2 \rangle_\text{D}(\epsilon) - \left[\langle h \rangle_\text{D}(\epsilon) \right]^2 \right\} \,,
\end{aligned}
\tag{3.99}
$$

where $\langle h^2 \rangle_D$ is the mean square number of HH contacts in the D state. The last line in Eq. (3.99) indicates that the second derivative of ΔG is equal to the variance of the number of HH contacts in the D state divided by kT.

It is obvious from Eq's (3.95) and (3.99) that both $m(\epsilon)$ and $\partial m(\epsilon)/\partial \epsilon$ are nonnegative. Hence all ΔG vs. ϵ curves have positive slopes and are concave upward (Fig. 3.21). When there is little or no attraction between H monomers ($\epsilon \to 0$), the favored D conformations are open and have small h. On the other hand, when the HH attraction is very strong ($\epsilon \to -\infty$), the favored D conformations are the most compact, with large $h \approx h_N - 1$. broader distribution, i. e., a larger variance, of h is only possible at intermediate values of ϵ. Hence by Eq. (3.99) ΔG vs ϵ has its largest curvature at intermediate ϵ (see Fig. 3.21).

It is useful to express $m(\epsilon)$ as a power series in $e^{\epsilon/kT}$. Differentiation of Eq. (3.94) yields

$$
\begin{aligned}
m(\epsilon) &= 1 + e^{\epsilon/kT}\frac{g(h_N - 2)}{g(h_N - 1)} + e^{2\epsilon/kT}\left\{2\frac{g(h_N - 3)}{g(h_N - 1)} - \left[\frac{g(h_N - 2)}{g(h_N - 1)}\right]^2\right\} \\
&\quad + O(e^{3\epsilon/kT}) .
\end{aligned}
\tag{3.100}
$$

When the HH attraction is very strong ($\epsilon \to -\infty$), the m value for all sequences with $g(h_N - 1) > 0$ reduces to unity [reduces to j for the general case of Eq. (3.92)]. When the HH attraction is sufficiently strong, Eq. (3.100) shows that the m value is essentially determined by the ratio of the number of conformations at the $h = h_N - 1$ and $h = h_N - 2$ levels. This feature is illustrated by the examples discussed in the text (Fig. 3.23). At the other extreme, when there is little or no HH attraction ($\epsilon \to 0$), the m value tends to be large. This follows from Eq. (3.95) and the fact that $\langle h \rangle_D$ is small for small $|\epsilon|$. The exact value of $m(\epsilon)$ at $\epsilon = 0$ is sequence dependent.

Effects of Mutations

When a mutation changes sequence S to sequence S', the change in stability and m value are, respectively,

$$\Delta\Delta G(\epsilon) \equiv \Delta G(\epsilon)' - \Delta G(\epsilon) , \tag{3.101}$$

and

$$\Delta m(\epsilon) \equiv m(\epsilon)' - m(\epsilon) = \Delta h_{\mathrm{N}} + \left[\langle h\rangle_{\mathrm{D}}(\epsilon)\right] - \left[\langle h\rangle_{\mathrm{D}}(\epsilon)\right]' ,$$
$$\Delta h_{\mathrm{N}} \equiv h_{\mathrm{N}}' - h_{\mathrm{N}} , \tag{3.102}$$

where the unprimed and primed terms correspond to the wild-type sequence S and the mutated sequence S' respectively. Note that when the HH attraction is very strong,

$$\lim_{\epsilon \to -\infty} \Delta m(\epsilon) = 0 \tag{3.103}$$

by Eq. (3.100), implying that no mutation can lead to a change in m value at infinite $|\epsilon|$. It follows that for large $|\epsilon|$ the distribution of $\Delta m(\epsilon)$ over all possible mutations should peak narrowly around zero.

The ϵ-dependence of $\Delta m(\epsilon)$ is illustrated by Fig. 3.25, which shows the distribution of $\Delta m(\epsilon)$ at different ϵ for mutations of $n = 16$ sequences. Because of Eq. (3.103), $\Delta m(\epsilon)$ approaches zero for all mutations at large ϵ (three bottom panels of Fig. 3.6 for $\epsilon = -10kT$).

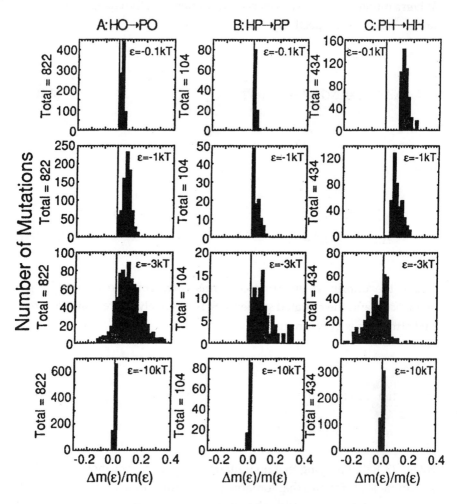

Figure 3.25: Distribution of fractional change $\Delta m(\epsilon)/m(\epsilon)$ as a function of HH contact free energy ϵ (for $\epsilon = -0.1, -1, -3$, and $-10kT$).

References

[1] D. O. V. Alonso and K. A. Dill (1991). *Biochemistry* **30**, 5974.

[2] C. B. Anfinsen and H. A. Scheraga (1975). *Adv. Protein Chem.* **29**, 205.

[3] E. V. Anufrieva, T. M. Birshtein, T. N. Nekrasova, O. B. Ptitsyn, and T. V. Sheveleva (1968). *J. Polymer Sci.* **C16**, 3519.

[4] M. N. Barber, B. W. Ninham (1970). *Random and Restricted Walks: Theory and Applications* (Gordon & Breach, New York).

[5] W. J. Becktel, W. A. Baase, R. Wetzel, and L.J. Perry (1986). *Biophys. J.* **49**, 109a.

[6] T. M. Birshtein and O. B. Ptitsyn (1966). *Conformations of Macromolecules*, translated from the Russian edition by S. N. Timasheff and M. J. Timasheff, in *Higher Polymer Vol. 22*, H. Mark, ed. (Interscience, New York).

[7] J. D. Bryngelson and P. G. Wolynes (1987). *Proc. Natl. Acad. Sci. USA.* **84**, 7524.

[8] J. D. Bryngelson and P. G. Wolynes (1989). *J. Phys. Chem.* **93**, 6902.

[9] W. Burchard (1986). In *Cyclic Polymers*, J. A. Semlyen, ed., pp. 43-84. (Elsevier, New York).

[10] H. S. Chan and K. A. Dill (1989). *J. Chem. Phys.* **90**, 492 [see also *J. Chem. Phys.* (1992) **96**, 3361].

[11] H. S. Chan and K. A. Dill (1989). *Macromolecules* **22**, 4559.

[12] H. S. Chan and K. A. Dill (1990). *J. Chem. Phys.* **92**, 3118.

[13] H. S. Chan and K. A. Dill (1990). *Proc. Natl. Acad. Sci. USA.* **87**, 6388.

[14] H. S. Chan and K. A. Dill (1991). *J. Chem. Phys.* **95**, 3775.

[15] H. S. Chan and K. A. Dill (1991). *Annu. Rev. Biophys. Biophys. Chem.* **20,** 447-490.

[16] H. S. Chan, M. R. Wattenbarger, D. F. Evans, V. A. Bloomfield, and K. A. Dill (1991). *J. Chem. Phys.* **94,** 8542 [see also *J. Chem. Phys.* (1992) **96,** 3361].

[17] P. Y. Chou and G. D. Fasman (1974). *Biochemistry* **13,** 211; 222.

[18] B. Chu, I. H. Park, Q. -W. Wang, and C. Wu (1987). *Macromolecules* **20,** 2833.

[19] B. Chu and Z. Wang (1988). *Macromolecules* **21,** 2283.

[20] B. Chu and Z. Wang (1989). *Macromolecules* **22,** 380.

[21] B. Chu, R. Xu, Z. Wang and J. Zuo (1988). *J. Appl. Cryst.* **21,** 707.

[22] B. Chu, R. Xu, and J. Zuo (1988). *Macromolecules* **21,** 273.

[23] D. G. Covell and R. L. Jernigan (1990). *Biochemistry* **29,** 3287.

[24] P. -G. de Gennes (1972). *Phys. Lett.* **A38,** 339.

[25] P. -G. de Gennes (1975). *Journ. de Phys. Lett.* **36,** L55.

[26] P.-G. de Gennes (1977). *Rivista del Nuovo Cimento* **7,** 363.

[27] P.-G. de Gennes (1978). *Journ. de Phys. Lett.* **39,** L299.

[28] P.-G. de Gennes (1979). *Scaling Concepts in Polymer Physics* (Cornell University Press, Ithaca NY).

[29] B. Derrida (1981). *Phys. Rev.* **B24,** 2613.

[30] J. des Cloizeaux (1974). *Phys. Rev.* **A10,** 1665.

[31] J. des Cloizeaux (1975). *Jour. de Phys.* **36,** 281.

[32] J. des Cloizeaux (1980). *Jour. de Phys.* **41,** 223.

[33] K. A. Dill (1985). *Biochemistry* **24,** 1501.

[34] K. A. Dill (1990). *Biochemistry* **29,** 7133.

[35] K. A. Dill and D. O. V. Alonso (1988). In *Colloquium Mosbach der Gessellschaft für Biologische Chemie: Protein Structure and Protein Engineering* Vol. 39, ed. R. Huber, E. L. Winnacker, pp. 51-58 Berlin: Springer-Verlag.

[36] K. A. Dill, D. O. V. Alonso, and K. Hutchinson (1989). *Biochemistry* **28**, 5439.

[37] K. A. Dill and D. Shortle (1991). *Annu. Rev. Biochem.* **60**, 795.

[38] C. Domb (1963). *J. Chem. Phys.* **38**, 2957.

[39] C. Domb (1969). *Adv. Chem. Phys.* **15**, 229.

[40] C. Domb (1974). *Polymer* **15**, 259.

[41] C. Domb, J. Gillis and G. Wilmers (1965). *Proc. Phys. Soc.* **85**, 625.

[42] J. F. Douglas and K. F. Freed (1985). *Macromolecules* **18**, 2445.

[43] S. F.Edwards (1965). *Proc. Phys. Soc.***85**, 613.

[44] S. F. Edwards and P. W. Anderson (1975). *J. Phys.* **F5**, 965.

[45] R. P. Feynman (1948). *Rev. Mod. Phys.* **20**, 367.

[46] R. P. Feynman and A. R. Hibbs (1965). *Quantum Mechanics and Path Integrals* (McGraw-Hill, New York).

[47] A. V. Finkelstein and E. I. Shakhnovich (1989). *Biopolymers* **28**, 1681.

[48] M. E. Fisher (1966). *J. Chem. Phys.* **44**, 616.

[49] M. E. Fisher (1968). *J. Phys. Soc. Japan (Suppl.)* **26**, 44.

[50] M. E. Fisher and B. J. Hiley (1961). *J. Chem. Phys.* **34**, 1253.

[51] M. Fixman (1955). *J. Chem. Phys.* **23**, 1656.

[52] P. J. Flory (1942) *J. Chem. Phys.* **10**, 51.

[53] P. J. Flory (1949). *J. Chem. Phys.* **17**, 303.

[54] P. J. Flory (1953). *Principles of Polymer Chemistry* (Cornell University Press, Ithaca NY).

[55] P. J. Flory (1969). *Statistical Mechanics of Chain Molecules* (Wiley, New York).

[56] P. J. Flory and S. Fisk (1966). *J. Chem. Phys.* **44,** 2243.

[57] P. J. Flory and J. A. Semlyen (1966). *J. Am. Chem. Soc.* **88,** 3209.

[58] P. J. Flory U. W. Suter and M. Mutter (1976) *J. Am. Chem. Soc.* **98,** 5733; 5745.

[59] K. F. Freed (1972). *Adv. Chem. Phys.* **22,** 1.

[60] K. F. Freed (1987). *Renormalization Group Theory of Macromolecules* (Wiley, New York).

[61] M. Gabay and T. Garel (1978). *Jour. de Phys. Lett.* **39,** L123.

[62] T. Garel and H. Orland (1988). *Europhys. Lett.* **6,** 307; 597.

[63] I. M. Gel'fand and A. M. Yaglom (1960). *J. Math. Phys.* **1,** 48.

[64] S. Geyler, T. Pakula and J. Reiter (1990). *J. Chem. Phys.* **92,** 2676.

[65] C. Ghelis and J. Yon (1982). *Protein Folding* (Academic Press, New York).

[66] S. J. Gill and I. Wadso (1976). *Proc. Natl. Acad. Sci. USA* **73,** 2955.

[67] Y. Goto and A. L. Fink (1989). *Biochemistry* **28,** 945.

[68] Y. Goto and A. L. Fink (1990). *J. Mol. Biol.* **214,** 803.

[69] H. R. Guy (1985). *Biophys. J.* **47,** 61.

[70] J. P. Hansen and I. R. McDonald (1986). *Theory of Simple Liquids* (Academic Press, New York).

[71] W. Heisenberg (1928). *Zeits. für Phys.* **49,** 619.

[72] J. J. Hermans and J. Th. G. Overbeek (1948). *Recl. Trav. Chim. Pays-Bas* **67,** 761.

[73] B. J. Hiley and M. F. Sykes (1961). *J. Chem. Phys.* **34,** 1531.

[74] T. L. Hill (1956). *Statistical Mechanics* (McGraw-Hill, New York).

[75] M. L. Huggins (1942). *J. Phys. Chem.* **46**, 151.

[76] M. L. Huggins (1942). *Ann. N.Y. Acad. Sci.* **43**, 1.

[77] E. Ising (1925). *Zeit. für Phys.* **31**, 253.

[78] H. Jacobson and W. H. Stockmayer (1950). *J. Chem. Phys.* **18**, 1600.

[79] W. Kabsch and C. Sander (1983). *Biopolymers* **22**, 2577.

[80] A. L. Kholodenko and K. F. Freed (1984). *J. Phys.* **A17**, 2703.

[81] H. A. Kramers and G. H. Wannier (1941). *Phys. Rev.* **60**, 252; 263.

[82] K. F. Lau and K. A. Dill (1989). *Macromolecules* **22**, 3986.

[83] K. F. Lau and K. A. Dill (1990). *Proc. Natl. Acad. Sci. USA* **87**, 638.

[84] J. C. LeGuillou and J. Zinn-Justin (1977). *Phys. Rev. Lett.* **39**, 95.

[85] I. M. Lifshitz (1969). *Sov. Phys. JETP* **28**, 1280.

[86] I. M. Lifshitz, A. Grosberg, A. Yu., and A. R. Khokhlov (1978). *Rev. Mod. Phys.* **50**, 683.

[87] S. Lifson and B. H. Zimm (1963). *Biopolymers* **1**, 15.

[88] K. U. Linderstrm-Lang (1924). *C. R. Trav. Lab. Carlsberg* **15**, 70.

[89] J. L. Martin, M. F. Sykes, and F. T. Hioe (1967). *J. Chem. Phys.* **46**, 3478.

[90] M. Matsumura, W. J. Becktel, M. Levitt, and B. W. Matthews (1989). *Proc. Natl. Acad. Sci. USA* **86**, 6562.

[91] J. E. Mayer and M. G. Mayer (1940). *Statistical Mechanics* (Wiley, New York).

[92] J. Mazur (1969). *Adv. Chem. Phys.* **15**, 261.

[93] D. S. McKenzie (1973). *J. Phys.* **A6**, 338.

[94] D. S. McKenzie (1976). *Phys. Rep.* **27**, 35.

[95] M. Mezard, G. Parisi, N. Sourlas, G. Toulouse, and M. Virasoro (1984). *Jour. de Phys.* **45**, 843.

[96] R. Miller, C. Danko, M. J. Fasolka, A. C. Balazs, H.S. Chan and K. A. Dill (1992). *J. Chem. Phys.* **96**, 768.

[97] M. A. Moore (1977). *J. Phys.* **A10**, 305.

[98] M. Muthukumar and B. Nickel (1984). *J. Chem. Phys.* **80**, 5839.

[99] Y. Nozaki and C. Tanford (1963). *J. Biol. Chem.* **238**, 4074.

[100] Y. Nozaki and C. Tanford (1970). *J Biol. Chem.* **245**, 1648.

[101] Y. Nozaki and C. Tanford (1971). *J Biol. Chem.* **246**, 2211.

[102] D. H. Ohlendorf, B. C. Finzel, P. C. Weber, and F. R. Salemme (1987). In *Protein Engineering*, ed. D. L. Oxender, C. F. Fox, pp. 165-173. (Alan Liss, New York).

[103] Y. Oono (1979). *J. Phys. Soc. Japan* **47**, 683.

[104] Y. Oono (1985). *Adv. Chem. Phys.* **61**, 301.

[105] Y. Oono and K. F. Freed (1981). *J. Chem. Phys.* **75**, 993; 1009.

[106] Y. Oono and T. Ohta (1981). *Phys. Lett.* **A85**, 480.

[107] Y. Oono, T. Ohta, and K. F. Freed (1981). *J. Chem. Phys.* **74**, 6458.

[108] T. A. Orofino and P. J. Flory (1957). *J. Chem. Phys.* **26**, 1067.

[109] W. J. C. Orr (1947). *Trans. Faraday Soc.* **43**, 12.

[110] C. N. Pace, G. R. Grimsley, J. A. Thompson, andB. J. Barnett (1988). *J. Biol. Chem.* **263**, 11820.

[111] A. A. Pakula and R. T. Sauer (1990). *Nature* **344**, 363-364.

[112] T. Pakula (1987). *Macromolecules* **20**, 679.

[113] T. Pakula and S. Geyler (1987). *Macromolecules* **20**, 2909.

[114] G. Parisi (1980). *J. Phys.* **A13**, 1101; 1887; L115.

[115] I. H. Park, Q. -W. Wang, and B. Chu (1987). *Macromolecules* **20,** 1965.

[116] R. K. Pathria (1980). *Statistical Mechanics* (Pergamon, New York).

[117] L. Peller (1959). *J. Phys. Chem.* **63,** 1194; 1199.

[118] L. J. Perry and R. Wetzel (1984). *Science* **226,** 555.

[119] D. C. Poland and H. A. Scheraga (1970). *Theory of Helix-Coil Transitions in Biopolymers, Statistical Mechanical Theory of Order-Disorder Transitions in Biological Macromolecules* (Academic Press, New York).

[120] C. B. Post and B. H. Zimm (1979). *Biopolymers* **18,** 1487.

[121] P. L. Privalov (1979). *Adv. Protein Chem.* **33,** 167.

[122] P. L. Privalov, Yu V. Griko, S. Yu Venyaminov, and V. P. Kutyshenko (1986). *J. Mol. Biol.* **190,** 487.

[123] O. B. Ptitsyn and Ye Yu. Eizner (1965). *Biofizika* (USSR) **10,** 3.

[124] O. B. Ptitsyn, A. K. Kron, and Ye Yu. Eizner (1968). *J. Polymer Sci.* **C16,** 3509.

[125] N. Qian and T. J. Sejnowski (1988). *J. Mol. Biol.* **202,** 865.

[126] G. N. Ramachandran, C. Ramakrishnan, and V. Sasisekharan (1963). *J. Mol. Biol.* **7,** 95.

[127] P. Ramond (1981). *Field Theory, A Modern Primer* (Benjamin/Cummings, Reading MA).

[128] G. S. Rushbrooke and J. Eve (1959). *J. Chem. Phys.* **31,** 1333.

[129] I. C. Sanchez (1979). *Macromolecules* **12,** 980.

[130] J. A. Schellman (1978). *Biopolymers* **17,** 1305.

[131] H. A. Scheraga (1980). In *Protein Folding,* R. Jaenicke, ed., pp. 261-288 (Elsevier, Amsterdam).

[132] H. A. Scheraga (1983). *Biopolymers* **22,** 1.

[133] S. I. Segawa and M. Sugihara (1984). *Biopolymers* **23,** 2473; 2489.

[134] J. A. Semylen (1976). *Adv. Polymer Sci.* **21**, 41.

[135] E. I. Shakhnovich and A. V. Finkelstein (1989). *Biopolymers* **28**, 1667.

[136] E. I. Shakhnovich and A. M. Gutin (1989). *Biophys. Chem.* **34**, 187.

[137] E. I. Shakhnovich and A. M. Gutin (1990). *Nature* **346**, 773.

[138] E. I. Shakhnovich and A. M. Gutin (1990). *J. Chem. Phys.* **93**, 5967.

[139] D. Shortle and A. K. Meeker (1986). *Proteins: Structure, Function, and Genetics* **1**, 81-89.

[140] D. Shortle, W. E. Stites and A. K. Meeker (1990). *Biochemistry* **29**, 8033-8041.

[141] D. Stigter and K. A. Dill (1990). *Biochemistry* **29**, 1262.

[142] D. Stigter, D. O. V. Alonso, and K. A. Dill (1991). *Proc. Natl. Acad. Sci. USA* **88**, 4176.

[143] W. H. Stockmayer (1960). *Die Makromolekulare Chemie* **35**, 54.

[144] S. T. Sun, I. Nishio, G. Swislow, and T. Tanaka (1980). *J. Chem. Phys.* **73**, 5971.

[145] U. W. Suter, M. Mutter, and P. J. Flory (1976). *J. Am. Chem. Soc.* **98**, 5740.

[146] G. Swislow, S. T. Sun, I. Nishio, and T. Tanaka (1980). *Phys. Rev. Lett.* **44**, 796.

[147] M. F. Sykes (1961). *J. Math. Phys.* **2**, 52.

[148] M. F. Sykes (1963). *J. Chem. Phys.* **39**, 410.

[149] C. Tanford (1968). *Adv. Protein Chem.* **23**, 121.

[150] J. M. Thornton (1981). *J. Mol. Biol.* **151**, 261.

[151] Y. Tsunashima and M. Kurata (1986). *J. Chem. Phys.* **84**, 6432.

[152] P. Vidakovic and F. Rondelez (1984). *Macromolecules* **17**, 418.

[153] M. V. Volkenstein (1963). *Configurational Statistics of Polymer Chains*, translated from the Russian edition by S. N. Timasheff, M. J. Timasheff, (in *Higher Polymer* Vol. 17, ed. H. Mark). New York: Interscience.

[154] F. T. Wall and F. T. Hioe (1970). *J. Phys. Chem.* **74,** 4410.

[155] F. T. Wall and S. G. Whittington (1969). *J. Phys. Chem.* **73,** 3953.

[156] M. C. Wang and G. E. Uhlenbeck (1945). *Rev. Mod. Phys.* **17,** 323.

[157] M. R. Wattenbarger, H. S. Chan, D. F. Evans, V. A. Bloomfield, and K. A. Dill (1990). *J. Chem. Phys.* **93,** 8343.

[158] J. A. Wells and D. B. Powers (1986). *J. Biol. Chem.* **261,** 6564.

[159] R. Wetzel (1987). *Protein Engineering* **1,** 79.

[160] C. Williams, F. Brochard, and H. L. Frisch (1981). *Annu. Rev. Phys. Chem.* **32,** 433.

[161] K. G. Wilson and J. Kogut (1974). *Phys. Rep.* **12,** 75.

[162] H. Yamakawa (1971). *Modern Theory of Polymer Solutions.* New York: Harper & Row.

[163] B. H. Zimm (1946). *J. Chem. Phys.* **14,** 164.

[164] B. H. Zimm and J. K. Bragg (1959). *J. Chem. Phys.* **31,** 526.

Finding Electron Transfer Pathways

J. J. Regan, J. N. Betts, D. N. Beratan, and
J. N. Onuchic

Introduction

Electron transfer reactions are among the simplest and best studied chemical reactions [25, 18, 10, 12, 8, 9, 26]. Biology makes use of these reactions in many places, notably in the photosynthetic apparatus which traps the energy of sunlight. We would like to understand the dynamics of these reactions, and the ways in which the biological catalysts of electron transfer - specific protein molecules - are capable of controlling the flow of electrons. The availability of genetic engineering and powerful synthetic chemistry techniques suggests that the true test of our understanding will be in our ability to design new molecules with certain desired properties.

The challenge of designing new biocatalysts requires an understanding of enzymatic reactions at the molecular level. The conceptual challenge of achieving a quantitative understanding of catalysis is also one of the most important and fundamental problems in molecular biophysics. Nearly all of the biological chemical reactions with which we are familiar (e.g. DNA replication, photosynthetic light harvesting, oxidative phosphorylation, etc.) would not occur without the presence of specific catalysts.

The design of biocatalysts, or enzymes, poses a unique set of challenges compared to the design of small molecule catalysts. First, our understanding of the molecular *details* of biochemical reactions is in a more primitive state than our understanding of small molecule chemistry. Second, it is not yet clear *a priori* whether the surrounding protein superstructure plays an active or passive role in a given biocatalytic reaction. This paper discusses a model for the simplest biocatalytic reaction, electron transfer. These reactions are chosen because of 1) the ubiquity of electron transfer in biocatalytic energy harvesting and conversion pathways *and* 2) the potential for exploiting these reactions to produce tailored catalysts. In order for biological electron transfer systems to be controlled and designed, one needs to develop a quantitative understanding of the molecular structure/property relationships for these reactions.

Conventional enzymatic catalysts usually function by lowering the activation barrier for the atomic motions involved in a given chemical reaction. Catalytic reactions carried out by the electron transfer enzymes are different: The electron transfer proteins can catalyze reactions *either* by changing activation barriers or by changing the *electronic interaction* between an electron donor and an acceptor.

The ability of electron transfer catalysts to modify rates of reactions by changing *orbital interactions* between two species opens a wide range of new design strategies for producing tailored biocatalysts. An example of this orbital effect is seen in the bacterial photosynthetic reaction center, which captures the energy of the excitation delivered to it with unit quantum yield via a cascade of electron transfer reactions. These reactions have been shown [25, 18, 10, 12, 8, 9, 26, 15, 16] to be fairly insensitive to the activation parameters, but extremely sensitive to the orbital interaction between the electron donors and acceptors (determined by the donor-acceptor (D-A) separation distance, energetics, and structure of the intervening medium). In addition to the biological literature, a rich chemical literature [1, 11] devoted to the orbital dependence of chemical electron transfer reaction rates prompts us to investigate the potential for *designing* new biocatalytic and biomimetic materials based on electron transfer reactions.

To focus the discussion of the D-A coupling mechanism, it is useful to consider the concept of a *physical tunneling pathway*. For a single physical pathway there are exact and perturbation theory methods to calculate the coupling arising from that physical pathway. The tunneling matrix element is calculated from the D (A) localized state at the appropriate nuclear configuration [7, 24, 22, 19, 20]. Numerical strategies (both exact and perturbation methods) usually write the decay of the wave function as a product of decays per bond (or delocalized group). Within a perturbation theory calculation, the per bond decay depends only on the tunneling energy and on the nature of the particular bonds.

This method (applied to lowest order) neglects scattering corrections to the wave function propagation in the protein. Scattering pathways are the distinct combinations of bonds that give rise to a large number of additional paths within the same physical pathway. For example, a physical pathway consisting of bonds 1, 2, 3, 4, $\cdots$ has the direct pathway 1-2-3-4$\cdots$ and the scattering pathways 1-2-3-2-3-4 $\cdots$, etc. The scattering pathways can be treated [21, 13, 14] exactly in the electronic coupling calculation for a one dimensional physical pathway by correcting the self energy of each orbital on the path. Exact methods, particularly green's function [21, 13, 14] approaches, often write the coupling as a product as well. In this case, the terms in the product explicitly include these scattering corrections by introducing self-energy terms into the site energies of bridge units in the pathways. The effect of side groups appended to the physical pathway can also be included by making self energy corrections [21]. (This simple description is valid when the physical pathways are independent - not intersecting.) This approach is adequate for the present survey purposes.

The problem is more complicated if the pathways are not independent because loops are present. In this case, the utility of the pathway approach is questionable [21]. Methods that attempt numerical solution of the Schrodinger equation do not usually break the problem into products of decays across a collection of groups. However, our initial concern is a qualitative description of the pathways. The goal is to identify important physical pathways, *not* to make firm quantitative estimates of the tunneling matrix elements.

The Model of Electron Transfer

The model [2, 23] used characterizes an electron transfer pathway as a series of bonds (bonding and anti-bonding orbitals) between a given donor orbital D and a given acceptor orbital A. An electron transfers from one orbital to another orbital with a rate proportional to the coupling between the two orbitals. A molecule (a general term taken here to mean anything from an individual amino acid to a multi-protein complex in the presence of solvent molecules) is viewed as a set of such orbitals, and pathway determination becomes a problem of looking for chains of coupled bonds connecting donor and acceptor orbitals.

In perturbation theory, which is called the "non-adiabatic" limit of electron transfer theory - non-adiabatic because it is the opposite from the Born-Oppenheimer approximation - the rate of electron transfer from D to A is given by

$$k_{\text{ET}} = \frac{2\pi}{\hbar} \ \ |T_{\text{DA}}|^2 (F.C.) \tag{4.1}$$

where $(F.C.)$ is the Frank-Condon factor associated with the atomic motions which provide the "density of states" in Fermi's golden rule. T_{DA}

is the electronic D-A coupling,

$$T_{DA} \propto \prod_n \epsilon_n = \prod_i \epsilon_i^{(C)} \prod_j \epsilon_j^{(H)} \prod_k \epsilon_k^{(S)}. \tag{4.2}$$

The form of this overall coupling suggests the three general classes of decay between neighboring bonds recognized in the model. The decay factor $\epsilon^{(C)}$ couples two bonds associated with a common atom (covalent chain). The factor $\epsilon^{(H)}$ couples the bonding orbital holding a hydrogen in place and a nearby lone pair orbital (i.e. the orbitals lined up in a hydrogen bond). The factor $\epsilon^{(S)}$ couples two orbitals that are not associated with a common atom. Whereas the first two interactions are considered bonded interactions, this last interaction is called a non-bonded interaction, or through-space jump.

In order for the nonadiabatic approximation that leads to the electron-transfer rate given by Eq. (4.1) to be valid, the electronic frequency, $T_{DA}/\hbar$, must be slow compared to that of the relevant nuclear motion. In long-distance electron transfer, the tunneling matrix elements are so small that this approximation is most likely adequate.

The form of decay factors [Eq. (4.2)] used most frequently in the implementation to be described are

$$\begin{aligned}
\epsilon^{(C)} &= 0.6, \\
\epsilon^{(H)} &= 0.6 \exp\left[-1.7(r-1.4)\right], \\
\epsilon^{(S)} &= 0.3 \exp\left[-1.7(r-1.4)\right],
\end{aligned} \tag{4.3}$$

where r is the distance in angstroms between the centers of the two orbitals involved in the interaction. These are only typical decay factor functions, for the three general classes of interactions suggested above. In the implementation, decay factors can be defined which depend on (for example) the atoms involved in the bond, the local environment, the relative orientation of the bonding orbitals, etc. The functional form of the decay factors will match within a class, but the actual ϵ values computed may differ from orbital pair to orbital pair within the same class.

This model of electron transfer readily accommodates the identification of pathways and their coupling in a molecule. The set of interacting orbitals comprising the molecule maps directly onto a graph with bidirectional, weighted edges (a graph with these properties is usually called a network). In the network, the vertices are the bonding orbitals, and the edges represent the interactions between the orbitals. The weight of an edge between two orbitals is given by the ϵ decay factor associated with the coupling between the orbitals. A given orbital will usually couple to at least two other orbitals via a bonded interaction, and bond to any number of other orbitals via a non-bonded interaction. From Eq's. (4.1) and (4.2) the rate of transfer mediated by a pathway between any two orbitals is proportional to the product of the decay factors between the individual orbitals on the pathway. Since all the decay factors are between 0 and 1, the more steps one takes, the slower the rate. The maximum coupling pathway will be the one via orbital interactions that have a maximal product of decay factors.

This is isomorphic to the "shortest path" problem of graph theory, where vertices represent cities and edge weights represent road lengths (or travel times) between cities. A traveler is interested in the shortest route between two cities. In the protein, an electron will tunnel through the pathway between two orbitals along a path with a maximal product of decay factors.

The Implementation of the Model

The search for pathways separates into two problems; the mapping of a molecule onto a network, and the analysis of the network. What follows is a description of the implementation of these tasks by a set of computer programs that will collectively be called *Pathways*.

Creation of the Network

The network has to be constructed from some representation of the molecule. The predominant representation at this time is a file in the Brookhaven Protein Data Bank (PDB) format. This format offers atomic coordinates and limited covalent bond connection information, based on inter-atomic distances. Additionally, one needs to specify bonding orbital interaction decay factors, as in Eq. (4.3). When assigning an interaction to a given pair of orbitals, *Pathways* looks up the pair in a table (read from a file) to find the decay factor (and note that there will generally be three forms for the decay factor, associated with the three interaction classes discussed above). Orbital pairs are defined in terms of the types of the two orbitals involved, and in turn an individual orbital type is defined in terms of the atom types associated with it (two in the case of a bond, one in the case of a lone pair). An atom type signifies at least the elemental species of the atom, but can also represent something about the environment of the atom (so that one can, for example, draw a distinction between a nitrogen in a protein backbone and a nitrogen in a histidine sidechain bound to an iron in a heme). Atom types can be specified in the atom name field of the PDB atomic record. Alternatively, *Pathways* can read the molecular description files associated with commercially available molecular modeling packages (e.g., Biosym or BioDesign) which have atom type descriptors (distinct from the atom name) that these packages use to assign potential functions for energy minimization and molecular dynamics. A user can exploit these type names in defining orbitals; this is preferable to using the atom names that come with a PDB file simply because the standard PDB atom names contain no atomic hybridization information.

The first task of *Pathways* is to use the covalent connection information in the PDB file to establish the initial network vertices of covalent bonding orbitals. It is important to emphasize that the vertices of the network are the bonds (not the atoms) and the edges are the interactions between bonds (not between atoms).

The second task of *Pathways* is to find the orbitals not accounted for in the typical PDB molecular description. First, the missing hydrogens must be found, or more precisely, the covalent bonding orbitals attaching the hydrogens must be added to the network. Secondly, lone pair orbitals must also be identified and added. Both of these classes of orbitals are terminal in the sense that they are not part of a chain of covalent bonds, but they form hydrogen bonds and can take part in nonbonded interactions. The identification of these orbitals consists of a search through the molecule for the proper types of atoms in the proper covalent bonding configurations.

After this is done, the vertices of the network have been identified, and all covalent bonding orbitals and lone pair orbitals are represented. Now one can add the edges, corresponding to interactions between bonds. The program could simply examine each orbital one at time and scan the rest of the molecule for orbitals that it could interact with, but in practice *all* the bonded interactions are added first, before any non-bonded interactions are considered.

The first edges to add are those representing covalent bond interactions through a shared atom, i.e those of the $\epsilon^{(C)}$ class mentioned above. This process is guided by the covalent connection information in the molecular data file, and the results of the hydrogen and lone pair searches.

Next, one adds the remaining bonded interactions; the interactions associated with hydrogen bonds. Such an interaction is between the bonding orbital located between the donor and its covalently bound hydrogen, and the lone pair orbital between the hydrogen and the acceptor. The pairs of orbital types that can take part in such an interaction are in tables read by *Pathways*, along with the decay factors. The decay factor assigned, in addition to being a function depending on the atom types (donor, hydrogen, acceptor) involved in the bond, will also depend on the relative positioning of the donor, hydrogen and the acceptor. The possibility of a bifurcated hydrogen bond can be included in the model, simply by not limiting the number of hydrogen bonds a given hydrogen atom can be involved in.

Finally, the edges representing the non-bonded interactions (hereafter called *jumps*) are added. The jump coupling is proportional to the spatial overlap between the orbitals involved. In this simple model, the overlap is approximated by considering only the distance between and relative orientation of the (spherically asymmetric) orbitals involved. Naturally, an overlap is present between *all* orbitals in the molecule, thus an N atom molecule with k bonding orbitals per atom has $kN(kN - 1)/2$ possible jumps (disregarding bonded interactions). The representation of each of these interactions by an edge in the network would greatly increase the network's complexity. However, for the purpose of estimating pathway coupling between orbitals, it is not necessary to add all possible jumps. The majority of orbital pairs have a chain of bonded interactions connecting them, and in most cases the transfer rate associated with this chain greatly exceeds the rate associated with a direct jump between the pair because the jump's potential barrier is so high. Such jumps should not be included in the network, since they only increase search times (in the analysis to be performed later) without having an effect on the results. Thus the process of adding edges representing jumps to the network is best described as a process of deciding which jumps do not matter.

There are two criteria a jump must meet to be included. One is global, the other is local. The global criterion is that the jump must be no more than a certain user specified distance. That is equivalent to saying the coupling must be at least some minimum value. The local criterion demands that the jump coupling between two orbitals must be at least as good as some (user specified) percentage of the coupling provided by the best pathway already present between them. This local criterion in the process of elimination is why all the bonded interactions are added first, rather than taking one orbital at a time and establishing all its interactions.

It should be pointed out that jumps can actually eliminate other jumps in the same way that bonded interactions can eliminate jumps. When establishing jumps from a given orbital, the neighboring orbitals are considered in order from the nearest to the furthest. A jump established to a close orbital could create a pathway to one of the orbitals further out that would disqualify a jump directly to that distant orbital. Thus, to keep the network as uncluttered as possible, not only is it important to establish the bonded interactions first, it is also important to consider the jumps in order of increasing distance.

Analysis of the Network

One's interests may involve questions about the effect of adding, deleting or replacing a residue in a protein, or the effect of attaching transition metal atoms at certain locations, etc., but all this information is incorporated in the network construction. The creation of the network is actually a much more difficult task than its subsequent analysis. Once the network is complete, atom types, potentials, physical distances, etc. no longer matter. The network stands apart from the molecule as an abstracted set of connected vertices, and can be searched using any of the path search algorithms developed for graphs (see Ref. [17], for example).

One might be interested in the maximum coupling pathway between two orbitals, the set of all pathways between two orbitals, the fastest path between a given donor orbital and all other orbitals in the molecule, etc., but all of these questions reduce to finding a path between two network vertices. The following pseudo-code example shows a simple data structure used to store the network, the initialization required to set up a search, and a call to perform a search:

program *path_search*
declare integer
  max_orbitals, max_interactions;
  start, goal, num_orbitals, num_interactions(max_orbitals),
  interacting_orbital(max_interactions,max_orbitals);
declare real
  min_coupling, coupling(max_interactions, max_orbitals);
declare logical
  visited(max_orbitals);

```
begin
load_network(num_interactions, num_orbitals, coupling, interacting_orbital);
do i := 1,num_orbitals;
  visited(i) := false;
enddo
read start, goal, min_coupling;
search(start,1.0);
end.
```

The procedure *search* is a depth first search, which means it branches out from the starting vertex, extending a single search path as far as possible until either the path coupling falls below the minimum required by the user, or until it hits the goal (in which case the path is reported). In both cases, the algorithm then backs up a step and tries another approach, until all possible paths better than *min_coupling* have been reported.

```
procedure search (integer parent,real coupling_from_start)
declare integer i, child;
declare real c;
begin
if parent = goal then
  write the path and the coupling_from_start;
else
  visited (parent):=true;
  doi := 1,num_interactions (parent);
   child:= interacting_orbital (i,parent);
    if not visited( child) then
     c := coupling_from_start * coupling (i, parent);
     if c ≥ min_coupling then
       search (child, c );
     endif
    endif
  enddo
  visited(parent) := false;
endif
```

This algorithm reports the paths as it finds them. Note that c is exactly the product [Eq. (4.3)] of all the decay factors from the start orbital to the orbital represented by *child*. To find the fastest path with this algorithm one has to call it with a low enough *min_coupling* to recover *some* paths; the fastest of these paths will be the fastest overall path. Such guesswork is not needed however, since in the actual implementation another algorithm (attributed to Dijkstra [17]] can be used to directly (i.e. quickly) find the pathway with the best coupling, and then some appropriate fraction of this amount can be passed as *min_coupling* to the depth first search. Unlike the depth first search, this second algorithm cannot be used to find families of alternative paths - it can only find the best one. However, it can be used in conjunction with a network decomposition technique to find alternative paths (wherein one breaks the network into sub-networks, and finds the best paths in the sub-networks, etc.) This second search technique is faster than the depth first search when the average number of edges per vertex in the network becomes high, and is available as an alternative to the depth first search.

Given the ability to determine paths between points, one can ask what the maximum coupling pathways are to every acceptor from a given donor. The listing of paths provided would give detailed information as to where an electron could go from a given donor site, but a more useful way to assimilate this information is to image the molecule with the atoms colored as a function of their best path coupling to the donor. Since coupling drops off by an order of magnitude in going through three bonds, the logarithm of the coupling, rather than the coupling itself, should be mapped to the colors.

A second useful coloring scheme, called a *hot-cold map*, highlights differences between actual couplings from acceptor sites to a given donor and a characteristic coupling function defined *for that donor*. The coupling provided (by the best paths) to acceptors from a donor fall off roughly exponentially with direct donor-acceptor distance r. The computed couplings can be used to fit the parameters A and β of a model coupling $\epsilon(r) = A \exp(-(\beta/2)r)$. $\beta/2$ can be interpreted as a characteristic distance decay for the given donor in the given molecule. The computed coupling to an acceptor at a given distance can be compared to this model; an acceptor with higher (lower) coupling than the model for its distance is considered hot (cold). The actual number used to color an acceptor is:

$$\log\left(\frac{\prod_n \epsilon_n}{A\exp(-(\beta/2)r)}\right) \tag{4.4}$$

Positive values are considered hot, while negative values are cold. All these coloring schemes all very compelling on a color monitor, but alas not so informative when reproduced in black and white.

Example Analysis: Cytochrome c

The following simple analysis was performed on the well studied electron transfer protein, cytochrome c, to demonstrate the basic usage of it Pathways.

Figure 4.1 shows the results of 34 searches through the molecule to establish jumps. The searches were performed using the absolute jump length limit given on the horizontal axis (the global criteria), with a common local criteria that for a jump to be established it had to provide at least 10% of the coupling provided by the best pathway already in existence between the donor and acceptor in question. Thus for example, with a jump limit of 0.45 nm, there were 13890 potential donor acceptor pairs within that distance of one another, but of these 10598 had pathways between them that would render the coupling provided by a jump negligible (under the 10% criterion), thus only 3292 jumps were added to the network. This was the criterion used to construct the network used for the results to follow.

Figure 4.2a shows the two highest coupling electron transfer pathways between the positively charged nitrogen at the end of the lysine sidechain

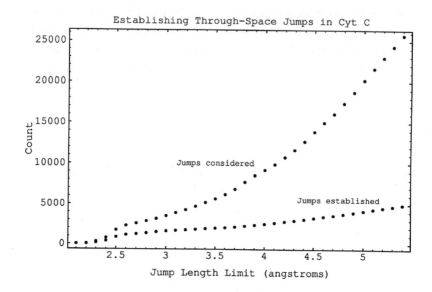

Figure 4.1: A jump may be established if a donor and acceptor are within a certain distance of one another (top line), and is actually established (bottom line) if the jump provides at least a certain percentage of the coupling already provided by the best path between the donor and acceptor in question. Here, the minimal percentage was 10%.

in position 72 and the iron in the heme. The direct distance from the nitrogen to the iron is 13 angstroms, and the through-space decay factor [Eq. (4.3)] for that distance is 1.63×10^{-9}. However, the total decay factors [Eq. (4.2)] computed for the top two paths were 1.81×10^{-5} and 1.45×10^{-5}. Each of these paths contains one jump, with the remaining links being covalently bonded interactions. The path with the highest coupling is the one which starts with a jump from Lys^{72} to Ile^{81}.

For comparison, the same pathway search was repeated on a network constructed with jumps prohibited. The result is shown in Fig. 4.2b, where the two top paths have total decay factors of 2.97×10^{-6} and 1.30×10^{-6}. In this case both paths utilize the hydrogen bonds connecting Thr^{78} to Ile^{75} and connecting Ile^{75} to Pro^{71}, but then they branch at Thr^{78} to follow

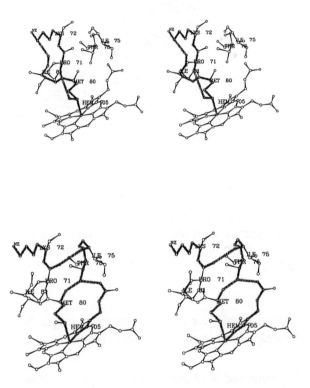

Figure 4.2: The relevant section of cytochrome c with covalent interactions shown in black. (a) The two maximum coupling pathways between the nitrogen at the end of the LYS72 sidechain and the iron in the heme. Each path has a jump. The path with the highest coupling is the one which starts with a jump from LYS72 to ILE81. (b) The results of the same pathway search on a network where no jumps were permitted. The four sections of the paths which do not follow covalent bonds are making use of hydrogen bonds.

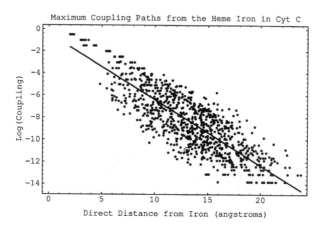

Figure 4.3: Each dot represents the coupling of the maximum coupling pathway from all possible donors (heavy atoms) in the protein to the heme iron. The fitted line offers a model coupling of 0.6190 exp(-(1.186/2) R) where R is the distance from the iron. The circled dot at R=13 is the nitrogen discussed in Fig. 4.2. Hot spots are by definition above the line, while cold are below.

different hydrogen bonds, for a total of three hydrogen bonds in both paths. The remaining links are again covalent interactions.

A search was performed to obtain the best coupling pathways from the iron to all possible acceptors in the molecule (or equivalently, from all possible donors to the iron), and the results were used to construct a hot-cold map. In Fig. 4.3 the logarithm of the maximum path coupling is plotted against distance from the iron for all possible donors. The model coupling computed (from these values) for this protein is $\epsilon(r) = 0.6190 \exp(-(1.186/2)r)$, which is represented by the fitted line. This line is taken as a model for how a wave function decays out from the iron, assuming a protein presenting a homogeneous environment for electron transfer, and is used only to give meaning to the words hot and cold. In some sense, it represents an average decay as a function of distance from the donor. The circle locates the nitrogen of the previous pathway example. Because of its position well below the line, it is considered a cold spot; the coupling provided is an order of magnitude below the model. This is confirmed by visualizing the hot-cold map.

Conclusion

This paper describes the software implementation of a simple model of electron transfer in proteins. The software has successfully predicted relative electron transfer rates in several transition metal labeled proteins [4, 5].

The current version of the *Pathways* software (implemented in FORTRAN) is available from the authors. A new C version is in development; it will include the effects of multiple interfering and scattering pathways [21] to calculate *net* couplings from a donor to an acceptor rather than couplings along individual pathways. Also, the software is being coupled to molecular dynamics code so that the temperature dependence of electron transfer pathways can be explored.

As a concrete example of the success of *Pathways* we conclude this manuscript presenting some experimental results obtained by the Gray [27] group at Caltech. Intramolecular electron transfer rates from Fe^{2+} to $Ru(bpy)_2(im)(His^X)^{3+}$ in modified cytochrome c were measured by time-resolved absorption spectroscopy. The histidines [X = 33 (horse heart), 39 (*Candida krusei*), 62 (genetically engineered $Asn^{62} \rightarrow$ His *Saccharomyces cerevisiae*), and 72 (semisynthetic $Lys^{72} \rightarrow$ His horse heart)] were modified to give $Ru(bpy)_2(im)(His^X)$ - protein derivatives. The reorganization energy λ, and therefore the $(F.C.)$ factor in Eq. (4.1) is basically the same in these four ET reactions. However, the electronic-coupling strengths show a great deal of variability for the different derivatives. Figure 4.4 shows the results for the maximum electron transfer rates k_{max} (i.e. the rate for the maximum value of $(F.C.)$). Semiclassical theory [16] predicts that k_{max}

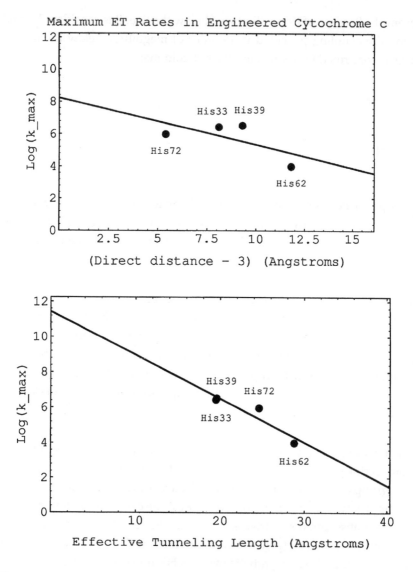

Figure 4.4: (a) Plot of k_{max} versus D-A distance minus 0.3 nm (van der Waals contact) for four cytochrome c derivatives (labeled by the location of the probe). No exponential fit is reasonable for these data, and the best fit suggests a rate near 10^8 s^{-1} at d = 0.3 nm (edge-edge contact) while physical values should be around 10^{12} s^{-1}. (b) Plot of k_{max} versus the tunneling length, $\sigma l = n_{eff} \times 0.14$ nm, where n_{eff} is the effective number of bonds in the pathway. The y intercept occurs around 10^{12} s^{-1} as expected.

values will fall exponentially with distance between donor and acceptor. Figure 4.4a shows a best exponential fit to the data, which is not only a poor fit, but suggests a rate at the direct D-A contact distance of 0.3 nm of 1.7×10^8 s^{-1}, or about four orders of magnitude below what would be expected for a direct contact rate. *Pathways* predicts that k_{max} correlates with the effective tunneling length of the pathway (i.e. the average covalent bond length of 0.14 nm times the effective number of bonds n_{eff}, where n_{eff} = the total pathway coupling divided by $\epsilon^{(C)}$). Fig. 4.4b shows that the effective tunneling length computed by *Pathways* is in agreement with the experimental data, a correlation which is not explained by the conventional theory. The disagreement becomes enormous for His[72], where conventional theory predicts a rate which is wrong by five orders of magnitude. This particular modified protein was specially designed using the hot-cold spot maps.

Acknowledgments

JNO is a Beckman Young Investigator. Work in San Diego was funded by the National Science Foundation - Grant No. MCB-9018768, and by the Public Health System Training Grant - Grant No. 5T32 GM08326-03 (JJR). Work in Pittsburgh was funded by the Department of Energy, Advanced Industrial Concepts Division.

References

[1] T. A. Albright, J. K. Burdett, and M.-H. Whangbo (1985). *Orbital Interactions in Chemistry* (Wiley, New York).

[2] D. N. Beratan, J. N. Onuchic, and J. J. Hopfield (1987). *J. Chem. Phys.* **86,** 4488.

[3] D. N. Beratan, J. N. Betts, and J. N. Onuchic (1992). *J. Phys. Chem.* **96,** 2852.

[4] D. N. Beratan, J. N. Onuchic, J. N. Betts, B. E. Bowler, and H. B. Gray (1990). *J. Am. Chem. Soc.* **112,** 7915.

[5] D. N. Beratan, J. N. Betts, and J. N. Onuchic (1991). *Science* **252,** 1285.

[6] J. N. Betts, D. N. Beratan, and J. N. Onuchic (1992). *J. Am. Chem. Soc.* **114,** 4043.

[7] W. Bialek, W. J. Bruno, J. Joseph, and J. N. Onuchic (1989). *Photosyn. Res.* **22,** 15.

[8] B. E. Bowler, A. L. Raphael, J. N. Onuchic, and H. B. Gray (1990). *Prog. Inorg. Chem.: Bioinorganic Chem.* **38,** 259.

[9] B. Chance, D. C. Devault, H. Frauenfelder, R. A. Marcus, J. R. Scheriffer, and N. Sutin, eds. (1979). *Tunneling in Biological Systems* (Academic Press, New York).

[10] W. A. Cramer and D. B. Knaff (1990). *Energy transduction in biological membranes* (Springer-Verlag, New York).

[11] L. Eberson (1987) *Electron Transfer Reactions in Organic Chemistry* Chapter X (Springer-Verlag, New York).

[12] G. Feher, J. P. Allen, M. Y. Okamura, and D. C. Rees (1989). *Nature* **339,** 111.

[13] A. A. S. da Gama (1990). *J. Theor. Biol.* **142,** 251.

[14] C. Goldman (1991). *Phys. Rev. A* **43,** 4500.

[15] M. R. Gunner, D. E Robertson, and P. L. Dutton (1986). *J. Phys. Chem.* **90,** 3783.

[16] J. J. Hopfield (1974). *Proc. Natl. Acad. Sci. USA* **7,** 3640.

[17] E. Horowitz and S. Sahni (1978). *Fundamentals of Computer Algorithms* (Computer Science Press, Potomac, MD).

[18] R. A. Marcus and N. Sutin (1985). *Biochem. Biophys. Acta* **811,** 265.

[19] K. V. Mikkelsen and M. A. Ratner (1988). *Chem. Rev.* **87,** 113.

[20] J. N. Onuchic, D. N. Beratan, and J. J. Hopfield (1986). *J. Phys Chem.* **90,** 3707.

[21] J. N. Onuchic, P. C. P. Andrade, and D. N. Beratan (1991). *J. Chem. Phys.* **95,** 1131.

[22] J. N. Onuchic and P. G. Wolynes (1989). *J. Phys. Chem.* **92,** 6495.

[23] J. N. Onuchic, D. N. Beratan (1990) *J. Chem. Phys.* **92,** 722.

[24] J. R. Riemers and N. S. Hush (1989). *Chem. Phys.* **134,** 323.

[25] H. Sigel and A. Sigel, eds. (1991) *Electron Transfer Reactions in Biological Systems: Metal Ions in Biological Systems 27* (Marcel Dekker, New York).

[26] H. Taube (1970). *Electron Transfer Reactions of Complex Ions in Solution* (Academic Press, New York).

[27] D. S. Wuttke, M. J. Bjerrum, J. R. Winkler, and H. B. Gray (1992). *Science* **256,** 1007.

Physical Constraints and Optimal Signal Processing in Bacterial Chemotaxis

Leonid Kruglyak

Introduction

This chapter is about how bacteria process sensory information. You will notice that it is not based on the lectures given by Steve Block at the 1991 NEC summer school. Instead, the chapter is based on the lecture I gave at the 1992 summer school. My interest in chemotaxis was inspired in part by Steve's lectures and I hope that this chapter will convey at least some of the excitement about the subject that was so clear in them.

Bacteria provide us with an opportunity to study sensory processing in its simplest form. They can detect light, chemicals, temperature, magnetic fields, and other stimuli [2, 12]. All the processing is performed by a single

197

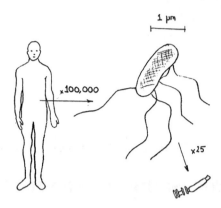

Figure 5.1: The size of *E. Coli* compared to that of man, also showing the flagella. Redrawn from Ref. [7].

cell. Nonetheless, bacteria can respond to sensory stimuli at threshold levels that approach fundamental physical limits on detectability set by the environment [9, 12]. Such near-optimal processing has also been found in a number of sensory tasks performed by higher organisms [10, 11, 12], as reviewed in Bialek's contributions to these proceedings. There have been suggestions that performance approaching physical limits can serve as a design principle for understanding sensory systems. Bacteria provide a good starting point for an exploration of these ideas.

There are several advantages to working with bacteria. Their environment, or at least those features of it that seem relevant to the bacteria, is relatively simple, and therefore the environmental constraints can be clearly defined. Bacteria also perform relatively simple tasks. There is a growing body of excellent experiments on bacteria, both qualitative and quantitative, and new experiments can quickly test theoretical predictions. Experiments and analysis are facilitated by the availability of many mutant strains. As a result, as I hope to convince you in this chapter, quantitative, predictive analysis of bacteria's behavior is possible and provides a model for the kind of analysis we would like to be able to carry out for higher organisms.

The chapter is organized as follows. I first cover some general facts about the best studied bacterium, *E. Coli* - how they sense chemicals, how they propel themselves, how they behave - and how we know all this. I then discuss the nature of the physical constraints they must overcome and some of the solutions that are used to overcome them. These constraints affect virtually every task the bacterium needs to perform: propulsion, signal detection, and information processing. The main goal of this chapter is to describe these constraints and the solutions bacteria use to enable them to function in their presence. Finally, then, I present some recent work that attempts to provide a quantitative explanation of the chemotactic strategy of *E. Coli*.

What we know about E. Coli

There are a few basic facts about bacteria that underlie the problems they face. To be specific, let me concentrate on the bacterium *Escherichia Coli (E. Coli)*. *E. Coli* is quite small, by our standards, measuring about one micron in diameter (see Fig. 5.1). It lives in a medium with roughly the physical properties (density, viscosity, and so on) of water. It also detects chemical stimuli present at very low concentrations, down to below $10^{-7}M$ [12]. The detection of chemical gradients modulates the motion of the bacterium, which is itself quite an interesting business. The short review in this section hardly does justice to the work on the phenomenology of bacterial motility and chemosensing, nor is the set of citations anywhere near complete. The reader is encouraged to delve into the references, which include both general reviews and some original classics.

Chemoreception

The first thing to know about the chemical sense of *E. Coli* is that they do in fact sense chemicals. This is not an obvious fact; once it was observed that bacteria respond to changes in the concentration of various substances a debate ensued as to whether they could actually sense the chemicals or were merely responding to the resulting changes in their own metabolic activity. This debate was settled emphatically in favor of chemoreception by Adler in 1969 [1], although it is known that other species of bacteria do in fact use metabolic cues [2]. In a classic paper Adler demonstrated by several different experiments that bacteria must have receptors for different chemical substances. In particular, he showed the following:

- Bacteria are not attracted to certain compounds that they do metabolize.

- They are attracted to some compounds that they are incapable of metabolizing.

- Structurally related attractors (that would bind to the same receptor) compete-the presence of one affects bacteria's ability to sense another.

- Structurally unrelated attractants do not compete, which they would be expected to do in a metabolism-based system.

- Finally, mutants exist that cannot sense specific chemicals that they still metabolize, and sense other substances.

On the basis of these findings Adler concluded that *E. Coli* possess separate chemoreceptors for a variety of substances including sugars and amino acids.

Chemotactic behavior

It has been known since the early work of Pfeffer that bacteria will swim toward regions of high concentration of an attractant [17]. The motion of bacteria in concentration gradients was first analyzed in detail by Berg and Brown using three-dimensional tracking [8]. They found that bacteria swim in more or less straight-line segments (called runs) interrupted by

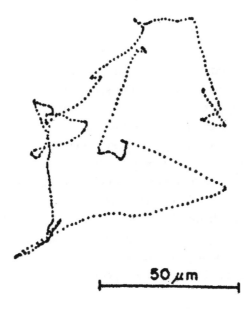

Figure 5.2: Random motions executed by *E. Coli* in the absence of external stimuli. Reproduced from Ref. [7].

tumbles - randomizing events during which the bacterium picks out a new run direction at random. This behavior is illustrated in Fig. 5.2. The durations of both runs and tumbles are exponentially distributed, indicating a Poisson process. The average run duration is about one second, while the average tumble duration is several times shorter. This is the situation in the absence of concentration gradients. How are bacteria able to respond to concentration changes? Berg and Brown found that the tumble rate drops for runs up a concentration gradient of attractant. This results in longer average run lengths up the gradient, biasing the random walk the bacterium executes in the absence of stimuli and introducing a drift up the gradient. The drift velocity reaches about ten percent of the bacterium's swimming velocity.

Propulsion

Many single celled organisms swim by waving long filaments called "flagella;" a sperm swims by the same scheme. The flagellum has all along its length molecules which can consume chemcial energy and generate local forces, and if one takes a cross-section of the flagellum one sees a correspondingly complex structure. Bacteria appear to be doing something similar, although the flagella are much thinner. In fact a bacterial flagellum is a polymer of a single protein, flagellin, and measures about 130 Angstroms in diameter; there is no indication that flagellin can catalyze any chemical reaction which could provide the energy for flagellar betaing [5]. The mystery was solved with the demonstration that *E. Coli* swim by *rotating* their flagella [4, 5]; because the flagella are thin and helical this rotation gives the appearance of a beating motion.

The flagellum is rotated by a rotary motor to which it is attached at the cell body [5]. The motor is proton-driven; how it works is still a subject of investigation. The rotary motion of flagella was confirmed by experiments on tethered bacteria-it is possible to attach one end of a flagellum to a glass slide using an antibody and then watch the cell body rotate. Rotations in both directions, as well as changes in direction, are observed [5]. An intact *E. Coli* cell has several flagella, as seen schematically in Fig. 5.1. During counterclockwise (CCW) rotation the individual flagella form a bundle and rotate together, propelling the bacterium forward. This occurs during a run. When the direction of rotation switches to clockwise (CW) the bundle destabilizes and flies apart, and the bacterium tumbles, picking out a new orientation at random [7].

Physical Constraints

The main theme of this lecture is that the bacterial strategy for chemosensing is determined by the physical constraints of its environment. In this section we review each of these constraints in turn.

Hydrodynamics

The hydrodynamic constraints on the motion of bacteria were pointed out by Purcell in one of the classic papers in biophysics [18]. He pointed out that bacteria live in a hydrodynamic regime very alien to us-that of low Reynolds number. Reynolds number measures the relative importance of inertial to viscous forces and is defined by

$$R \equiv \frac{av\rho}{\eta},\tag{5.1}$$

where a is the size of the object moving through a fluid, v is its velocity, ρ is the density of the fluid, and η is the viscosity. For a man swimming in water, $R \sim 10^5$, and inertial forces dominate. For a bacterium one micron in diameter and swimming at about $30\mu m/\text{sec}$, it is less than 10^{-4} - inertial forces are completely irrelevant. Purcell points out that if a bacterium swimming at full speed were to suddenly stop rotating its flagella, it would coast to a stop in a tenth of an Angstrom. Of course the bug wouldn't actually stop but rather would execute Brownian motion, and this will become important below; the point is that any net, deterministic forward motion would be stopped by viscous forces almost instantly.

An interesting feature of life at low Reynolds number is that swimming by reciprocal motion is impossible [18]. More precisely, any time-reversal symmetric pattern of motion will not propel the object, nor will any pattern which can be continuously deformed into a time-reversal symmetric one. In order to swim in a sustained fashion, an object must execute the same set of motions over and over. Such a set of motions could be moving an oar through the fluid and bringing it back to the original position, or using a piston first to suck in water and then to expel it. Neither of these schemes will work at low Reynolds number, because time plays no role in the equations describing low Reynolds number motion-all the time-dependent terms in the Navier-Stokes equations are inertial [18]. Only configuration is important, so if a movement is reversed along the same trajectory, no net displacement can result. It does not matter how quickly or slowly the movement is carried out nor whether or not it's interrupted-it is perfectly reversible. Microorganisms have found several solutions to this problem that allow them to swim; of necessity all involve nonreciprocal motion. In *E. Coli* the solution is to rotate a helical filament, which obviously violates T-reversal. From a more mechanistic view, rotation of a helix can apply a net propulsive force because the drag forces for transverse and longitudinal motion of a cylinder are different.

Although not so relevant to the case of bacteria, there are some elegant mathematical issues regarding low-Reynolds number propulsion. Specifically, if only configuration is important, the possibility of propulsion is related to whether the organism in fact is translated after executing one full cycle of its tour through configuration space. The problem is that any description of the configuration can be re-parameterized, and of course the propulsive translation must be parameterization invariant. Shapere and Wilczek have given this problem a gauge theory formulation, in which the re-parameterization invariance is central to the theory [22].

Diffusion

Many of the constraints on chemcial sensing which are imposed by bacteria's environment were originally examined by Berg and Purcell [9]. The first of these is due to diffusion. All the particles that the bacteria are interested in capturing are diffusing through the fluid. It turns out that as a result of diffusion bacteria cannot increase their intake of a substance significantly by swimming or stirring. Suppose a bacterium tries to increase its intake by either moving its whole body (swimming) or some appendage (stirring). The characteristic time it takes to move a particle through a distance L via this mechanism is $\tau_s \approx L/v$, where v is the velocity of the movement. At the same time, the particles are diffusing away, leaving the region of size L in roughly $\tau_d \approx L^2/D_m$, where D_m is the diffusion coefficient of the molecules in question. The ratio of the two times is a measure of the effectiveness of movement for transport. For typical molecules that bacteria are interested in, D_m is about 10^{-5} cm^2/sec, L is roughly the bacterium's size, 10^{-4} cm, and typical velocities are between 10^{-2} and 10^{-3} cm/sec. Hence $\tau_d/\tau_s \ll 1$-molecules diffuse away long before the movement has had any effect. Bacteria are passive absorbers of molecules that reach their surface by diffusion. They move solely in order to reach areas of higher attractant concentration.

We can then follow Berg and Purcell and look at diffusive intake [9]. In steady state chemical concentration c obeys the diffusion equation

$$D_m \nabla^2 c = 0. \tag{5.2}$$

This equation is identical in form to the Laplace equation for the electrostatic potential ϕ in the absence of charges,

$$\nabla^2 \phi = 0. \tag{5.3}$$

Continuing the analogy, the diffusive current entering a closed surface S,

$$J = -D_m \int_S \vec{\nabla} c \cdot d\vec{S}, \tag{5.4}$$

is equivalent to the charge on a conducting surface S,

$$Q = -\frac{1}{4\pi} \int_S \vec{\nabla} \phi \cdot d\vec{S}. \tag{5.5}$$

Hence the diffusive current into the surface of an object is given by

$$J = 4\pi C D_m c_\infty, \tag{5.6}$$

where C is the capacitance of a conductor with the same geometry and c_∞ is the concentration far from the object. For a sphere of radius a the current is $J_s = 4\pi a D_m c_\infty$. This is the maximum possible absorption for a sphere-the entire surface is absorbing. Suppose that instead the sphere is covered with N disc receptors of radius s. Then the current is given by [9]

$$J = \frac{4\pi D_m c_\infty N s a}{Ns + \pi a} = J_s \frac{Ns}{Ns + \pi a}. \tag{5.7}$$

The current is at half of the value for a totally absorbing sphere, $J = J_s/2$, when $N = \pi a/s$. In this case the ratio of the total area covered by receptors to the surface area of the entire sphere is $N\pi s^2/4\pi a^2 \approx s/a \sim 0.001$ for a typical bacterial chemoreceptor. We come to the surprising conclusion that a bacterium need only cover 0.1 percent of its area with receptors for one kind of molecules to reach half the intake achieved by covering the whole surface. Therefore a bacterium can have receptors for many different kinds of molecules without a big sacrifice in the intake of any of them. This of course is very important in the life of the bug, and explains some mysteries in the kinetics of receptor binding [9].

Concentration noise

Chemotactic bacteria would like to base their response on the best possible measurement of the concentration. This task is complicated by noise inherent in measuring low concentrations over a small region of space in a short time. We use the following simple argument to illustrate this; see also Refs. [9] or [12] for more details. Clearly a bacterium can measure the concentration only by looking at the fluid in its immediate vicinity - it cannot sense remotely. Suppose it is able to count every molecule of an attractant within a volume R^3, where R is roughly the bacterium's size. This is the best estimate of the concentration the bacterium can obtain in a single measurement; it is subject to counting noise $\Delta c \sim \sqrt{\bar{c}/R^3}$, where $\bar{c}$ is the mean attractant concentration. The noise can be reduced by integrating over time, but successive time measurements are correlated due to the finite time it takes molecules to diffuse in and out of the sampling volume. The correlation time is roughly $\tau_c \sim R^2/D_m$. Hence if the bacterium integrates for a time T its measurement is subject to noise

$$\frac{\Delta c}{\bar{c}} \sim \sqrt{\frac{1}{R^3 \bar{c}}} \cdot \sqrt{\frac{\tau_c}{T}} \sim \frac{1}{\sqrt{T D_m R \bar{c}}}, \tag{5.8}$$

allowing a measurement of concentration of $10^{-7} M$ to better than 1 percent in under a second.

Bacteria are really interested in sensing concentration changes rather than absolute concentration. So the problem is to measure a concentration gradient $\vec{\nabla} c$ in a time T. To simplify the argument assume that the gradient is constant in the $\hat{x}$ direction and the bacterium is swimming along it at velocity v. Then two measurements of the concentration separated by time T will show a concentration difference

$$\delta c \sim \frac{d\bar{c}}{dx} v T \tag{5.9}$$

due to the gradient. For reliable detection this difference must exceed the noise fluctuations in concentration, $\delta c > \Delta c$. From this requirement we can obtain a lower limit on integration time,

$$T_{int} > \left[D_m R \bar{c} \left(\frac{v}{\bar{c}} \frac{d\bar{c}}{dx} \right)^2 \right]^{-1/3}. \tag{5.10}$$

Putting in some order of magnitude numbers [9], ($D_m \sim 10^{-5} \, \text{cm}^2/\text{sec}$, $R \sim 10^{-4} \, \text{cm}$, $\bar{c} \sim 1 \, \text{mM}$, $v \sim 10^{-3} \, \text{cm/sec}$, $\frac{1}{\bar{c}} \frac{d\bar{c}}{dx} \sim 1 \, \text{cm}^{-1}$), we find $T_{int} \sim 1/8$ sec. This is only a rough estimate, but it does show that bacteria need to integrate for a significant fraction of a second in order to reach observed performance levels.

Brownian motion

Because of their small size *E. Coli* are affected by Brownian motion resulting from collisions with fluid particles. The main effect is that bacteria cannot hold a straight course-their direction of motion is subject to rotational diffusion. We can obtain a simple estimate for the time it takes the motion to decorrelate from its initial direction [6, 9]. Consider the bacterium to be a sphere of radius R swimming in water. The rotational diffusion coefficient for a sphere is given by [6]

$$D = \frac{k_B T}{8\pi\eta R^3},$$

(5.11)

where T is the temperature, η is the viscosity of the fluid, and k_B is the Boltzmann constant. If we take $T = 300°K$ (room temperature), $\eta = 10^{-2}$ poise, and $R = 10^{-4}$ cm, we find $D = 0.165\,\text{rad}^2/\text{sec}$. This means that the mean deviation in angle will reach $90°$ in about two seconds. The rotational diffusion coefficient for *E. Coli* is probably somewhat lower (it is not spherical and the medium it lives in is more viscous than water) but the key point is that a bacterium cannot integrate concentration changes for longer than several seconds because it "forgets" its orientation on that time scale.

Optimal Signal Processing

We have seen that physical constraints place severe restrictions on the computational strategies of E. Coli. They must integrate for a significant fraction of a second and cannot integrate for longer than several seconds. Can a more quantitative analysis of the strategy in the face of the constraints lead to more specific predictions? In particular, can we understand the experimentally observed chemotactic response? In the following sections we use design arguments to answer these questions in the (tentative) affirmative.

Quantitative measurements of the chemotactic response

Recently the chemotactic response of E. Coli has been measured for tethered cells [13, 14, 21]. Cells are attached (tethered) to a slide by one flagellum. They rotate clockwise (CW) or counterclockwise (CCW). In free-swimming cells CCW rotation of the flagellar bundle corresponds to runs and CW rotation corresponds to tumbles. This identification is made for the tethered cells, although the correspondence is not precise-for example the statistics of runs and tumbles differ from those of CCW and CW rotations. The experiments measure the probability that a bacterium is spinning counterclockwise (CCW) as a function of time while a given time-dependent chemical stimulus is present in the medium. The impulse response of the CCW vs. CW bias (the change in the bias in response to a brief attractant pulse) shows that E. Coli pass the concentration through a band-pass filter (Fig. 5.3). The positive lobe of the filter extends over one second and the negative lobe over four seconds; i.e., a bacterium integrates the concentration one second into the past and then subtracts the result of integrating over the previous four seconds. The two filter lobes are of equal area, showing that E. Coli respond to a low-pass version

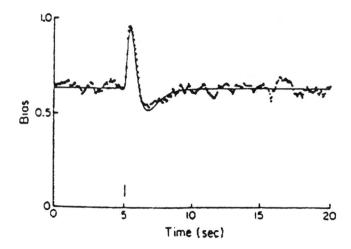

Figure 5.3: The CCW/CW bias response of tethered bacteria to brief pulses of attractant. Reproduced from Ref. [21].

of the derivative of the concentration, with a time scale that lies within the bounds on integration time discussed earlier. Other experiments show that for small concentration changes the response is linear [21], although larger changes can lead to response saturation.

Goals of the theoretical approach

Suppose we were to build a device that measured attractant concentration and moved to to regions of higher concentration while subject to the same noise and locomotion constraints that bacteria face. What design would allow such a device to perform its task optimally? How does the best design compare to the strategy actually utilized by bacteria? Below we describe a simple model for bacteria's behavior, propose an optimization principle, and compare the resulting strategy to the experimental results.

We work in the regime of slow concentration gradients that are essentially uniform over the length of one run. In this regime the strategy is determined by the physical constraints discussed earlier and not by short-scale variations in concentration. Whether or not these are the conditions actually encountered by *E. Coli* in their natural environment is a question that can only be answered by experimental investigations of the typical gradients they swim in. If it turns out that short-scale variations are important, a detailed analysis involving the statistics of natural concentration variations will need to be carried out.

A model of the chemotactic response

In the absence of stimuli bacteria tumble in a Poisson process with a fixed rate; the rate is lowered when the concentration as sensed by the bacterium is increasing. A simple model for this behavior is that bacteria use a concentration-dependent decision variable y in the following fashion: When $y \leq 0$ tumbles occur with rate r_0, while when $y > 0$ tumbles are suppressed entirely. Following the experimental evidence for linear filtering, we take $y(t) = \int dt' F(t')[\dot{c}(t - t') + \dot{\eta}(t - t')]$, where $c(t)$ is the "true" (e.g., experimentally imposed) concentration encountered by the bacterium; $\eta(t)$ is the noise in the bacterium's estimate of the concentration, and $F(t)$ is the linear filter through which the rate of change of the concentration is passed. Intuitively, the filter F allows the bacterium to reduce the noise level by integrating concentration changes over a period of time. It is reasonable to assume that the concentration noise is Gaussian; it is also spectrally white on the time scale of interest, since the correlation time of the noise is determined by the time (~ 1 msec) it takes for attractant molecules to diffuse in an out of a volume roughly the size of a bacterium. We further assume that concentration changes are small compared to the noise since the problem of chemotaxis is hardest in the low signal to noise limit, and hence the optimal strategy will be determined by this regime. With these assumptions the tumble rate to lowest order in $\dot{c}$ is given by

$$r(t) = \bar{r} \left[1 - \frac{2}{\sqrt{2\pi N \int dt \dot{F}^2(t)}} \int dt' F(t') \dot{c}(t - t') \right], \qquad (5.12)$$

where $\bar{r} = r_0/2$ and N is the spectral density of the concentration noise. Note that the rate does not depend on the overall scale of F. We vary F and $\bar{r}$ to achieve optimal performance within the model.

Defining optimality

We must first define what we mean by "optimal." We compute the average displacement $\langle \vec{x} \rangle$ during a run and its variance $\langle x^2 \rangle$. To lowest order in the concentration gradient the displacement will be proportional to the gradient, $\langle \vec{x} \rangle = \alpha \vec{\nabla} c$. Recall that we are working in the low signal to noise limit and hence $\vec{\nabla} c$ is small, and also that we take the gradient to be essentially uniform over the length of a run. In the random walk executed by a bacterium $\langle \vec{x} \rangle$ acts as the drift velocity and $\langle x^2 \rangle$ as the diffusion constant. We recall that for diffusion with drift $v_d/C_D = \mathcal{F}/k_B T$, where v_d is the drift velocity, C_D is the diffusion constant, $\mathcal{F}$ is the applied force, k_B is the Boltzmann constant, and T is the temperature [6]. In our analogy $\vec{\nabla} c$ plays the role of the applied force, and hence the inverse temperature $\beta \equiv 1/k_B T$ is given by

$$\beta = \frac{\alpha}{\langle x^2 \rangle}. \tag{5.13}$$

If we start with an ensemble of particles (e.g. bacteria) that move according to these dynamics the equilibrium distribution of the particles in space will be the Boltzmann distribution, $P(\vec{x}) = \exp(\beta c(\vec{x}))/Z$, where Z is the normalization. It is possible to derive this distribution more rigorously [16]. We see that if we maximize β - the equivalent of minimizing the temperature in a physical system - the probability of finding bacteria will be greatest in the regions of highest concentration, which is the desired result. Hence our optimization principle is to maximize β.

Computing β

We now need to compute $\langle \vec{x} \rangle$ and $\langle x^2 \rangle$ for a run. The displacement for a single run of duration t will be

$$\vec{x} = v_0 \int_0^t dt' \hat{v}(t'), \tag{5.14}$$

where $\hat{v}(t)$ is the direction of bacterium's trajectory at time t and v_0 is bacterium's (constant) speed. To obtain $\langle \vec{x} \rangle$ over one run we need to average Eq. (5.14) over the distribution of run times and over the possible trajectories $\hat{v}(t)$ generated by rotational diffusion. The distribution of run times for a Poisson process with a time-dependent rate is given by

$$P(t) = r(t) \exp(-\int_0^t dt' r(t')), \tag{5.15}$$

where $r(t)$ is given by Eq. (5.12). Therefore we have

$$\langle \vec{x} \rangle = v_0 \langle \int_0^\infty dt P(t) \int_0^t dt' \hat{v}(t') \rangle, \tag{5.16}$$

where $\langle ... \rangle$ is the average over trajectories. In Eq. (5.12) $\dot{c}$ is the time derivative of the concentration as seen by the bacterium and can be written $\dot{c} = \vec{v} \cdot \vec{\nabla} c$. We take uniform the gradient to be in the $\hat{z}$ direction without loss of generality and expand Eq. (5.16) to lowest order in $\vec{\nabla} c$. We then carry out the average over trajectories to find

$$\langle \vec{x} \rangle = v_0^2 \vec{\nabla} c \frac{1}{(\bar{r} + D)} \frac{1}{\sqrt{2\pi N}} \frac{1}{\sqrt{\int \dot{F}^2}} \int_0^\infty dt F(t) e^{-(\bar{r}+2D)t}. \tag{5.17}$$

Recall that N is the spectral density of the concentration noise, and D is the bacterium's rotational diffusion coefficient measured in radians^2/sec. The factor of e^{-2Dt} arises from the decay of correlations between the current direction of swimming and the direction t seconds ago. Note that both rotational diffusion and tumbling combine to disorder the trajectory and make contributions from earlier times less important. A similar calculation gives

$$\langle x^2 \rangle = 2v_0^2 \frac{1}{\bar{r}(\bar{r} + 2D)}. \tag{5.18}$$

Substituting these results into Eq. (5.13) gives β in terms of F:

$$\beta \propto \bar{r} \frac{1}{\sqrt{\int \dot{F}^2}} \int_0^\infty dt\, F(t) e^{-(\bar{r}+2D)t}, \tag{5.19}$$

where constant factors that do not affect optimization are not shown.

Response optimization

We find the form of the optimal filter by carrying out functional maximization of β with respect to F. This leads to a differential equation for F that we solve with the conditions that $\dot{F}^2$ must vanish at infinity (so that $\int \dot{F}^2$ is finite) and that $F(t) = 0$ for $t \leq 0$ (since F must be causal). Then the solution for $t > 0$ is

$$F(t) = 1 - e^{-(\bar{r}+2D)t}. \tag{5.20}$$

We can now compute β:

$$\beta = \frac{\bar{r}}{4(\bar{r}+2D)^{\frac{3}{2}}}. \tag{5.21}$$

Maximizing with respect to $\bar{r}$ we find $\bar{r}_{opt} = 4D$.

We are not quite done. We have now found the filter F in Eq. (5.12) for the rate. However, in that expression the integral extends infinitely far back in time. The bacterium is only interested in the integral since the last tumble - all contributions from earlier times are uncorrelated with its current direction and hence only serve to degrade its estimate of the concentration gradient. We assume that only the contribution since the last tumble is used - the filter is "zeroed" at every tumble. Hence while the filter actually used to make decisions is F, responses from different trials will differ due to the variations in the time since the previous tumble. Many such responses are averaged to obtain the impulse response. Hence the average tumbling rate we expect to observe is given (again to lowest order in $\dot{c}$) by

$$ r(t) = \bar{r} \left[1 - \sqrt{\frac{2}{3\pi ND}} \int_0^\infty d\tau \bar{r} e^{-\bar{r}\tau} \int_0^\tau dt' F(t') \dot{c}(t - t') \right], \qquad (5.22) $$

where τ is the time since the last tumble and we average over the distribution of run times. Integrating over τ we find

$$ r(t) = \bar{r} \left[1 - \sqrt{\frac{2}{3\pi ND}} \int_0^\infty dt' e^{-\bar{r}t'} F(t') \dot{c}(t - t') \right]. \qquad (5.23) $$

The experimentally measured response is an average of responses to pulses of different magnitude and even of different attractants [21]. Therefore only the shape of the response is meaningful, not the overall scale. The overall scale becomes important in the discussion of the response gain below. Also, the experiments measure the response to concentration and not to its time derivative. We must take the derivative of the predicted filter in order to obtain the predicted response to concentration. This

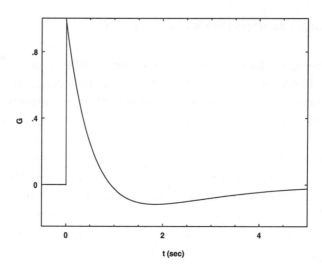

Figure 5.4: The predicted impulse response $G(t)$, from Eq. (5.24).

response is given by

$$G(t) = 10De^{-10Dt} - 4De^{-4Dt}. \tag{5.24}$$

We plot $G(t)$ in Fig. 5.4, using $D = 0.165$ radians^2/sec from above.

Comparison with experiments

The response can be compared to the measured response shown in Fig. 5.3. We note that both filters are band pass, although the negative lobe the predicted filter is somewhat longer. The time scale for forgetting old information is $1/\bar{r}$ [see Eq. (5.23)] which is about 1.5 seconds since $\bar{r} = 4D = 0.66\,\mathrm{sec}^{-1}$. We see that forgetting is determined by rotational diffusion as expected. Note that the tumble rate we predict is slightly longer than the observed rates of roughly one a second. Several explanations are possible. One is that the disagreement is due to the difference between free-swimming and tethered cells; as we point out below, experiments on free-swimming cells are crucial for a precise test of the theory. One of the effects not considered here is that consecutive runs are weakly correlated in direction [8]. This would make the effective run length somewhat longer than the actual mean run length; since we assume successive runs to be uncorrelated we predict a longer run length. Mean run lengths are also known to increase substantially in certain attractants [8]. The predicted response rises more sharply than the observed response. This difference arises from the distinction between rate and bias: the experiment measures the response of the *probability* of CCW vs. CW rotation while we predict the response of the *rate* of tumbling. To obtain the probability of being in one state vs. the other, we must integrate changes in the rate over some time period. Hence the probability will be a low-pass filtered version of the rate. Experiments on free-swimming cells are necessary in order to make a direct comparison with the prediction.

In addition to the filter shape we can look at the overall gain of the response. The gain measured experimentally is the change in the run/tumble bias divided by the time rate of change of fractional receptor occupancy, $\mathcal{G} \equiv \Delta b / \dot{o}$. Hence we need to compute Δb and $\dot{o}$. We formulate the problem of chemotaxis in terms of concentration changes rather than receptor occupancy changes, but when concentration changes are slow and the mean concentration is such that $o \approx 0.5$ (both reasonable assumptions in the regime we study), occupancy changes are simply related to concentration changes by $\dot{o} \approx \dot{c}/4c$. If $\dot{c}$ is small we can assume that c is roughly constant, in which case a constant $\dot{c}$ corresponds to a constant $\dot{o}$. A constant $\dot{c}$ will cause a constant change in r. Hence we can look at the steady-state bias, $b = r_t/(r_t + r)$, where r_t is the rate for switching from tumble to run. From Eq. (5.23) we can then compute Δb and find

$$\mathcal{G} = \frac{4}{5} \frac{r_t}{(r_t + \bar{r})^2} \dot{c} \sqrt{\frac{6}{\pi N D}}. \qquad (5.25)$$

We can get an estimate of the concentration noise N as before [9]. The local fluctuations in concentration will have a variance of order the number of molecules in the volume occupied by the bacterium. N is roughly the variance times the correlation time of the noise, with the latter corresponding to the time required by the molecules to diffuse across the volume. To get concentration rather than number fluctuations, we need to divide by volume squared. Hence $N \sim c/R \, D_m$, where D_m is the diffusion constant of the molecules. Substituting into Eq. (5.25) we obtain

$$\mathcal{G} = \frac{4}{5} \frac{r_t}{(r_t + \bar{r})^2} \sqrt{\frac{6c \, R \, D_m}{\pi D}}. \tag{5.26}$$

Representative values for the quantities in this expression are $c = 6 \times 10^{14}$ molecules/cm^3, $R = 10^{-4}$ cm, and $D_m = 10^{-5}$ cm^2/sec. Using the optimum value for $\bar{r}$ and taking $r_t = 10$ sec^{-1} (a typical value for free-swimming cells) we find $\mathcal{G} \approx 210$ sec. This value is to be compared with an observed gain of 114 sec [21]. The sensitivity of E. Coli is within a factor of two of the optimal sensitivity predicted by our model.

Discussion

We have largely succeeded in predicting a bacterium's response to nutrient concentration changes from physical constraints and a simple optimization principle. Not only the qualitative features of the response but also some quantitative details - the mean tumble rate, the gain, and the response time scales - are not far from the experimental measurements. Note that all predictions are made with no free parameters. Experiments on the responses of free-swimming cells are crucial to further test the theory. The prediction suggests several interesting questions that can be explored experimentally. One is whether bacteria can maintain the optimal filter over a variety of conditions. Does the filter change if the fluid viscosity (and hence the rotational diffusion coefficient) is altered? If a change is observed, does it occur adaptively within an existing generation of bacteria or arise evolutionarily after several generations are grown in the new medium? In environments where the concentration gradients vary on shorter length scales, can the bacterium adapt to make use of this *a priori* information? Another test involves other species of chemotactic bacteria - are they also optimal in accordance with their media and physical characteristics? Of course, we need not restrict ourselves to chemotaxis and we should be able to analyze the processing of other stimuli within the same framework.

Conclusion

We know of a number of sensory systems that perform at or near the limits set by basic physical constraints [10, 12]. One common example is photon counting in vertebrate vision - we can detect just a few photons entering the eye in low light conditions [15, 20]. Another amazing example is bat echolocation - bats can detect temporal differences down to 10 *nanoseconds* in echo return times and perform as ideal detectors in the presence of noise [23, 24]. Perhaps it should not be surprising that bacteria also exhibit signal processing that approaches physical limits - while bacteria are the simplest living organisms, they also have had the longest time to evolve, and do so at a faster rate than more complex organisms.

If an organism's response to sensory stimuli is limited by physical constraints imposed by the environment, that must mean that the processing circuitry is essentially perfect and does not introduce additional errors. This fact can place tight constraints on the circuitry design. Such design arguments can be used to understand why sensory processing works the way it does. Design arguments have been applied with success to such problems compound eye design in insects [25, 26], bipolar cell responses in the salamander retina [19], and ganglion cell responses in the vertebrate retina [3]. Aspects of these problems are reviewed by Atick and Bialek in their contributions to these proceedings. The fact that we have succeeded in explaining some aspects of bacterial chemotactic response using design arguments lends further credence to the idea that such arguments can provide a powerful and generic approach to understanding sensory systems.

We have greatly extended the range of organisms where this approach has been used and have been able to predict behavioral strategy starting from basic physical constraints of the environment. This is of course the ultimate goal for higher organisms; whether or not it is reachable and by what means remains to be seen, but it is encouraging that progress can be made at least for the simplest sensory system.

Acknowledgements

The work on optimal filtering was done in collaboration with Bill Bialek. I thank Bill Bruno, Fred Rieke, Allan Schweitzer, Mark Schnitzer and Steve Block for stimulating discussions and useful comments and the Institute for Advanced Study for its hospitality. Work at the Institute was supported in part by a grant from the Seaver Institute.

References

[1] J. Adler (1969). Chemoreceptors in bacteria, *Science* **166**, 1588-1597.

[2] J. P. Armitage (1992). Behavioral responses in bacteria, *Ann. Rev. Physiol.* **54**, 683-714.

[3] J. J. Atick and A. N. Redlich (1990). Towards a theory of early visual processing, *Neural Comp.* **2**, 308-320.

[4] H. C. Berg and R. A. Anderson (1973). Bacteria swim by rotating their flagellar filaments, *Nature* **245**, 380-392.

[5] H. C. Berg (1975). Bacterial behaviour, *Nature* **254**, 389-392.

[6] H. C. Berg (1983). *Random Walks in Biology* (Princeton University Press, Princeton NJ).

[7] H. C. Berg (1988). A physicist looks at bacterial chemotaxis, in *Cold Spring Harbor Symp. Quant. Biol.* **53**.

[8] H. C. Berg and D. A. Brown (1972). Chemotaxis in *Escherichia coli* analysed by three-dimensional tracking, *Nature* **239**, 500-504.

[9] H. C. Berg and E. M. Purcell (1977). Physics of chemoreception, *Biophys. J.* **20**, 193-219.

[10] W. Bialek (1987). Physical limits to sensation and perception, *Ann. Rev. Biophys. Biophys. Chem.* **16**, 455-478.

[11] W. Bialek (1990). Theoretical physics meets experimental neurobiology, in *1989 Lectures in Complex Systems, SFI Studies in the Sciences of Complexity, Lect. Vol. II,* E. Jen, ed., pp. 513-595 (Addison-Wesley, Menlo Park CA).

[12] S. Block (1992). Biophysical principles of sensory transduction, in *Sensory Transduction* (Rockefeller University Press, New York).

[13] S. M. Block, J. E. Segall, and H. C. Berg (1982). Impulse response in bacterial chemotaxis, *Cell* **31**, 215-226.

[14] S. M. Block, J. E. Segall, and H. C. Berg (1983). Adaptation kinetics in bacterial chemotaxis, *J. Bacteriol.* **154**, 312-323.

[15] S. Hecht, S. Shlaer, and M. Pirenne (1942). Energy, quanta and vision, *J. Gen. Physiol.* **25**, 819-840.

[16] L. Kruglyak and W. Bialek (1992). Optimal signal processing in bacterial chemotaxis, in preparation.

[17] W. Pfeffer (1884). Locomotorische richtungsbewegung durch chemische reize, *Untersuch. Bot. Inst. Tubingen*, **1**, 363-482.

[18] E. M. Purcell (1977). Life at low reynolds number, *Am. J. Phys.* **45**, 3-11.

[19] F. Rieke, W. G. Owen, and W. Bialek (1991). Optimal filtering in the salamander retina, in *Advances in Neural Information Processing Systems 3,* D. Touretzky and J. Moody, eds., pp. 377-383 (Morgan Kaufmann, San Mateo CA).

[20] B. Sakitt (1972). Counting every quantum, *J. Physiol.* **223**, 131-150.

[21] J. E. Segall, S. M. Block, and H. C. Berg (1986). Temporal comparisons in bacterial chemotaxis. *Proc. Nat. Acad. Sci. USA* **83**, 8987-8991.

[22] A. Shapere and F. Wilczek (1989). Geometry of self-propulsion at low Reynolds number, *J. Fluid Mech.* **198**, 557-585.

[23] J. A. Simmons (1989). A view of the world through the bats ear: The formation of acoustic images in echolocation, *Cognition* **33**, 155-199.

[24] J. A. Simmons, M. Ferragamo, C. F. Moss, S. B. Stevenson, and R. A. Altes (1990). Discrimination of jittered sonar echoes by the echolocating bat, *Eptesicus fuscus*: The shape of target images in echolocation, *J. Comp. Physiol. A* **167**, 589-616.

[25] A. W. Snyder (1977). Acuity of compound eyes: Physical limitations and design, *J. Comp. Physiol.* **116**, 161-182.

[26] A. W. Snyder, D. G. Stavenga, and S. B. Laughlin (1977). Spatial information capacity of compound eyes, *J. Comp. Physiol.* **116**, 183-207.

Could Information Theory Provide An Ecological Theory of Sensory Processing?

Joseph J. Atick

Introduction

This manuscript explores the use of Information theory [55] as a basis for a first principles approach to neural computing. The relevance of this theory to the nervous system ultimately derives from the fact that the nervous system possesses a multitude of subsystems that acquire, process and communicate information. This is especially true in the sensory pathways. One could use Information theory to assess efficiency of information representation in many of these pathways, and this has led to the design and analysis of new experiments in simpler neural systems [17, 64]. The results of these studies support the general idea that efficiency of information representation in the nervous system has evolutionary advantages ([1, 2, 3, 5, 6, 7, 8, 9, 16, 26, 43, 44, 45, 58, 61], see also [10] and references therein). We shall consider the hypothesis that much of the processing in the early levels of sensory pathways is in fact geared towards building efficient representations of sensory stimuli in an animal's environment.

The above efficiency principle, formulated as an optimization problem, can be used as a *design* principle to predict neural processing: Starting with the natural representation of environmental signals as sampled by the array of sensory cells, one can try to find the recodings needed to improve efficiency subject to identifiable biological hardware constraints. The several stages of processing required to cast incoming data into the optimal form can then be compared to the stages of neural processing observed in sensory pathways. This principle has been shown to successfully predict retinal processing in space-time and color [1, 2, 3, 5, 6] and there are encouraging signs that it could be equally successful in predicting some of the cortical computation strategies ([10, 12, 27],Atick and Redlich, to appear). The approach just described can be termed "ecological", since it attempts to predict neural processing from physical properties of the stimulus environment. It is clear, that essential to the success of this program is a quantitative knowledge of (statistical) properties of natural signals. Several studies on properties of natural stimuli are currently underway.

The organization of the manuscript is as follows. We start with a brief review of information theory cast in a language suited for our subsequent analysis. We then speculate on why efficiency of information representation could be an organizing principle underlying sensory processing, and formulate this principle as an optimization problem and discuss how in general it might be solved. Finally we analyze in detail some biological systems where information theory has been shown to predict the observed neural processing. Contrast-coding of the LMC cells in the blow-fly compound eye [40, 42], and the spatio-temporal [1, 2, 3] and color coding [5, 6] of the mammalian retina. Our discussion on retinal processing is self-contained since we have included a brief review of the relevant experimental facts on retinal coding in space, time and color.

Information Theory: A Quick Primer.

Information theory evolved in the 1940s and 1950s in response to the need of electrical engineers to design practical communication devices. The theory, however, despite its practical origins, is a deep mathematical theory [55] concerned with the more basic aspects of the communication process. In fact, it is a framework for investigating fundamental issues such as efficiency of information representation and its limitations in reliable communication. The practical utility of this theory stems from its multitude of powerful theorems that are used to compute optimal efficiency bounds for any given communication process. Physicists might find these bounds reminiscent of the bounds set by the laws of thermodynamics on the performance of heat engines, and these ideal bounds serve as benchmarks to guide the design of better information systems.

In this section we give a brief review of Information theory. This review is not intended to be a full account of the theory. In fact, it focuses primarily on one aspect of Information theory, namely the effect of statistical regularities on efficiency of information representation. Among the other important aspects it ignores, is the role of noise and the reliability of representation. However, this account is adequate to enable the reader with no prior knowledge of Information theory to follow its subsequent applications to neural computing. Readers interested in further details are encouraged to consult the literature.

Information sources and channels

In Information theory any device, system or process that generates messages as its output is generically referred to as an *information source*. Although each source has its own representation that it uses to put out messages, generally speaking sources represent their messages as combinations of symbols selected from the list of all possible symbols they are capable of producing. The latter are often called the *source symbols* or *alphabets*, or the *representation elements*. The choice of alphabet (the set of symbols) and the way they are used to construct messages constitutes a representation or a code --- source coding.

For example, a book in English can be thought of as the output of an information source --- English language --- whose alphabet is $A, \cdots, Z, +$ blank. Similarly, a neuron or a layer of neurons can act as an information source whose symbols are the different neuronal response levels. Finally, an information source that is discussed often in this manuscript is the visual environment, where the symbols are the different gray levels of light pixels in the image mosaic. For simplicity, we introduce Information theory for the discrete case, where there is a countable number, N, of symbols that can be produced by the information source. In written English $N = 27$, while in an 8-bit gray scale imaging, $N = 2^8 = 256$.

An important fact about "natural" information sources is that they never produce messages which are random combinations of their alphabets. Instead, their messages tend to possess regularities or what is known as *statistical structure*. In other words, the way symbols are put together to form messages obeys certain statistical rules that are source specific. To begin with, information sources do not utilize elments of their alphabet with equal frequency. In long sequences of written English for example, E occurs at the rate of once in every ten letters while Z occurs only once in a thousand [51]. In totally random sequences the frequency of occurrence would be once in every 27 for all the alphabets. The frequency of occurrence of source symbols is captured by the probability distribution $\{P(m), m = 1, \cdots, N\}$.

More importantly, the selection of a symbol in a message is influenced by previous selections *i. e.*, symbols in a message are not statistically independent, instead there are intersymbol dependencies or correlations. Again, in English when a T occurs somewhere in a text it is very likely it will be followed by an H while it is never followed by a P or a Q. This statistical influence can be quite significant and can extend up to many symbols. Mathematically, it is captured by conditional probabilities or equivalently by joint probabilities among symbols. For messages of length l symbols the joint probabilities are denoted by $\{P(m_1, \cdots, m_l)\}$.

We model real information sources as stochastic systems [49] that generate sequences of symbols subject to some statistical rules [29, 37]. Since our knowledge of the statistical regularities of natural information sources is somewhat limited at this time, the rules we impose on our models represent only a subset of all regularities real information sources might possess. This is not necessarily a handicap since at any given stage in a sensory pathway, especially at the early stages, we suspect *only* incomplete knowledge of statistical regularities of stimulus source is available to neurons anyway. For example, we shall argue that the retina only makes use of the (two-point) pixel-pixel correlation function. Thus an approximate model of natural scenes that generates luminosity pixels subject only to the constraint of fixed pixel-pixel correlation function will be sufficient for studying the retina. Of course, to predict the processing in the cortex, knowledge of more complex regularities is necessary.

Finally, another basic concept in this theory is the concept of an *information channel*, which is the medium through which messages from sources are transmitted or stored. Just like an information source a channel possesses a set of alphabets, called *channel symbols*, which are used to carry the messages. The problem of mapping source symbols into channel symbols is referred to as the channel coding problem [28]. For the sake of brevity in the present manuscript we ignore all differences between source and channel coding and deal only with the generic problem of information representation regardless of where the coding is happening. This is justifiable especially since we are focusing on discrete noiseless Information theory.

Efficiency of information representation

As mentioned earlier, one of the main concerns of noiseless Information theory is quantifying efficiency of information representation. Intuitively, inefficiency can be attributed to the fact that information sources are constrained to obey statistical rules in constructing their messages. These rules build some degree of redundancy where, for example, many pieces in a message are *a priori* predictable from other pieces and from knowledge of the statistical structure. Also, the presence of constraints implies that information sources do not utilize their alphabets to their fullest capacity. Hence, a representation that possesses any statistical regularities is in many ways wasteful or inefficient. In this section we find a quantitative measure for this inefficiency.

To begin with, Information theory attributes to each message in the ensemble M of all messages that can be produced by a source, a statistical quantity known as the *information* which is given by

$$I(w) \equiv -\log_2 P(w), \qquad\qquad\qquad (6.1)$$

where $P(w)$ is the probability of the message w normalized so that $\sum_{w=1}^{\mathcal{N}} P(w) = 1$, with $\mathcal{N}$ the total number of messages in the ensemble M. Since we use $\log_2$ the units of I are bits (or binary digits)/message. $I(w)$ is essentially a measure of "surprise" or *a priori* "unexpectedness" of a message. According to it, a message that occurs often $P(w) \sim 1$ has low surprise or information value $I(w) \sim 0$, while that which is unexpected has high information. This measure does conform to the usual editorial policy where rare events are given more attention than frequently occurring ones. However, we should emphasize that it ignores the semantic value of a message; in this theory, the unexpectedness of a message plays an important but distinct role from the meaning of the message.

Averaging Eq. (6.1) over all messages in the ensemble M defines

$$H(M) = \sum_{w=1}^{\mathcal{N}} P(w)\, I(w) = -\sum_{w=1}^{\mathcal{N}} P(w) \log_2 P(w). \tag{6.2}$$

which is known as the *entropy* or average information per message. As is shown below, $H(M)$ is the mathematical object one needs to construct a quantitative measure of efficiency. Its precise significance derives from the powerful theorems that were proven about it. For example, the source coding theorem (see for example [28]) shows that $H(M)$ is the *minimum* length in binary digits (bits) per source message that are needed on average to represent the outputs of the source. Immediately, this says that a representation is most efficient iff on average messages in the ensemble M are equal to $H(M)$ bits in length.

To see how $H(M)$ is used to define a quantitative measure of efficiency, we investigate its dependence on what one intuitively perceives as the cause of inefficiency, namely the statistical structure. For concreteness, we consider a representation where each message w is built out of a combination of l symbols, then $P(w) = P(m_1, \cdots, m_l)$. We examine the value of $H(M)$ as a function of the statistical structure of the source keeping the N alphabets and the length l fixed. We show that $H(M)$ decreases the more statistical constraints the source has to obey in generating messages.

Consider first the case of a source that uses a representation where the symbols are statistically independent, i. e., the only statistical structure is that given by $\{P(m_i)\}$. In that case $P(m_1, \cdots, m_l) = P(m_1) \cdot P(m_2) \cdots P(m_l)$ and the entropy $H(M)$ can be written as a sum over the individual *symbol* (or *pixel*) entropies, $H(i)$,

$$H(M) = -\sum_{i=1}^{l} \sum_{m_i=1}^{N} P(m_i) \log_2 P(m_i) \equiv \sum_{i=1}^{l} H(i). \tag{6.3}$$

In general, however, the symbols are not statistically independent, so $P(m_1, \cdots, m_l)$ does not factorize into a product and the total entropy does not equal the sum of symbol entropies. Instead it satisfies

$$H(M) \leq \sum_{i=1}^{l} H(i), \qquad (6.4)$$

with equality if the symbols are statistically independent. This means that statistical influence among symbols lowers $H(M)$ or the amount of information carried by those symbols, which is intuitive since in this case many of the symbols redundantly carry the same information. To see how the proof goes consider the simple case of two symbols. Define the matrix $D_{ij} = P(m_i)P(m_j) - P(m_i, m_j)$, then using the fundamental inequality $x \geq \ln(1 + x)$ applied to $x = D_{ij}/P(m_i, m_j)$ we have the inequality $D_{ij}/P(m_i, m_j) \geq \ln(1 + D_{ij}/P(m_i, m_j))$. Multiplying this by $P(m_i, m_j)$ on both sides and summating on i and j remembering that $P(m_i) = \sum_j P(m_i, m_j)$ and $\sum_i P(m_i) = 1$ one arrives at $H(1) + H(2) \geq H(1, 2)$. Generalizing this proof to arbitrary number of symbols is straightforward.

The upper bound on $H(M)$ in Eq. (6.4) is not the absolute maximum since one can still look for the distribution $\{P(m_i)\}$ that maximizes the symbol entropy $H(i) = -\sum_{m_i=1}^{N} P(m_i) \log_2 P(m_i)$. Again, it is not hard to show that the maximum occurs when $\{P(m_i) = 1/N, \quad \forall \, m_i\}$, or when the alphabets are utilized with equal frequency, as anticipated. The maximum this gives is

$$\max_{s.s.} (\, H(M) \,) = \max_{\{P(m_i)\}} \left(\sum_{i=1}^{l} H(i) \right) = l \log_2 N \equiv C, \qquad (6.5)$$

where the first maximization is over the full statistical structure, and the second is over the distribution $\{P(m_i)\}$. Thus, maximum entropy is achieved by a source that represents its messages such that no statistical regularities exist among the symbols. A representation with no statistical structure is one where the receiver's knowledge about what to expect is minimal and thus on average a message when received conveys maximum amount of "surprise" or equivalently maximum amount of information $H(M)$.

The last equality in Eq. (6.5) defines another important information theoretic quantity, namely the *capacity* C of the representation or the channel, which is the absolute maximum information that l symbols selected from a list of N distinct alphabets could ever carry. Notice that $C = l\log_2 N = \log_2 N^l$, is the logarithm of the total number of messages, N^l, that the representation can carry. It can also be interpreted as the actual length of messages in binary digits. In English, $C = \log_2 27 = 4.73$ bits/alphabet, while the capacity of an 8-bit gray scale 256×256 pixel screen is $8 \times 256 \times 256$.

We are now ready to define a measure of efficiency: for any source with $H(M)$ using a representation of capacity C one useful measure of efficiency is

$$\mathcal{R} = 1 - H(M)/C, \qquad (6.6)$$

which is called the Shannon redundancy. Since $H(M) \le C$, $0 \le \mathcal{R} \le 1$ with $\mathcal{R} = 0$ being the most efficient where $C = H(M)$. This measure has two interpretations. First, thinking of $H(M)$ as the *actual* amount of information transmitted and C as the *maximum* amount that could be transmitted, efficiency calls for using a channel where the transmitted rate $H(M)$ is as close as possible to the maximum rate C. Alternatively, since C is the average length of a message in bits and $H(M)$ is the smallest average length that can ever be achieved by any representation (source coding theorem), efficiency calls for finding a representation where the actual length C is as close as possible to minimum allowed $H(M)$.

In general, to improve efficiency one recodes the output of the source into a representation that uses C as close to $H(M)$ as possible. This data compression is achieved by discarding the structure that is *a priori* predictable from the messages (the statistical structure) leaving only the so-called "textual" or non-predictable information. In principle a coding strategy that takes advantage of all statistical regularities can compress the representation down to its minimal size *i. e.*, can allow the use of $C = H(M)$. In practice, it might prove computationally prohibitive to achieve the optimal compression. In general one tries to find a compromise between the complexity of the representation and its efficiency, for example by ignoring certain aspects of the statistical structure and concentrating on those regularities that are simple to disentangle and discard in recoding. Also in real information systems noise is always there. In that case it is not advantageous to eliminate the redundancy completely since it is redundancy after all that distinguishes what is signal from what is noise. Information theory formulated for noisy channels can be used to find the best compromise. In our analysis of real neural coding in chapter five we use an effective approach to handle the noise without the need for developing the complicated machinery of Information theory in the presence of noise. The more general approach for handling noise in early sensory processing can be found in [1, 2].

The cost of inefficiency

To illustrate the cost of inefficiency, it is helpful to start with an example. Consider the DNA of a fictional creature whose bases, A, T, C, G are assumed to occur with probabilities listed in the second column of Table 1; never mind the fact that they violate Chargaff's rule. The problem is to find a coding that will store long sequences of this DNA on a computer disk economically. Since Table 1 does not supply any knowledge of statistical structure beyond base probabilities, we have to treat this information source approximately as if no statistical influence among the bases existed, and deal with each symbol as an independent message. Then the entropy of this DNA is $H(M) = \frac{1}{2} \times 1 + \frac{1}{4} \times 2 + \frac{1}{8} \times 3 + \frac{1}{8} \times 3 = \frac{7}{4}$ bits/base. This means that there exists a code that can represent this DNA's sequences with as few as $\frac{7}{4}$ bits/base. If we code the four bases into $00, 11, 01, 10,$ then

Symbol	$P(i)$	Code1	Code 2
A	$\frac{1}{2}$	00	0
T	$\frac{1}{4}$	01	10
C	$\frac{1}{8}$	10	110
G	$\frac{1}{8}$	11	111

Table 6.1: The probability distribution of the bases A, T, C, G of the DNA of a ficitonal creature and the two simple binary codes discussed in the text.

the average length (or capacity) used is 2 bits/base which is greater than $H(M)$. However, if we code in the fashion illustrated in the third column of Table 1, then the average length is $\frac{1}{2} \times 1 + \frac{1}{4} \times 2 + \frac{1}{8} \times 3 + \frac{1}{8} \times 3 = \frac{7}{4}$, which is exactly the entropy of the source and thus the most efficient code possible given base probabilities only.

One might think that since the bases in code 2 are not of equal length that decoding sequences would be difficult. This is not true; the code by construction has a trivial decoding algorithm. In any sequence, a zero signals the end of a coded base; with one exception, where one does not encounter zero for three consecutive digits, in that case the base is G and the next digit is part of the next coded base. There is a general procedure for constructing these minimal redundancy codes, known as Huffman coding, which generalizes this trivial example to arbitrarily complicated real problems [28].

Notice that code 2 is on average $\frac{1}{4}$ bits/base shorter than than code 1. This saving may seem insignificant; however, imagine the situation where this creature has 10^9 bases in its DNA. Then code 1 effectively requires an additional $\frac{1}{4} \times 10^9 = 250$ Mega bits or ~ 62 Megabytes to store the same information. Further savings in storage space could be achieved using a code that can discard other statistical regularities that this DNA might have, such as correlations among the bases. Of course, this comes at the cost of increasing the complexity of the code.

The above example leads into the general question of what is the cost of inefficiency. In man-made systems, inefficiency usually means more storage space, more expenditure of transmission power, longer transmission times or in general larger bandwidths or dynamic range to transmit or store the same amount of information.

In biological systems the consequences of inefficiency are not as clear and they are most likely animal dependent. What one needs is a way to translate the information theoretic cost into a biologically significant cost to an animal. We suspect that most areas in the nervous system of many species could not afford to be inefficient, since invariably neurons have a limited response range (capacity or dynamic range) especially in comparison with the wide range of stimuli the animal encounters [11, 56]. Pooling its dynamic range resources, we believe the brain possesses a relatively limited number of states with which to build representations of the great multitude of objects and events in its information rich environment. Under such circumstances, an efficient representation can allow the brain to extract more information about its environment without the need to evolve to larger sizes.

In addition to savings in dynamic range, efficient representations could potentially facilitate certain cognitive tasks, such as associative learning [10] and pattern recognition. Actually, in higher animals we feel it is more likely that cognitive benefits are the driving force towards efficient representations. These issues are discussed in further detail below.

Types of inefficiencies

There are two types of inefficiencies that one encounters in information systems. As we shall see shortly, both types can influence the computational strategies of real sensory neurons. Both were alluded to in our discussion above, here for future reference we exhibit them more explicitly. To do that, we rewrite $\mathcal{R} = \frac{1}{C}(C - H(M))$ in the following equivalent form

$$\mathcal{R} = \frac{1}{C}\left(C - \sum_{i=1}^{l} H(i)\right) + \frac{1}{C}\left(\sum_{i=1}^{l} H(i) - H(M)\right), \qquad (6.7)$$

where we have added and subtracted $\frac{1}{C}\sum_{i=1}^{l} H(i)$ to the definition of $\mathcal{R}$.

The two terms in the brackets in Eq. (6.7) explicitly quantify the contribution of the two forms of redundancy to $\mathcal{R}$. First if the alphabets are used with equal frequency then $\sum_{i=1}^{l} H(i) = l \times \log_2 N = C$, and the first term in the bracket drops out. In general, however, $\sum_{i=1}^{l} H(i) < C$, and this term contributes positively to the redundancy. Second, if there are no intersymbol dependencies, then the total entropy $H(M)$ equals $\sum_{i=1}^{l} H(i)$ exactly and the second term vanishes. Typically, however, there are statistical relations among the symbols in which case $\sum_{i=1}^{l} H(i) > H(M)$, and hence the second term contributes a positive amount to the redundancy. In a system where there is absolutely no redundancy $C = \sum_{i=1}^{l} H(i) = H(M)$ to make $\mathcal{R} = 0$.

To get a feel for the relative significance of the two types of inefficiencies, consider written English. There $C = \log_2 27 = 4.76$ bits/letter, while $H(i)$ computed using the well known probabilities of different alphabets [51] is 4.03 bits/letter, which gives a redundancy of about only 15%. In general inefficiency due to unequal use of alphabets is minor. The major source of redundancy comes from statistical correlations among symbols. For English, an estimate of $H(M)$ was first done by Shannon [54] using a method that takes into account statistical correlations among the alphabets. He found that the entropy is around 1.4 bits/letter. From $C = 4.76$, $H(M) = 1.4$ and $H(i) = 4.03$ we can see that redundancy due to intersymbol correlations in English is about 55%, making the total redundancy of written English close to 70%; for estimates of redundancy in other western languages see Ref. [14]. The situation is very similar in many natural sensory information sources.

Minimum-redundancy vs. minimum-entropy codes

The expression for $\mathcal{R}$ in Eq. (6.7) also makes explicit two classes of codes that we will refer to in our discussions on neural coding. A code that minimizes the full $\mathcal{R}$ is known as *minimum redundancy* code, while that which minimizes the part of $\mathcal{R}$ due to intersymbol correlations is known as *minimum entropy* code (or factorial code); Elegant examples of factorial codes can be found in Refs. [10, 33, 65, 66]. Minimum entropy codes minimize the difference $\sum_{i=1}^{l} H(i) - H(M)$. In the limit $\sum_{i=1}^{l} H(i) = H(M)$, they produce a representation where the symbols are statistically independent, so the probability of any message is given by the product of the probabilities of the symbols making up the message. If one insists on no loss of information, then minimum entropy codes minimize $\sum_{i=1}^{l} H(i)$ subject to the constraint of fixed total entropy $H(M)$. We should emphasize that these codes are not by themselves redundancy reducing. In fact, from Eq. (6.7) we can see that these codes preserve the total redundancy by transforming redundancy due to correlations to redundancy due to unequal use of symbols.

The interest in minimum redundancy codes in engineering is clear; they allow the use of smaller dynamic range or smaller capacity. The reason minimum entropy codes are also interesting is that usually after minimizing $\sum_{i=1}^{l} H(i)$ one can find trivial transformations to fit the coded messages into a channel with a smaller C, and thus they can be viewed as a convenient first step for achieving minimum redundancy codes. A simple example of this type of two stage coding is given below.

In sensory pathways, we expect minimum entropy codes to play an important role for two reasons: just as in engineering, minimum entropy codes are excellent first steps towards redundancy reduction. This is especially true for natural stimuli where the most significant part of the redundancy is coming from intersymbol dependency. Second, minimum entropy codes could have an intrinsic cognitive advantage beyond the fact that they enable the nervous system to use smaller dynamic range. Both issues are elaborated on in the next chapter.

Information Theory and Sensory Ecology?

The neural networks in the sensory pathways of animals are well adapted to processing signals from the "natural" environment. One fact about these special stimuli is that they are never random, instead they tend to possess statistical regularities. For example, in natural images, due to the morphological consistency of objects, nearby pixels are very similar in their visual appearance. The luminosity profile in these images changes gradually in space and only abruptly at edges or borders. Similarly in time and color, where there is continuity and smoothness. This means that in natural images there is a high degree of spatio-temporal and chromatic correlations among pixels, and hence a pixel by pixel representation of natural scenes, which is the representation formed by the photoreceptor mosaic, is inefficient. This fact was well known to engineers in the television industry as far back as the fifties. In fact, the statistical studies on television signals that they conducted indicate that redundancy could run well in excess of ninety percent in natural images [31, 32, 38, 53]. The situation is expected to be similar for most other senses.

Given that natural stimuli come in a highly inefficient form, there are several reasons why the nervous system might invest some of its resources to recode incoming signals to improve efficiency. We present three potential benefits of efficient representations. The first is an advantage of strict redundancy reduction, while the other two are advantages of both redundancy reduced and minimum entropy representations. At this stage we cannot tell which of the two strategies, redundancy reduction or minimum entropy, is more fundamental in the nervous system. However since they are closely related we will continue to treat both on an equal footing under the banner of efficiency. These benefits, however, are not mutually exclusive and do not exhaust all potential advantages of efficiency. This discussion is somewhat heuristic; we hope to present a more mathematical analysis of the material in this section elsewhere. Armed with this motivation, however, we formulate an optimization principle for the representation of sensory data.

Information bottleneck

It is possible that at some point along a sensory pathway there exists what may be termed an *information bottleneck*. This means that somewhere there exists a restriction on the rate of data flow into the higher levels of a pathway. This could arise from a limited bandwidth or dynamic range of a neural link, which is not unlikely given that neurons invariably possess limited response range [11, 56]. Alternatively, the limitation could be due to a computational bottleneck in the deeper levels of the sensory pathway which restricts the number of bits of data per second that can be analyzed in the object recognition process. An example of such limitation might be the "attention bottleneck" which is suspected to occur somewhere between area $V4$ and the inferotemporal cortex IT [63]. Actually it is very unlikely that the bottleneck is abrupt. It is most likely happening through a gradual constriction of data flow.

Studies on the speed of visual perception [60] and reading speeds [39], consistently give numbers around $40 - 50$ bits/sec for the perceptual capacity of the visual pathway in humans. This number can be interpreted as the maximum rate of visual information that can be processed by the deep layers of the visual pathway and is in a sense a measure of the bottleneck. On the other hand the rate at which visual data is collected by the photoreceptor mosaic is known to exceed 5×10^6 bits/sec [35] In order to fit the huge range of incoming signals into the limited capacity anticipated at higher levels a sensory pathway might have to perform a series of data compressions. One strategy for data compression in neural systems is redundancy reduction [7, 8].

Of course, if the animal's needs are very specific then it could develop specialized feature detectors -- bug detectors -- very early on in its pathways that are tuned for objects and patterns that are critical for its survival. Such detectors will cut down on the data rate since they discard almost everything they do not detect. In higher animals, where the needs are not very specific and where flexibility to changing environment is critical, a better strategy is one which recodes to improve efficiency without discarding a lot of information early on. In reality, a combination of the two mechanisms are in place. After all, an animal chooses a sensory sampling unit --- acuity limit or resolution --- below which it discards all data. Other strategies include noise filtering and generalization.

Associative learning

Barlow [10] argued that the way the nervous system represents objects and events in the environment might have dramatic implications for an animal's ability to perform associative learning. The idea is that for an animal to learn a new association between any two events, m_1 and m_2, the brain should have knowledge of the prior probability of occurrence or the *a priori* coincidence rate of m_1 and m_2. Without this information the animal cannot tell whether event m_1 has become a good predictor of m_2 or whether the joint occurrence of m_1 and m_2 (or m_1 followed by m_2) is consistent with the random coincidence rate. What the animal needs is knowledge of the prior joint probability $P(m_1, m_2)$. Similarly for associations among any number of events.

However, knowledge of the prior probability of joint events in the environment is not easy to achieve. In general, there is a huge number of events and conjunctions among them. By any reasonable estimate, knowledge of the prior probabilities of all these conjunctions would require storage of an exponentially large set of numbers that far exceeds any estimate of brain storage resources. The only way out seems to be if the representation of events and objects in the brain is very special. In fact, if the representation is such that the elements are statistically independent --- of course, until the association to be learned occurs --- then the probability of any combination of them can be obtained very simply from individual probabilities, since in that case $P(m_1, \cdots, m_n) = P(m_1) \cdots P(m_n)$. Thus for any N events the N^n probabilities $\{P(m_1, \cdots, m_n)\}$ can be computed from knowledge of the N individual probabilities $\{P(m_i); m_i = 1, \cdots, N\}$.

So the fact that the brain is finite in its resources suggests that a minimum entropy representation of the world might be necessary for it to perform a cognitive task essential for survival, namely associative learning.

Pattern recognition

The ultimate goal of any sensory pathway is pattern recognition: for its survival, an animal needs to acquire from its senses knowledge of the location and identity of all objects in its immediate environment. A third possible explanation for why a sensory pathway might choose to preprocess incoming signals to improve their efficiency is that efficiency might facilitate the pattern recognition process [9, 65, 66].

Consider for instance the visual pathway: In the incoming representation, pixels are highly correlated and thus have low information value. A large number of pixels is needed to define any feature. An efficient representation, on the other hand, decomposes images in terms of elements that are statistically independent and thus necessarily more informative elements. These elements are the features or the "vocabulary" from which natural images can be assembled most economically. It is possible that these building blocks, arrived at by pure statistical considerations, are closer to the patterns and objects an animal needs to recognize in its environment and hence a representation that uses them could simplify the subsequent pattern recognition process.

Independent of whether the visual system takes advantage of efficiency for pattern recognition, it is of interest to find what the features in efficient representations of natural images turn out to be. This is a concrete proposal, since starting with a data-base of natural images, one can look for transformations that drive $\mathcal{R}$ or some variant of it down. One promising approach for doing this is to use neural networks which can be trained using unsupervised learning algorithms that incrementally improve efficiency of representation as the network is exposed to more examples of natural images. Some unsupervised learning algorithms that achieve this in some simple settings have appeared in [4, 12, 30, 34, 50, 52].

An optimization problem

Here we formulate the principle of coding to improve efficiency as an optimization problem. For concreteness, we focus on visual processing. We make the hypothesis that the visual system is concerned with building a minimum entropy representation of the natural world. Since by a simple transformation we can also achieve minimum redundancy, the results of this section are equally relevant to minimum redundancy coding. What this means is that the visual system has to map the photoreceptor signals, which are highly correlated, to a representation where the elements are statistically independent. It is unlikely that any system could achieve this in one recoding. It is more likely that it would have to work in an iterative scheme that tries to improve efficiency by successively eliminating more complex forms of correlations. For instance, we shall see that if at the first stage one insists on eliminating only second order statistics ignoring all the higher order regularities, one arrives at filters with properties that are close to those observed in the retina. It is then conceivable that the elimination of more complex statistical structures could lead to processing similar to what is found in the primary visual cortex.

To begin with, let $\{L_i, i = 1, \cdots, n\}$ denote the activities of the n neurons in the input layer and $\{O_i, i = 1, \cdots, l\}$ the corresponding activities in the output layer. (l is not necessarily equal to n). The response of the output neurons is assumed to be some general function of the input activities:

$$O_i = K_i(L_1, \cdots, L_n), \quad \forall i. \tag{6.8}$$

The input and output layer could be any two consecutive stages along the visual pathway. The question is then how the recoding functions $\{K_i\}$ should be chosen in order to achieve the desired statistical independence.

We have seen above that a recoding which minimizes the sum over pixel entropies $\sum_{i=1}^{l} H(O_i)$ to its absolute minimum while keeping the total entropy fixed, achieves statistical independence. In general, one may not be able to find the $\{K_i\}$ that achieves the absolute minimum. For this reason, we define a fitness or energy functional, $E\{K_i\}$, that grades different recodings, $\{K_i\}$, according to how well they minimize the sum of pixel entropies without loss of information. A recoding is considered to yield an improved representation if it possesses a smaller value for E. The simplest energy functional for statistical independence is

$$E\{K_i\} = \sum_{i=1}^{l} H(O_i) - 2\rho \left[H(O_1, \cdots, O_l) - H(L_1, \cdots, L_n) \right], \tag{6.9}$$

where ρ is a parameter penalizing information loss. It can also be treated as a Lagrange multiplier in which case it enforces the constraint $H(O_1, \cdots, O_l) = H(L_1, \cdots, L_n)$ exactly. Any hardware constraint can be added with the appropriate Lagrange multiplier.

The optimal recoding can be found by solving the variational equations:

$$\frac{\delta E\{K_i\}}{\delta K_i} = 0. \tag{6.10}$$

In general, these equations are hard to solve if $\{K_i\}$ is allowed to be any arbitrary function. Anyway, it is not clear that biology could implement recodings by arbitrary functions. A better approach would be to find the optimal solution for a restricted class of functions that are implementable by realistic layers of neurons. For example, since the retina to a good approximation performs a linear transform on the photoreceptor signals, one could solve Eq. (6.10) for the class of linear functions.

Actually, an interesting simplification occurs when $\{K_i\}$ is restricted to the class of linear one to one ($l = n$) recodings, $i.\,e.$, if $O_i = \sum_{j=1}^{n} K_{ij} L_j$, $\forall i$. By a change of variables, keeping in mind that $P(O_1, \cdots, O_n)$ transforms as a density it is not hard to show that $H(O_1, \cdots, O_n) - H(L_1, \cdots, L_n) = \log \det K$ independent of the statistical structure of natural scenes, where K stands for the matrix K_{ij}. The only knowledge of the statistics resides in the pixel entropies $\{H(O_i)\}$. We will solve Eq. (6.10) explicitly for this special class of codes. But first we discuss the statistics of natural scenes which are needed to compute $\sum_{i=1}^{l} H(O_i)$.

Statistics of Natural Scenes

Unfortunately only little is known at the quantitative level about the statistical properties of natural scenes. Some of that knowledge has come from the early work on the statistics of television images [31, 32, 38, 53] and from the more recent measurements of the pixel-pixel correlation function of natural scenes by Field [26, 27]. Thus our model of natural scenes will have to be approximate.

The two-dimensional pixel-pixel correlation function, or alternatively the spatial autocorrelator, is defined as

$$R(x_1, x_2) = \langle L(x_1)L(x_2)\rangle, \tag{6.11}$$

where the brackets denote ensemble averaging over scenes or average over one large scene assuming ergodicity [49]. $L(x_1)$, $L(x_2)$ are the light levels above the mean level at two spatial points x_1 and x_2. Actually, by homogeneity of natural scenes the autocorrelator is only a function of the relative distance, $X \equiv x_1 - x_2$: $R(X)$. One can thus define the *spatial power spectrum* which is the Fourier transform of the autocorrelator

$$R(f) = \int dX \, e^{i f \cdot X} R(X). \tag{6.12}$$

For an ergodic system [49], the power spectrum $R(f)$ is simply given by $L(f)L(-f)$, and therefore it is only necessary to take the Fourier transform of a scene $L(x)$ in order to compute the power spectrum.

This is what Field did, where he found that for natural scenes

$$R(f) \sim \frac{1}{|f|^2}, \tag{6.13}$$

which corresponds to a scale invariant autocorrelator: under a global rescaling of the spatial coordinates $x \to \alpha x$ the autocorrelator $R(\alpha x) \to R(x)$. Although this scale invariant spatial power spectrum is by no means a complete characterization of natural scenes, it is the simplest regularity they possess.

The model of natural scenes that we adopt is one where the pixels $(L(x_1), \cdots L(x_n))$ making up an image are chosen with a Gaussian probability distribution of the form

$$P(L) = [(2\pi)^n \det(R)]^{-1/2} \exp\left[-\frac{1}{2} L \cdot R^{-1} \cdot L\right]. \tag{6.14}$$

In writing this expression R stands for the matrix $R_{ij} \equiv < L(x_i)L(x_j) >$ and is given by the Fourier transform of Eq. (6.13); L is the vector

$(L(\mathrm{x}_1), \cdots, L(\mathrm{x}_n))$. The distribution in Eq. (6.14) is the one that gives maximum total entropy $H(L)$ consistent with the autocorrelator being R. In other words it is the distribution that incorporates no knowledge beyond what is specified by the autocorrelator, and hence is the one that most honestly reflects what we know about natural scenes.

LMC Gain Control in the Blowfly Compound Eye

The number of examples of neural systems where a computational strategy to improve efficiency has been demonstrated is growing. In this manuscript, we only have space to discuss in detail two examples. These two illustrate coding strategies designed to deal with the two types of inefficiencies described above. Our first example, discussed in this section, illustrates a coding scheme from the fly compound eye that eliminates inefficiency due to unequal use of neural response levels [40].

The large monopolar cells (LMC) in the blowfly compound eye have been studied extensively over the last two decades (for reviews see [41, 57]). They are interneurons known to respond to contrast signals. These neurons, just like all other neurons, face a serious coding problem since they have a strictly limited dynamic range, *i. e.*, they possess only a small number of distinguishable response levels. The question is how should the LMC choose its gain (or contrast sensitivity) so as to most efficiently represent the different contrast levels. We are working at high luminosity so we can ignore the role of noise and treat the problem with the tools of noiseless Information theory.

Clearly, the problem is that if the LMC sets its sensitivity too high, inputs very often would saturate the response and much of the information about high contrast inputs would be lost. While if the sensitivity is too low, the information about low contrast inputs would be lost. In both cases, the different output levels would be far from being equally utilized. In the first case, the higher output states are used much more often than the lower ones, while in the second case large parts of the output at the high end remain under-utilized. To achieve an efficient encoding, the LMC must choose its gain such that all response levels are used with equal frequency.

This problem was first analyzed information theoretically by Laughlin [40]; here, we paraphrase his analysis. The first step in trying to discover the optimal code is to find out what are the statistical regularities of the input. In this case we only need to know the probability distribution of contrast signals occurring in the natural environment of the fly. Laughlin [40] measured it from samples of horizontal scans of dry woodland and lakeside vegetation.

Let us denote the input contrast signal by c, and use o to represent any one of the output or response levels, measured in some appropriate quantization units. The probability distribution for the input is $P(c)$ and it looks something like what is shown in Fig. 6.1A adapted from Laughlin [40]. The contrast signal is defined as $(I - I_0)/I_0$ where I is the intensity of a given pixel while I_0 is the average intensity within some visual window. This definition gives a contrast that cannot be smaller than -1. The neural transfer function or the neural gain g defines a mapping from the input c to the output $o = g(c)$. To achieve optimal coding, the function g should be chosen such that the probability distribution of the output, $P(o)$, is constant for all output states o, i.e., $P(o) = \alpha$ for some constant α. Since the transform from the input c to the output o can be thought of as a change of variables, and since the probabilities transform as densities, then

$$P(o)\, do = P(c)\, dc. \tag{6.15}$$

Setting $P(o) = \alpha$, we can integrate the resulting equation to find the transformation on the input needed to equalize the output probabilities:

$$o = g(c) \;=\; \frac{1}{\alpha} \int^{c} dc'\, P(c'),$$

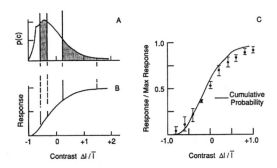

Figure 6.1: Probability distribution of contrasts, A, in the fly environment from the measurements of Laughlin [40]. The contrast-response predicted by information theory is the cumulative probability map in B. C is a comparison between the predicted response and that actually measured by Laughlin [40] in the LMC.

$$\frac{o}{o_{max}} = \int_{-1}^{c} dc' P(c'),$$ (6.16)

which can be recognized as the cumulative probability map. Notice that the constant α is given by $1/o_{max}$ and that $o \in [0, o_{max}]$. Also, the sensitivity of the cell, defined as do/dc, in this coding scheme is simply $P(c)$. So the neuron is hyper-sensitive around the most probable input contrast, with its sensitivity dropping to zero as the signal c becomes improbable (see Fig. 6.1B).

Laughlin compared this predicted neural coding strategy to that found in experiments where he measured the response of light-adapted LMC to sudden increments or decrements of light about the steady background level. The results of the comparison are shown in Fig. 6.1C. The solid curve is the cumulative probability computed from measured contrast probability distribution in the fly environment, while the dots with error bars are actual measurements of contrast response in the the LMC. Actually, the dots represent the average of repeated responses to the same stimulus. The agreement is clearly very good.

Gain control in a layer of n neurons

In this section we generalize Laughlin's result to a layer of n neurons each receiving inputs from a spatial array of n sensory cells. If we denote the inputs from the sensory cells by $\{c_i, i = 1, \cdots, n\}$ and the response of the neurons by $\{o_i, i = 1, \cdots, n\}$, the question again is how to choose the gain function, defined by $o_i = g_i(c_1, \cdots, c_n)$, in order to use the neuronal output levels most efficiently.

The analog of Eq. (6.15) here is

$$P(o_1, \cdots, o_n)\, do_1 \cdots do_n = P(c_1, \cdots, c_n)\, dc_1 \cdots dc_n. \qquad (6.17)$$

In general, contrary to the one neuron case, Eq. (6.17) is not integrable when we set $P(o_1, \cdots, o_n) = \alpha$. However, suppose that the neurons before choosing their gain function, coded the signals into a minimum entropy representation, i. e., coded the input signals $\{c_i\}$ into $\{\gamma_i\}$ such that $P(\gamma_1, \cdots, \gamma_n) = P(\gamma_1) \cdots P(\gamma_n)$. Eq. (6.17) can then be easily integrated to derive the necessary gain control on the resulting signals. The later is simply given by the cumulative probability maps

$$\frac{o_i}{o_{max}} = \int_{-1}^{\gamma_i} d\gamma_i' P(\gamma_i'), \qquad (6.18)$$

for each neuron independently. Thus for this system, efficiency of output representation predicts a two stage coding shown in Fig. 6.2.

Figure 6.2: The problem is how to represent correlated signals most efficiently if the output neurons possess limited dynamic range. The simplest solution is shown as a two stage process in this figure. In the first stage input signals are decorrelated to provide a minimum entropy code which can be followed by a simple cumulative probability map that functions as the gain control. The later serves to fit the decorrelated signals into the limited dynamic range of each neuron independently.

To be more concrete we work out in detail the coding for the simple case of $n = 2$ and where the signals c_1 and c_2 are Gaussian signals with a correlator given by

$$\begin{pmatrix} < c_1 \, c_1 > & < c_1 \, c_2 > \\ < c_2 \, c_1 > & < c_2 \, c_2 > \end{pmatrix} = \begin{pmatrix} 1 & r \\ r & 1 \end{pmatrix}, \tag{6.19}$$

where r is a number < 1 characterizing the degree of overlap between the two channels, and the brackets denote ensemble averages. According to Fig. 6.2, the signals c_1, c_2 are first transformed to the decorrelated signals γ_+, γ_-

$$\gamma_+ = \frac{1}{\sqrt{2}} (c_1 + c_2),$$

$$\gamma_- = \frac{1}{\sqrt{2}} (c_1 - c_2). \tag{6.20}$$

The γ_+, γ_- signals also gaussian with variances $1 + r$ and $1 - r$ respectively. Thus the final transformation from γ_+, γ_- to o_+, o_- is given by a cumulative integral over a Gaussian which is simply related to the standard error functions. The net transformation for this system is

$$o_+ \quad \sim erf\left(\frac{1}{\sqrt{2}}\frac{c_1+c_2}{\sqrt{1+r}}\right) + \text{const.},$$

$$o_- \quad \sim erf\left(\frac{1}{\sqrt{2}}\frac{c_1-c_2}{\sqrt{1-r}}\right) + \text{const.}. \tag{6.21}$$

Notice in the regime where the contrast signals are small in comparison with the square root of the variance, the response linearizes, $o_\pm \sim (c_1 \pm c_2)/\sqrt{1 \pm r}$, and the only effect of the gain control is to normalize the signals by dividing by the square root of the variance.

In the next sections, we generalize the above minimum entropy code to the more realistic case of the array of retinal ganglion cells and will modify the coding to take into account the noise. However, we will ignore the gain control transform or the cumulative probability map and work purely within the linearized approximation.

Retinal Coding Strategies in Space-time and Color

The mammalian retina is a rather unique neural system. It is a network which is complex enough, so insight gained from understanding it promises to be useful in understanding other areas in the brain, yet it is still simple and isolated enough that quantitative experiments with clear outcomes can be performed. As such, the retina is ideal for developing and testing theoretical ideas on neural computations. In this section, we start by giving a brief review of relevant experimental facts about the retina. We then explore the efficiency principle discussed above in the context of the retina. This gives a predictive framework that seems to explain many of the experimental facts.

The retina is the thin tissue lining the back of the eyeball. As a neural network, it has feedforward architecture with three essential layers:

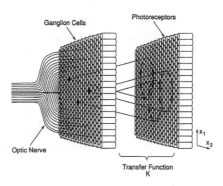

Figure 6.3: The retina as a black-box processor.

photoreceptors, bipolar cells, and ganglion cells. However, it also has important lateral connections and interneurons acting within a given layer. The photoreceptors' layer forms the input to this network, where photons from an image focused on the surface of the retina are captured and transduced into graded voltage signals. The output, on the other hand, is built of spike trains generated by ganglion cells, and it propagates down the optic nerve to the LGN and subsequently to the visual cortex.

Since, here, we are only interested in functional properties of the retina, neither detailed connectivity of its network nor properties of cells other than photoreceptors and ganglion cells are of interest to us. For further information about retinal organization the reader should consult reviews on the subject e.g. Refs. [20, 56, 59]. We simply think of the retina as a black-box processor whose input is the photoreceptors' activities and output is the ganglion cells' activities. This processor can be characterized by its *transfer function* which specifies how the output is related to the input, see Fig. 6.3.

The retinal transfer function is measured in single-cell recordings of ganglion cell outputs or inferred from psychophysical contrast sensitivity measurements subject to some plausible assumptions (see [56], and references therein).In single-cell experiments, one finds that after adaptation to the light level the output of any given ganglion cell, measured as the rate of spikes in spikes/sec, is to a good approximation given by a weighted sum of the photoreceptor activities over a small contiguous region on the surface of the retina known as the cell's *receptive field*, RF (Fig. 6.3). The linear cells in cat are often referred to as the X cells, while in monkey they are known as the parvocellular cells which constitute about 80% of the ganglion cells in the retina. In monkey, they are considered to be part of a pathway that extends into the deep layers and is believed to be concerned with detailed form recognition (see for example Van Essen and Anderson [60]).

The output of a ganglion cell whose RF is centered at x_i and at time t can be written as

$$O(x_i, t) = \int dx' dt' K(x_i, x'; t, t') L(x', t') \equiv K \cdot L, \qquad (6.22)$$

where $L(x', t')$ is the activity of the photoreceptor at location x' and at time t', while $K(x_i, x'; t, t')$ is the retinal kernel or retinal transfer function.

Without loss of generality, the kernel can always be re-expressed in terms of relative coordinates, $X \equiv (x_i - x')/2$, and average coordinates $(x_i + x')/2$: $K(\frac{x_i - x'}{2}, \frac{x_i + x'}{2}; t, t')$. However, in many species, K has a weak dependence on the average coordinates. In other words, the kernel changes gradually with eccentricity or with angular distance from center of gaze. Also, after adaptation it is known to be only a function of the temporal difference $T = t - t'$. Thus, to a first approximation one can assume translation invariance and retain only the dependence on the relative coordinates, $K(x_i, x'; t, t') = K(x_i - x'; t - t')$. This is convenient since it enables us to define the retinal filter, $K(f, w)$, simply by Fourier transforming K

$$K(f, w) = \int dX \, dT e^{-if \cdot X - iwT} K(X, T). \qquad (6.23)$$

This is the object that is actually measured in experiments. Furthermore, by rotational symmetry it is only a function of $(|\mathbf{f}|, w)$. In experiments, a luminosity grating , $L = I_0(1 + m\cos(fx)\cos(wt))$ is projected onto the RF of a cell and the minimum contrast $m_{f,w}$ needed to elicit a certain level of response, r_0, at that spatio-temporal frequency of stimulation is recorded. The recording is repeated for different values of (f, w). By linearity of the output:

$$K(|\mathbf{f}|, w) = \frac{r_0}{m_{|\mathbf{f}|,w}} \qquad (6.24)$$

Actually, there is a family of retinal filters, one for each adaptation or luminance level I_0. In Fig. 6.4, which is reproduced from the data of Refs. [25] and [24], we show two typical families of filters, one for the cat and one for monkey as a function of f and at a given low temporal frequency w. Actually, what is shown is $I_0 \times K$ which is called *contrast sensitivity*. A prominent feature in that figure is the transition from band-pass to low-pass filtering as I_0 is lowered. A similar transition is also observed as the temporal frequency of stimulation is increased for a given spatial frequency.

If a retinal filter at high luminance is Fourier transformed back into space, it looks like the curve in Fig. 6.5. This is a one dimensional slice in a two-dimensional rotationally invariant spatial profile, and it shows the familiar center-surround organization of ganglion cell RF: The cell effectively receives excitatory input (+ ve) from the photoreceptors in a small region around its RF center and inhibitory excitations (− ve) from the surround region. These cells are known as *on*-center cells. The other class of spatially opponent cells found in the retina, have an inhibitory center and an excitatory surround and are known as *off*-center cells. A similar organization exists in the temporal domain.

In retinas of species that possess color vision, such as most primates and shallow water fish, ganglion cells' RFs possess a more complicated center-surround organization. In these retinas, there are several types of

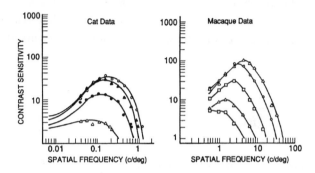

Figure 6.4: Measured contrast sensitivity. The data in the left figure is reproduced from cat, while that on the right is from monkey. In both cases, the luminance level I_0 decreases by one log unit each time we go to a lower curve.

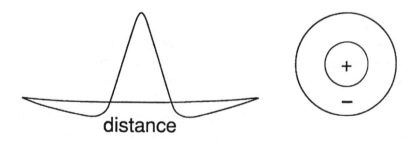

Figure 6.5: Retinal kernel at high adaptation level showing the opponent spatial organization of a ganglion cell's RF.

photoreceptors that possess different photosensitive pigments. Function-ally, the various pigments are identical except they differ in the location of their peak spectral sensitivity. In humans for example, the three types of pigments often referred to as B, G, and R for blue, green, and red respectively (or alternatively known as S, M and L for short, medium and long spectral wavelength respectively) absorb best light of spectral wavelength around 419 nm, 530 nm, and 558 nm respectively.

Corresponding to the diversity of photoreceptors there are several types of spatially opponent ganglion cells. These cells differ in the way the three photoreceptor types are used in the organization of their RF. In primate's retina, the most common on-center ganglion cells receive excitatory input dominantly from one type of photoreceptors in the center and inhibitory input from a different type in the surround: the two most common on-center cell types are +R in center and -G in the surround or +G in center and -R in surround [23], see Fig. 6.6A. Similarly for the off-center cells. These color coded cells are known as *single opponency* cells. There are other opponent cell types that involve blue cones. However, since blue cones are rare in the retina (nonexistent in fovea) these cells are also rare and hence will not be discussed here and in Ref. [22]

Single opponency cells are not found in retinas of all species that possess color vision. In fact, they represent one extreme in color coding. The other extreme is found in shallow-water fish which possess what are called *double opponency* cells [21]. As the name implies, these cells receive inputs of comparable strengths from two types of cones at every spatial location in their RF. For example, in one double opponent cell type found in goldfish retina, the RF has a center that receives excitatory R and inhibitory G stimulation while its surround receives inhibitory R and excitatory G inputs, Fig. 6.6B.

The fact that color coding is qualitatively dependent on the environment of the animal makes it an interesting dimension for testing ecological theories. A successful theory of the retina should not only explain the shape of the retinal kernel and its dependence on background luminance, it should

plus neurons that prefer reddish and greenish stimuli. Conversely, the cells that prefer reddish are distinct in their differences compared to those that prefer greenish, with the other...

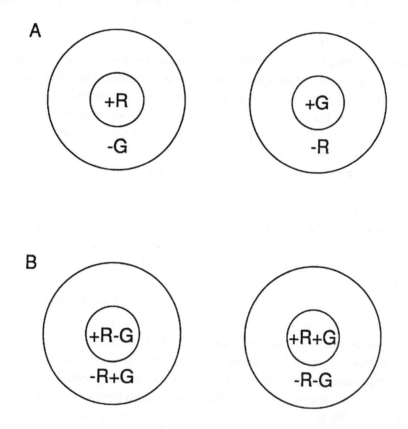

Figure 6.6: The two extremes of opponent color coding. The cell types found in primates are shown in A, while those found in shallow-water fish are shown in B.

also account for differences seen among species. In the theory of retinal processing presented here, differences in computation strategies among species are attributed to identifiable differences in the visual environment (information source differences). We start by examining the problem in the purely spatial domain, then show how to incorporate time. We also discuss something that we have ignored thus far, namely the role of the noise. Finally we introduce color. The problem in pure space-time was first considered by Atick and Redlich [1, 2], and in pure color domain by Buchsbaum and Gottschalk [18], while in the fully mixed dimension of space-time and color by Atick *et al.*[5, 6].

Theoretical Approach to the Retina: Spatial Processing

In chapter three, we have given several reasons why a sensory pathway, such as the visual pathway, might recode incoming signals from the natural environment into a more efficient representation. In this section, we show how to use this idea to predict retinal processing in the spatial domain. We work with the hypothesis that the retina's main goal is to build a minimum entropy representation *i. e.*, a representation where the elements are statistically independent or decorrelated (the same procedure followed by the appropriate gain control yields a redundancy reduced code, as we have seen). However, we limit the class of recodings to linear transformations. With this restriction we shall see that the retina can only eliminate pixel-pixel decorrelations; but since we will be assuming Gaussian signals, two-point decorrelation and statistical independence are equivalent.

Decorrelation in the absence of noise

In our previous discussion the problem of finding minimum entropy codes was formulated as a variational problem of some well defined energy functional Eq. (6.9) The solution to the variational equations Eq. (6.10) then gives the optimal transformation that best minimizes Eq. (6.9) Here, we explicitly solve Eq. (6.10) for the class of linear one to one mappings, where Eq. (6.9) takes the following simpler form

$$E\{K\} = \sum_{i=1}^{l} H(O_i) - 2\rho\left[H(O) - H(L)\right] \tag{6.25}$$

$$= \sum_{i=1}^{l} H(O_i) - \rho \log \det K^T \cdot K, \tag{6.26}$$

$$\tag{6.27}$$

where $O_i \equiv O(\mathsf{x}_i)$ is the response level of ganglion cell at location x_i, and we have used the bold-faced symbols to denote matrices and vectors; K denotes the matrix $K_{ij} \equiv K(\mathsf{x}_i - \mathsf{x}_j)$, and $O \equiv (O_1, \cdots, O_l)$; similarly for L. We have also used the fact that $H(O) - H(L) = \log \det K = \frac{1}{2} \log \det K^T \cdot K$ which is valid when O is related to L through a linear transformation.

To exhibit $E\{K\}$ more explicitly we need to compute the sum over pixel entropies $\sum_{i=1}^{l} H(O_i)$. Treating, the discrete response levels O_i as a continuous variable, the ith pixel entropy can be approximated by a simple integral:

$$H(O_i) \equiv - \sum_{O_i} P(O_i) \log P(O_i) \rightarrow - \int dO_i P(O_i) \log P(O_i), \tag{6.28}$$

which depends on the ith pixel probability $P(O_i)$. The latter is computable from the input probability P(L) since $O_i = \sum_{j=1}^{l} K_{ij} L_j$. $P(L)$ in Eq. (6.14) is a Gaussian of variance R, therefore $P(O)$ is also a gaussian but of variance $\tilde{R} \equiv K \cdot R \cdot K^T$. In Eq. (6.28) we only need the individual pixel probability $P(O_i)$ for every i, which is given by

$$P(O_i) = \int \prod_{j \neq i} dO_j P(O). \tag{6.29}$$

It is easy to show that

$$P(O_i) = \frac{1}{2\pi \tilde{R}_{ii}} \exp -\frac{1}{2\tilde{R}_{ii}} O_i^2 \qquad (6.30)$$

Again $\tilde{R}_{ii} = <O_i^2>$, the diagonal part of $\tilde{R}_{ij} = <O_i O_j>$.

Substituting the expression for $P(O_i)$ from Eq. (6.30) in Eq. (6.28), we find $H(O_i) = \log \tilde{R}_{ii}$, which when summed over all pixels yields

$$\sum_{i=1}^{l} H(O_i) = \log \prod_{i=1}^{l} \tilde{R}_{ii}. \qquad (6.31)$$

By translation invariance, all the $\tilde{R}_{ii}$ are equal, $\tilde{R}_{ii} = \langle O_0^2 \rangle$ for a pixel at some arbitrary location 0, thus $\sum_{i=1}^{l} H(O_i) = l \log(\langle O_0^2 \rangle)$. This can be substituted for the first term in Eq. (6.27), however there are a couple of mathematical steps that lead to an even simpler form of the energy functional.

Implicit in the approximation of the response levels by a continuous variable, is the fact that this variable has to be measured in some quantization units. This means that $\langle O_0^2 \rangle$, which in these units is related to the square of the number of distinct levels the system possesses, has to be greater than one.

Since necessarily $\langle O_0^2 \rangle \geq 1$, we can drop the logarithm from $\log(\langle O_0^2 \rangle)$ and minimize instead the simpler quantity $\langle O_0^2 \rangle$. However, by translation invariance, minimizing $\langle O_0^2 \rangle$ is equivalent to minimizing the explicitly invariant expression $\sum_i <O^2(x_i)> = \sum_i (K \cdot R \cdot K^T)_{ii} = Tr(K \cdot R \cdot K^T)$. The final energy function is then

$$E\{K\} = Tr(K \cdot R \cdot K^T) - \rho \log \det(K^T \cdot K). \qquad (6.32)$$

The advantage of this invariant form of E is that we can now go to Fourier space very easily:

$$E\{K\} = \int df |K(f)|^2 R(f) - \rho \int df \log |K(f)|^2. \qquad (6.33)$$

where we have used the identity $\log \det Q = Tr \log Q$ valid for any positive matrix Q.

The variational equations in frequency space, $\delta E\{K\}/\delta K(f) = 0$, are trivial in this case: the optimal solution is just

$$|K(f)|^2 = \frac{\rho}{R(f)}. \qquad (6.34)$$

This could have been guessed more easily by diagonalizing the the autocorrelator matrix of the output $\tilde{R}(x_i - x_j) \equiv \langle O(x_i)O(x_j)\rangle$. However, we have gone through the analysis systematically to illustrate the general procedure which will be useful for more complex codings. Since $R(f) = 1/|f|^2$ for natural scenes, the predicted kernel is simply $K(f) = \rho|f|$. On a log-log plot this gives a curve of slope one.

We can compare this simple prediction with retinal filters in the regime where the noise is not significant, namely in the regime of high luminance I_0 and at low frequencies. In Figs. 6.7A and 6.7C we have plotted some typical experimentally measured retinal filters at high luminance I_0. The data is taken from Refs. [24] and [36], respectively. In Figs. 6.7B and 6.7D, we show the ratio $\chi(f) = K_m(f)/K_p(f)$ where K_m and $K_p \sim |f|$ are the measured and predicted filters respectively. At low frequency, we can see that $\chi(f)$ is flat or that both filters have the same slope.

Another way to interpret the results in Figs. 6.7B and 6.7D is as follows. The power spectrum of the output is given by the square of the retinal

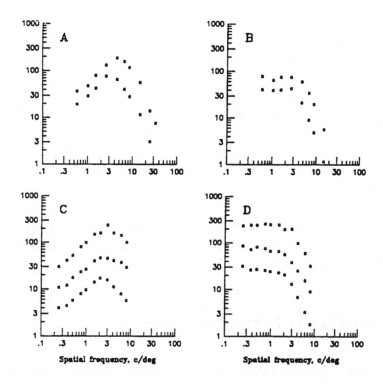

Figure 6.7: Retinal filers (A and C) at high mean luminosities, taken from the data of Refs. [24] and [36] respectively. B (D) is the data in A (C) multiplied by $1/|f|$ which is the amplitude spectrum of natural scenes. This gives the retinal ganglion cell's output amplitude spectrum. Notice the whitening of the output at low frequency. The ordinate units are arbitrary.

filter times the input power spectrum:

$$\langle O(f)O^*(f) = \langle (K(f)L(f))(K(f)L(f))^* \rangle = |K(f)|^2 R(f). \tag{6.35}$$

However, $R(f) \sim 1/|f|^2$. The output amplitude spectrum, which is the square root of the power spectrum, is then proportional to $\chi(f)$ which is what is plotted in Figs. 6.7B,D. Thus at low frequencies, the input spectrum $|f|^{-2}$ is converted into a flat spectrum at the retinal output: $\langle O(f)O^*(f) \rangle =$ constant. This *whitening* of the input by the retina continues up to the frequency where the kernel in Figs. 6.7A,C peaks. Beyond that the noise is no longer ignorable and the actual kernel deviates from the pure whitening kernel. This whitening is the statement in frequency space of decorrelation in regular space. Of course, since the whitening does not continue all the way to the system's cutoff the decorrelation in space is not perfect. In the next section we shall see that by incorporating a strategy for noise suppression in addition to decorrelation we arrive at filters that agree with what is measured not only over the entire range of visible spatial frequencies but also at all luminance levels.

Decorrelation in the presence of noise

The above agreement does support a strategy of decorrelation in the absence of noise. However, decorrelation cannot be the only goal in the presence of input noise such as photon (or quantum) noise which always exists. In that case, decorrelation alone would be a dangerous computational strategy as we now argue: If the retina were to whiten all the way up to the cutoff frequency or resolution limit, the kernel $K(f)$ would be proportional to $|f|$ up to that limit. This would imply a constant average squared response KRK^* to natural signals $L(x)$, which for $R \sim |f|^{-2}$ have large spatial power at low frequencies and low power at high frequencies. But this same $K(f) \sim |f|$ acting on input noise whose spatial power spectrum is approximately flat has a very undesirable effect, since it amplifies the noise at high frequencies where noise power, unlike signal power, is not becoming small. Therefore, even if input noise is not a major problem without decorrelation, after complete decorrelation (or whitening up to cutoff) it would become a problem. Also, if both noise and signal are decorrelated at the output, it is no longer possible to distinguish them. Thus, if decorrelation is a strategy, there must be some guarantee that no significant input noise is passed through the retina to the next stage. We believe this is why the retina stops whitening its input at a frequency far lower that the cutoff frequency.

Further evidence that the retina is concerned about not passing significant amounts of input noise is found in the fact that the ganglion cell kernel, as we have seen, makes a transition from band-pass to complete low-pass as the retina adapts to very low I_0. Since as I_0 decreases, the signal to noise ratio of the input signals decreases one expects low-pass filtering as a way of suppressing the noise, which is what the retina does.

Since here we are primarily interested in testing the predictions of minimum entropy coding (equivalently redundancy reduction), we take a somewhat simplified approach to the problem with noise. Instead of doing a full-fledged information theoretic analysis that unifies minimum entropy with noise suppression [5], we work in a formalism where the signal is first low-pass filtered to eliminate noise and the resulting signal is then decorrelated as before. The advantage of this modular approach is that it leads to a more intuitive picture of the various processing stages in the retina and it also gives parameters that have physical significance. Furthermore, the analysis is not as complicated as that in the unified formalism.

We start by going over the stages of signal processing that we assume precede the decorrelation stage. Figure 6.8 shows a schematic of those stages. First, images from natural scenes pass through the optical medium of the eye and in doing so their image quality is lowered. It is well known that this effect can be taken into account by multiplying the images by the optical *modulation transfer function* or MTF of the eye, a function of spatial frequency that is measurable in purely non-neural experiments

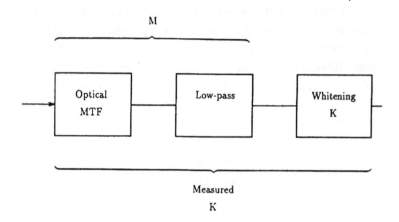

Figure 6.8: Schematic of the signal processing stages assumed to take place in the retina.

[19]. In fact, an exponential of the form $\exp(-(|f|/f_c)^\alpha)$, for some scale f_c characteristic of the animal (in primates $f_c \sim 22$ c/deg and $\alpha \sim 1.4$) is a good approximation to the optical MTF. The resulting image is then transduced by the photoreceptors and is low-pass filtered to eliminate input noise. Finally, we assume that it is decorrelated. In this model, the output-input relation schematically takes the form

$$O = K \cdot (M \cdot (L + n) + n_0), \tag{6.36}$$

where the dot denotes a convolution, $n(x)$ is the input noise (such as quantum noise) while $n_0(x_i)$ is some intrinsic noise level which models post-receptor synaptic noise. Finally, M is the filter that takes into account both the optical MTF as well as the low-pass filtering needed to eliminate noise. An explicit expression for M will be derived below.

With this model, the energy functional determining the decorrelation filter

$K(f)$ is

$$
\begin{aligned}
E\{K\} &= \int df\, |K(f)|^2 \left[M^2(f)(R(f) + N^2) + N_0^2\right] \\
&\quad - \rho \int df \log |K(f)|^2,
\end{aligned}
\tag{6.37}
$$

where $N^2(f) \equiv \langle |n(f)|^2 \rangle$ and $N_0^2(f) \equiv \langle |n_0(f)|^2 \rangle$ are the input and synaptic noise powers respectively. This energy functional is the same as that in Eq. (6.33) but with the variance $\tilde{R}(f)$ replaced by the variance of O in Eq. (6.36).

As before, the variational equations $\delta E/\delta K(f) = 0$ are easy to solve for $K(f)$. The predicted filter that should be compared with experimental measurements is this variational solution, K, times the filter M. We denote this by K_{expt}:

$$
|K_{expt}(f)| = |K(f)|\; M(f) = \frac{\sqrt{\rho}M(f)}{[M^2(f)\,(R(f) + N^2) + N_0^2]^{1/2}}.
\tag{6.38}
$$

An identical result can be obtained in space-time trivially by replacing the autocorrelator $R(f)$ and the filter $M(f)$ by their space-time analogs $R(f, w)$ and $M(f, w)$, respectively, with w the temporal frequency. However, we focus here on the purely spatial problem where we have Field's [26] measurement of the spatial autocorrelator $R(f)$.

Deriving the low-pass filter

In our explicit expression for K_{expt}, below, we shall use the following low-pass filter

$$
M(f) = \frac{1}{N} \left(\frac{1}{I_0} \frac{R(f)}{R(f) + N^2} \right)^{1/2} e^{-(|f|/f_c)^\alpha}.
\tag{6.39}
$$

The exponential term is the optical MTF while the first term is a low-pass filter that we derive next using Information theory. The reader who is not interested in the details of the derivation can skip this rather technical section without loss of continuity.

It is not clear in the retina what principle dictates the choice of the low-pass filter or how much of the details of the low-pass filter influence the final result. In the absence of any strong experimental hints, of the type which imply redundancy reduction, we shall try a simple information theoretic principle to derive an M: We insist that the filter M should be chosen such that the filtered signal $O' = M \cdot (L + n)$ carries as much information as possible about the *ideal* signal L subject to some constraint. To be more explicit, the amount of information carried by O', about L, is the mutual information $I(O', L)$ (see the appendix). However, as we discuss in the appendix $I(O', L) = H(O')$ − Noise Entropy (for L and n statistically independent gaussian variables), and thus if we maximize $I(O', L)$ keeping fixed the entropy $H(O')$ we achieve a form of noise suppression.

We can now formulate this as a variational principle. To simplify the calculation we assume Gaussian statistics for all the stochastic variables involved. The output-input relation including the appropriate quantization units, n_q, takes the form: $O' = M \cdot (L + n) + n_q$. A standard calculation leads to

$$I(O', L) = \int df \log \left(\frac{M^2(R + N^2) + N_q^2}{M^2 N^2 + N_q^2} \right). \tag{6.40}$$

Similarly, one finds for the entropy $H(O') = \int df \log(M^2(R + N^2) + N_q^2)$. The variational functional or energy for smoothing can then be written as $E\{M\} = -I(O', L) + \eta H(O')$. It is not difficult to show that the optimal noise suppressing solution $\delta E/\delta M = 0$ takes the form:

$$M = \left(\frac{N_q}{N}\right)\left(\frac{1}{\eta}\frac{R}{R+N^2} - 1\right)^{1/2}, \tag{6.41}$$

with the parameter $\eta \sim N_q^2 I_0$ in order to hold $H(O')$ fixed with mean luminance. Actually, below we will be working in the regime where the quantization units are much smaller than one, in this case we can safely drop the -1 term, since the $1/\eta$ term dominates. We can also ignore any overall factors in M that are independent of f. This then is the form that we exhibit in the first term in Eq. (6.39).

Analyzing the solution

Let us now analyze the form of the complete solution. In Fig. 6.9 we have plotted $K_{expt}(f)$ (curve a) for a typical set of parameters. We have also plotted the filter without noise $R(f)^{-1/2}$, (curve b), $M(f)$, and (curve c). There are two points to note: at low frequency the kernel $K_{expt}(f)$ (curve a) is identically performing decorrelation, and thus its shape in that regime is completely determined by the statistics of natural scenes: the physiological functions M and N drop out. At high frequencies, on the other hand, the kernel coincides with the function M, and the power spectrum of natural scenes R drops out.

We can also study the behavior of the kernel in Eq. (6.38) as a function of mean luminosity I_0. If one assumes that the dominant source of noise is quantum noise, then the dependence of the noise parameter on I_0 is simply $N^2 = I_0 N'^2$ where N' is a constant independent of I_0 and independent of frequency (flat spectrum). This gives an interesting result. At low frequency where K_{expt} goes like $1/\sqrt{R}$ and its I_0 dependence will be $K_{expt} \sim 1/I_0$ (recall $R \sim I_0^2$) and the system exhibits a Weber law behavior i. e., its contrast sensitivity $I_0 K_{expt}$ is independent of I_0. While in the other regime -- at high frequency -- where the kernel asymptotes M with $N^2 > R$

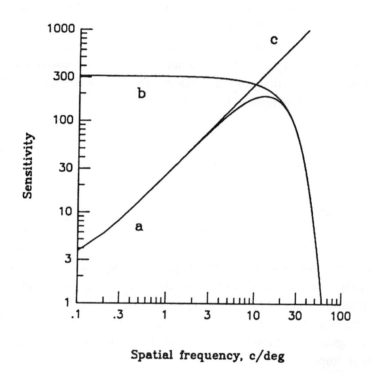

Figure 6.9: A typical predicted retinal filter (a) from Eq. (38), while curve b is $R(\underline{f})^{-1/2}$ which is the pure whitening filter in Eq. (34). Finally curve c is the low-pass filter M.

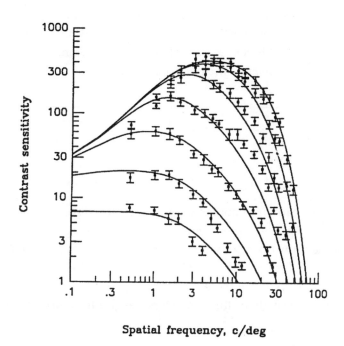

Spatial frequency, c/deg

Figure 6.10: Predicted retinal filters, Eq. (38), at different I_0 separated by one log unit, assuming that the dominant source of input noise is quantum noise ($N^2 \sim I_0$). No other parameters depend on I_0. The fixed parameters are $f_c = 22$ c/deg, $\alpha = 1.4$, $\rho = 2.7 \times 10^5$, $N' = 1.0$. The data [48] are psychophysical contrast sensitivity measurements.

then $K_{expt} \sim 1/I_0^{1/2}$ which is a De Vries-Rose behavior $I_0 K_{expt} \sim I_0^{1/2}$. This predicted transition from Weber to De Vries-Rose with increasing frequency is in agreement with what is generally found (see [36], Fig. 6.3).

Given the explicit expression in Eq. (6.38) and the choice of quantum noise for N we can generate a set of kernels as a function of I_0. The resulting family is shown for primates in Fig. 6.10. We need to emphasize that there are no free parameters here which depend on I_0. The only variables that needed to be fixed were the numbers f_c, α, ρ and N' and they are independent of I_0. Also we work in units of synaptic noise n_0, so the synaptic noise power N_0^2 is set to one. We have superimposed on this family the data from the experiments of van Ness and Bouman [48]on human psychophysical contrast sensitivity. It does not take much imagination to see that the agreement is very reasonable, especially keeping in mind that this is not a fit but a *parameter free* prediction.

Introducing Color

Images in nature carry information through their spectral compositions in addition to their spatio-temporal modulations. So an image is generally a function of the form $L(x, t, \lambda)$, where λ is the spectral wavelength. Many animals have evolved visual pathways capable of extracting this color information. In the retina of these species, images are first sampled in the spectral domain through the three cone types to give the output activities

$$P^a(\mathbf{x}, t) = \int d\lambda C^a(\lambda) L(\mathbf{x}, t, \lambda) + n(\mathbf{x}, t), \tag{6.42}$$

where the functions $C^a(\lambda)$ are the spectral sensitivity functions for the three photoreceptor types, $a = 1, 2, 3$ for R, G and B respectively. In Fig. 6.11A and 6.11B we show the spectral sensitivity curves for the cones in the retinas of primates and shallow water fish, respectively. (The two systems that form the two extremes in retinal color coding.)

Some experimental facts

One important feature to notice about the two sets of curves in Fig. 6.11 is the fact that the R and G spectral sensitivity curves overlap. The degree of overlap is more significant for primates' retina than for shallow water fish. To be quantitative, in the monkey *Macaca fascicularis* the separation

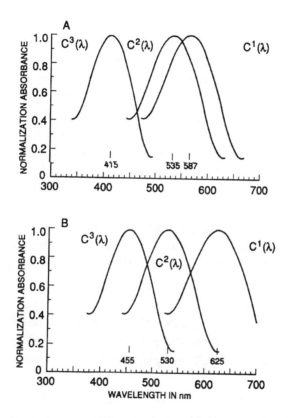

Figure 6.11: Spectral sensitivity curves for primates A, and shallow water fish, B. Notice that the overlap of C^1 and C^2 especially in the primates.

of the spectral peak sensitivities between R and G is about 30 nm while for goldfish the corresponding separation is about 90 nm. This difference between the two species is due to adaptation of cone pigments to different visual environments and will play an essential role in explaining the subsequent differences in the neural coding strategies.

In the case of primates, which are believed to have evolved in a forest like environment, one finds that the proximity of R and G cones can be explained by the fact that most of the information in a forest is squeezed in a narrow spectral band centered about 550 nm. Thus one needs to sample that region more densely if one is to resolve different objects found in that spectral band. On the other hand, under water light in the spectral band between 550 nm and 610 nm is heavily absorbed by water with the amount of absorption increasing dramatically with distance traveled. Thus if shallow water fish had adopted pigments around 568 just like primates, they would not have been able to see far under water. Shallow water fish instead evolved cones that sampled near the infrared, an area where the signal under water travels much farther before complete absorption. Additional discussion regarding the adaptation of the cone system of various species to the environment can be found in the excellent book of Lythgoe [46].

The fact that the cones sample in such an overlapping fashion introduces an additional source of correlations in the photoreceptor signals and thus an additional source of inefficiency that has to be eliminated [5, 8, 18].

In passing, we should mention that we limit our analysis to the two cone (R and G) system, since in primate retina these photoreceptors occur with equal density and are more abundant than the blue cones. In fact, the blue cones constitute only 15% of the total cone population in the entire retina while in the fovea they are virtually nonexistent. For a discussion of the role of blue see [5].

Theoretical formulations

We now generalize the analysis of the previous section to include color. The chromatic-spatio-temporal correlator is a matrix of the form $R^{ab}(f, w)$. Unfortunately, not much is known experimentally about the entries of this matrix. Thus, we are forced to make some assumptions. Although, it is possible to do the analysis entirely for the most general form of $R^{ab}(f, w)$ [5]. It is just as informative and much simpler to analyze the case where $R^{ab}(f, w)$ can be factorized into a pure spatio-temporal correlator times a 2×2 matrix describing the degree of overlap between the R and G systems. We will also only examine color coding under conditions of slow temporal stimulation or zero temporal frequency. In that case, we can replace the spatio-temporal correlator by by $I_0^2/|f|^2$. Thus we take

$$R^{ab}(f, 0) = \begin{pmatrix} 1 & r \\ r & 1 \end{pmatrix} \frac{I_0^2}{|f|^2} \qquad (6.43)$$

where $r < 1$ is a parameter describing the degree of overlap of R and G. We should emphasize, that we do not advocate that this is the form of R^{ab} necessarily found in nature. We have reduced R^{ab} to one degree of freedom in order to illustrate very simply the possibilities. More complex R^{ab}, in particular those where space and color are not decoupled, lead to quantitatively different but qualitatively similar solutions.

As before, the output O is related to the input P through

$$O = K \cdot (M \cdot (P + n) + n_0)6.8 \qquad (6.44)$$

where $n^a(x, t)$ is input noise including transduction and quantum noise, while $n_0(x, t)$ is noise (e.g. synaptic) added following the low-pass filter M. We have introduced bold face to denote in this section matrices in the 2×2 color space; also in Eq. (6.8) each   denotes a convolution in space. To see how the presence of two channels affect the spatial low-pass filtering, it is helpful to rotate in color space to the basis where the color matrix is diagonal. For the simple color matrix in Eq. (6.43), this is a $45°$ rotation,

$$U_{45} = \frac{1}{\sqrt{2}} \begin{pmatrix} 1 & 1 \\ -1 & 1 \end{pmatrix}, \tag{6.45}$$

to the luminance, G+R, and chromatic, G-R, channels [in vector notation, the red and green channels are denoted by R = (1,0) and G = (0,1)]. In the G±R basis, the total correlation matrix plus the contribution due to noise is

$$U_{45}\left(R(\mathrm{f}) + N^2\right) U_{45}^T = \frac{I_0^2}{|\mathrm{f}|^2}\begin{pmatrix} 1+r & 0 \\ 0 & 1-r \end{pmatrix} + N^2\begin{pmatrix} 1 & 0 \\ 0 & 1 \end{pmatrix} \tag{6.46}$$

where the noise, $\langle n^a n^b \rangle = \delta^{ab} N^2$, is assumed equal in both the R and G channels, for simplicity. Since in the G±R basis the two channels are decoupled, the spatial filters $M_\pm(\mathrm{f})$ are found by applying our single-channel result in Eq. (6.39). More specifically they are found by replacing $R(\mathrm{f})$ in Eq. (6.39) by

$$R_\pm(\mathrm{f}) = (1 \pm r)I_0^2/|\mathrm{f}|^2. \tag{6.47}$$

Notice that the two channels differ only in their effective S/N ratios: $(S/N)_\pm = \sqrt{(1 \pm r)}(I_0/N)$ which depend multiplicatively on the color eigenvalues $1 \pm r$. In the luminance channel, G+R, the signal to noise is increased above that in either the R or G channel alone, due to the summation over the R and G signals. The filter $M_+(\mathrm{f})$, therefore, passes relatively higher spatial frequencies, increasing spatial resolution, than without the R plus G summation. On the other hand, the chromatic channel, G-R, has lower S/N, proportional to $1 - r$, so its spatial filter $M_-(\mathrm{f})$ cuts out higher spatial frequencies, thus sacrificing resolution in favor of color discriminability. The complete filter is finally obtained by rotating from the G±R basis by 45 degrees back to the R and G basis

$$M^{ab}(\mathrm{f}) = \frac{1}{2} \begin{pmatrix} 1 & -1 \\ 1 & 1 \end{pmatrix} \begin{pmatrix} M_+(\mathrm{f}) & 0 \\ 0 & M_-(\mathrm{f}) \end{pmatrix} \begin{pmatrix} 1 & 1 \\ -1 & 1 \end{pmatrix}, \tag{6.48}$$

After filtering noise, the next step is to decorrelate the signal as if no noise existed as we did in the purely spatial problem. In this case this means that we have to find the kernel K that achieves diagonalization of the spatio-chromatic autocorrelator R, i. e., $K \cdot R \cdot K^T = D$ with D a diagonal matrix in space and color. In the purely spatial problem, we have insisted on translationally invariant, local set of retinal filters: the approximation where all retinal ganglion cells (in some local neighborhood, at least) have the same receptive fields, except translated on the retina, and these fields sum from only a nearby set of photoreceptor inputs. These assumptions force D to be proportional to unity. In generalizing this to include color, we note when D is proportional to the unit matrix, the mean squared outputs ($(KRK^T)^{aa}_{xx}$ for output O^a_x) of all ganglion cells are equal. This equalization provides efficient use of optic nerve cables (ganglion cell axons) if the set of cables for the cells in a local neighborhood all have similar information carrying capacity. We therefore continue to take D proportional to the identity matrix in the combined space-color system. Taking D proportional to the identity, however, leaves a symmetry, since one can still rotate by a 2×2 orthogonal matrix U^{ab}_θ i. e., $K(\mathrm{f}) \to U_\theta K(\mathrm{f})$, that leaves D proportional to the identity. Note that U^{ab}_θ is a constant matrix depending only on one number, the rotation angle; it satisfies $U_\theta U^T_\theta = 1$. The freedom to rotate by U_θ will be eliminated later by looking at how much information (basically S/N) is carried by each channel. We shall insist that no optic nerves are wasted carrying signals with very low S/N.

We are now ready to write down the prediction for $K^{ab}(\mathrm{f})$. To do that we go to the G±R basis where $M^{ab}(\mathrm{f})$ is diagonal in color space. $K^{ab}(\mathrm{f})$ can then be taken to be diagonal since there are no correlations in color in that basis: it consists of two functions $K_\pm(\mathrm{f})$ which are chosen to separately whiten the G±R channels. Since the complete frequency space correlators in the two channels after filtering by $M_\pm(f)$ are $M_\pm^2(f)(R_\pm(f) + N^2) + N_0^2$, the $K_\pm(f)$ are therefore

$$K_\pm(f) = \frac{\sqrt{\rho}}{\left[M_\pm^2(f)(R_\pm(f) + N^2) + N_0^2\right]^{1/2}} \qquad (6.49)$$

where N_0^2 is the power of the noise which is added following the filter $M^{ab}(f)$, see Eq. (6.8).

Now putting Eq. (6.49) together with Eq. (6.47) and Eq. (6.39), we obtain the complete retinal transfer function --- the one to be compared with experiment ---

$$K_{expt} = U_\theta \begin{pmatrix} K_+(f) & 0 \\ 0 & K_-(f) \end{pmatrix} \begin{pmatrix} M_+(f) & 0 \\ 0 & M_-(f) \end{pmatrix} \begin{pmatrix} 1 & 1 \\ -1 & 1 \end{pmatrix}. \qquad (6.50)$$

As a reminder, the rightmost matrix transforms the G, R inputs into the G$\pm$R basis. These signals are then separately filtered by $K_\pm M_\pm$. Finally, the rotation U_θ to be specified shortly, determines the mix of these two channels carried by individual retinal ganglion cells. We should emphasize that the outputs of the two color channels defined by Eq. (6.50) continue to be correlated for any choice of rotation angle θ.

Analyzing the color solutions

In this section we show how the diverse processing types such as those found in goldfish and primates are both given by Eq. (6.50) but for different values of the parameter r in the color correlation matrix.

For the case of goldfish, as mentioned earlier, one expects only small overlap between R and G responses and thus r is small. The diagonal channels G$\pm$R then have eigenvalues $1 \pm r$ of the same order: $(1 - r)/(1 +$

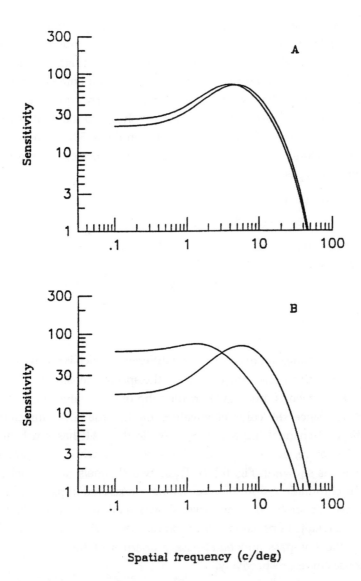

Figure 6.12: The luminance and chromatic channels for goldfish (A), and for primates (B). In both figures the curve that is more band-pass like is for the luminance G+R channel, while the other is for the G-R channel. Parameters used are $I_0/N = 5.0$, $\alpha = 1.4$, $f_c = 22.0$ c/deg, $N_0 = 1.0$ for both figures A and B. The only difference between A and B is that for A $r = .2$ while for B $r = .85$.

r) $\sim$ 1. This means both channels on average carry roughly the same amount of information and transmit signals of comparable S/N. Thus the filters $K_+(f)M_+(f)$ and $K_-(f)M_-(f)$ are very similar. In fact, they are both band-pass filters as shown in Fig. 6.12A for some typical set of parameters. Since these channels are already nearly equalized in S/N, there is no need to mix them by rotating with U_θ, so that matrix can be set to unity. Therefore, the complete solution Eq. (6.50) when acting on the input vectors R, G, gives two output channels corresponding to two ganglion cell types:

$$Z_1 = (G + R) \, K_+ M_+,$$
$$Z_2 = (G - R) \, K_- M_-. \tag{6.51}$$

If we Fourier transform these solutions to get their profiles in space, we arrive at the kernels $K^{ab}(x - x')$ shown in Fig. 6.13 for some typical set of parameters. The top row is one cell type acting on the R and G signals, and the bottom row is another cell type. These have features of double opponency cells.

Moving to primates, there is one crucial difference which is the expectation that r is closer to 1 since the overlap of the spectral sensitivity curves of the red and green is much greater: the ratio of eigenvalues $(1 - r)/(1 + r) \ll 1$. Since the color eigenvalues modify the S/N, this implies that the G-R channel has a low S/N while the G+R has much higher S/N. Therefore, $K_-(f)M_-(f)$ is a low-pass filter while $K_+(f)M_+(f)$ is band-pass as shown in Fig. 6.12B. These two channels can be identified with the chromatic and luminance channels measured in psychophysical experiments, respectively. The curves shown in Fig. 6.12B do qualitatively match the results of psychophysical contrast sensitivity experiments [47]: namely the low-pass and band-pass properties of the chromatic and luminance curves, respectively.

Although there is psychophysical evidence that indicates that color information in primate cortex is organized into luminance and chromatic channels under normal adaptation conditions [47], this is not how the primate retina transmits information down the optic nerve [23]. One reason that might explain why the primate retina chooses not to use the

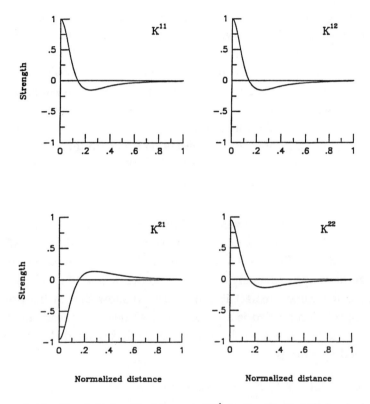

Figure 6.13: Predicted retinal kernel K^{ab} in the R and G basis in the goldfish regime $r = .2$ and for the same parameters as those in Fig. 6.12. These cells can be termed double opponency cells

G±R basis is that the representation of information in chromatic and luminance channels, has one undesirable consequence: If we compute the signal to noise ratio as a function of frequency in the chromatic channel, given by $(S/N)^2_- = K^2_- M^2_- R_- / [K^2_- (M^2_- N^2 + 1)]$, and compare it with the corresponding ratio in the luminance channel we find that the ratio $(S/N)_- / (S/N)_+ \ll 1$ because $(1 - r)/(1 + r) \ll 1$. So for primates, transmitting information in the luminance and chromatic basis would result in one channel with very low S/N, or equivalently one channel that does not carry much information. Transmitting information at low S/N down the optic nerve could be dangerous, especially since the optic nerve introduces intrinsic noise of its own; it also may be wasteful of optic nerve hardware. What we propose here is to use the remaining symmetry of multiplication by the rotation matrix U_θ, to mix the two channels so they carry the same amount of information i. e., such that they have the same S/N at each frequency. Keep in mind that this does not affect the decorrelated nature of the two signals.

In the case of primates, where the hierarchy in S/N between the two channels is large, the mixing of the two channels is significant. Mixing occurs in the goldfish case also, but there the two channels Z_1 and Z_2 already have approximately equal S/N so the degree of mixing is very small or ignorable. In fact it is not hard to show that in the primate angle of rotation needed is approximately 45 deg. This leads finally to the following solutions for the two optimally decorrelated channels with equalized S/N ratios

$$
\begin{aligned}
Z_1 &= (G + R) K_+ M_+ + (G - R) \\
&= R (K_+ M_+ + K_- M_-) + G(K_+ M_+ - K_- M_-), \\
Z_2 &= -(G - R) K_- M_- + (G + R) \\
&= R (K_+ M_+ - K_- M_-) + G (K_+ M_+ + K_- M_-).
\end{aligned}
\tag{6.52}
$$

Since for primates, $K_+(f)M_+(f)$ and $K_-(f)M_-(f)$ are very different, the end result is a dramatic mixing of space and color. For example, cell no. 1 at low frequency has $K_-(f)M_-(f) > K_+(f)M_+(f)$ so it performs an opponent R-G processing. As the frequency is increased, however, $K_-(f)M_-(f)$ becomes smaller than $K_+(f)M_+(f)$ and the cell makes a transition to a

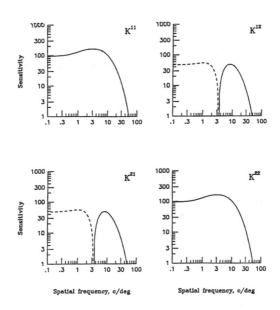

Figure 6.14: Predicted retinal filter $K^{ab}(\underline{f})$ in the R and G basis for the primate regime $r = .85$ and for the same parameters as those in Fig. 6.12. The solid (dashed) lines represent excitatory (inhibitory) responses. Notice that both cells Z_1 and Z_2 make a transition at some frequency from opponent color G-R or R-G to non-opponent G+R.

smoothing G+R type processing [23]. In Fig. 6.14, we show the filters in frequency space, in the R and G basis. These filters are in principle directly measurable in contrast sensitivity experiments. We view the zero crossing at some frequency as a generic prediction of this theory.

In Fig. 6.15 (dashed line), we show how the solutions look for a typical set of parameters after Fourier transforming back to space. We can see cell type 1 summates red mostly from its center and an opponent green mostly from its surround, while for type 2 the red and green are reversed. These cells can be termed single opponency cells, as seen in primates [23]. One might object that the segregation of the red and green in the center is not very dramatic. Actually, this is due to the simplified model we have taken. Complete segregation can be achieved if one allows the synaptic

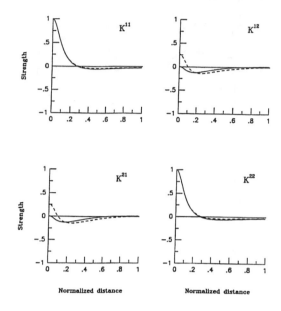

Figure 6.15: Predicted retinal kernel K^{ab} in the R and G basis in the primate regime $r = .85$ and for the same parameters as those in Fig. 6.12 – dashed curves. The solid curves use the same parameters except that the parametere N_0 was allowed to be different in the luminance and chromatic channels by a factor of two. This was done to illustrate that complete color segregation in the cell's center can easily be achieved.

noise parameter N_0, which was set to 1 for the dashed line, to be different for the two channels. In fact, a difference of $1/2$ between the two noises produces the solutions shown by the solid curves in Fig. 6.15.

We hope the results of this and the previous sectiosn have convinced the reader that the application of information theory to neural systems merits further investigation.

Acknowledgements

I would like to thank Z. Li and N. Redlich, L. Kruglyak and K. Miller for many hours of useful discussions, J. Kmiec for comments on the manuscript and L. Ferraro for interesting discussions. This work is supported in part by a grant from the Seaver Institute. This chapter is a slightly revised and updated version of an article which appeared in the journal *Network*.

Appendix

Another useful concept in information theory is that of mutual information of two events or variables, O and L defined as

$$I(O; L) \equiv H(O) + H(L) - H(O, L) \tag{6.53}$$

where $H(O, L)$ is the joint information $- \sum_{O,L} P(O, L) \log P(O, L)$. This quantity has some interesting properties. For example if the events O and L are completely statistically independent then $P(O, L) = P(O)P(L)$ and $H(O, L) = H(O) + H(L)$ making $I(O; L) = 0$. On the other hand, if the two events are completely dependent then $H(O, L) = H(L) = H(O)$ and $I(O; L)$ is the same as $H(O)$ (or equivalently $H(L)$). Thus, $I(O; L)$ in general is a measure of the interdependence of the two events. In fact, it can be thought of as the information carried by O about the event L. If $O = L + n$ where n is some additive noise and if all the variables are gaussian distributed with some variance, then $I(O; L) = H(O) - $ Noise Entropy, a fact that we needed in our analysis above.

References

[1] J.J. Atick and A.N. Redlich (1990a). Towards a theory of early visual processing, *Neural Comp.* **2**, 308-320.

[2] J. J Atick and A.N. Redlich (1990b). Quantitative tests of a theory of retinal processing: Contrast sensitivity curves. Inst. for Advanced Study Report no. IASSNS-HEP-90/51, submitted for publication.

[3] J.J. Atick and A.N. Redlich (1992a). What does the retina know about natural scenes?, *Neural Comp.* **4**, 196-210.

[4] J.J. Atick and A.N. Redlich (1992b). Convergent algorithm for sensory receptive field development. Inst. for Advanced Study report no. IASSNS-HEP-91/80. Submitted for publication.

[5] J.J. Atick, Z. Li and A.N. Redlich (1990). Color coding and its interaction with spatiotemporal processing in the retina. Inst. for Advanced Study report no. IASSNS-HEP-90/75.

[6] J.J. Atick, Z. Li and A.N. Redlich (1992). Understanding retinal color coding from first principles, *Neural Comp.* **4**, 559-572.

[7] F. Attneave (1954). Some informational aspects of visual perception, *Psych. Rev.* **61**, 183-193.

[8] H. B. Barlow (1961). Possible principles underlying the transformation of sensory messages, In *Sensory Communication*, W. A. Rosenblith, ed., pp. 217-234 (M.I.T. Press, Cambridge MA).

[9] H. B. Barlow (1985). Perception: what quantitative laws govern the acquisition of knowledge from the senses?, In *Functions of the brain*, C. W. Coen, ed. (Clarendon Press, Oxford).

[10] H. B. Barlow (1989). Unsupervised learning, *Neural Comp.* **1**, 295-311.

[11] H. B. Barlow, M. Hawken, T. P. Kaushal and A. J. Parker (1987). Human contrast discrimination and the contrast discrimination of cortical neurons, *J. Opt. Soc. Am. A,* **4,** 2366-2371.

[12] H. B. Barlow and P. Foldiak (1989). Adaptation and decorrelation in the cortex, in *The Computing Neuron* (Addison-Wesley, New York).

[13] H.B. Barlow, T.P. Kaushal, and G.J. Mitchison (1989). Finding minimum entropy codes, *Neural Comp.* **1,** 412-423.

[14] G.A. Barnard (1955). Statistical calculations of word entropies for four western languages, *I. R. E. Trans. Inf. Theory,* **IT-1,** 49-53.

[15] W. Bialek 1990. Theoretical physics meets experimental neurobiology, In *Lectures in Complex Systems, SFI Studies in the Sciences of Complexity, Lect. Vol. II,* E. Jen, ed., pp. 513-595 (Addison-Wesley, Menlo Park).

[16] W. Bialek, D.L. Ruderman and A. Zee (1991). Optimal sampling of natural images: A design principle for the visual system?, In *Advances in Neural information processing systems 3,* R. P. Lippmann, J. E. Moody, and D. S Touretzky, eds. pp. 363-369 (Morgan Kaufman, San Mateo CA).

[17] W. Bialek, F. Rieke, R.R. de Ruyter van Steveninck and D. Warland (1991). Reading a neural code, *Science* **252,** 1854-1857.

[18] G. Buchsbaum and A. Gottschalk (1983). Trichromacy, opponent colours coding and optimum colour information transmission in the retina, *Proc. R. Soc. Lond. Ser. B* **220,** 89-113.

[19] F.W. Campbell and R.W. Gubisch (1966). Optical quality of the human eye, *J. Physiol.,* **186,** 558-578.

[20] H. Davson (1980). *Physiology of the eye* (Academic Press, London).

[21] N.W. Daw (1968). Colour-coded ganglion cells in the goldfish retina: Extension of their receptive fields by means of new stimuli, *J. Physiol.* **197,** 567-592.

[22] F.M. De Monasteriom, E.P. McCrane, J.K. Newlander and S.J. Shein, (1985). Density profiles of blue-sensitive cones along the horizontal meridian of macaque retina, *Invest. Opthalmol. Vis. Sci.* **26,** 289-302.

[23] A.M. Derrington, J. Krauskopf and P. Lennie (1984). Chromatic mechanisms in lateral geniculate nucleus of macaque. *J. Physiol.* **357**, 241-265.

[24] R.L. De Valois, H. Morgan and D. M. Snodderly (1974). Psychophysical studies of monkey vision. III: Spatial luminance contrast sensitivity tests of macaque and human observers, *Vision Res.* **14**, 75-81.

[25] C. Enroth-Cugell and J.G. Robson (1966). The contrast sensitivity of retinal ganglion cells of the cat, *J. Physiol.* **187**, 517-552.

[26] D.J. Field (1987). Relations between the statistics of natural images and the response properties of cortical cells, *J. Opt. Soc. Am. A*, **4**, 2379-2394.

[27] D.J. Field (1989). What the statistics of natural images tell us about visual coding, *Proc. SPIE: Human vision, visual processing and digital display*, **1077**, 269-276.

[28] R. G. Gallager (1968). *Information theory and reliable communication* (John Wiley and Sons,New York).

[29] S. Geman and D. Geman (1984). Stochastic relaxation, Gibbs distributions, and the Bayesian restoration of images, *Trans. Patt. Anal. and Machine Intell.* **PAMI-6**, 721-741.

[30] M.C. Goodall (1960). Performance of stochastic net, *Nature* **185**, 557-558.

[31] G.G. Gouriet (1952). A method of measuring television picture detail, *Electronic Engineering* **24**, 308.

[32] G.W. Harrison (1952). Experiments with linear prediction in television, *Bell System Tech. J.* **31**, 764.

[33] H.G. Hentschel and H.B. Barlow (1991). Minimum-entropy coding with Hopfield networks, *Network* **2**, 135-148.

[34] G. E. Hinton and T.J. Sejnowski (1983). Optimal perceptual inference, *Proc. IEEE conference on computer vision and pattern recognition.* pp. 448-453.

[35] H. Jacobson (1951). The informational capacity of the human eye, *Science* **113**, 292-293.

[36] D. H. Kelly (1972). Adaptation effects on spatio-temporal sine-wave thresholds, *Vis. Res.* **12**, 89-101.

[37] D. Kersten (1990). Statistical limits to image understanding, In *Vision: Coding and Efficiency*, C. Blakemore, ed. (Cambridge University Press, Cambridge).

[38] E. R. Kretzmer (1952). Statistics of television signals, *Bell System Tech. J.* **31**, 751-763.

[39] H. H. Kornhuber (1973). Neural control of input into long term memory: limbic system and amnestic syndrome in man, In *Memory and transfer of information*, H. P. Zippel, ed., pp. 1-22 (Plenum Press, New York).

[40] S. B. Laughlin (1981). A simple coding procedure enhances a neuron's information capacity, *Z. Naturforsch* **36c**, 910-912.

[41] S. B. Laughlin (1987). Form and function in retinal processing, *Trends Neurosci.* **10**, 478-483.

[42] S. B. Laughlin (1989). Coding efficiency and design in visual processing, in *Facets of Vision*, D.G. Stavenga and R.C. Hardie, eds., pp. 213-234 (Springer, Berlin).

[43] R. Linsker (1988). Self-organization in a perceptual network. *Computer* **21**, 105-117.

[44] R. Linsker (1989a). An application of the principle of maximum information preservation to linear systems, In *Advances in Neural Information Processing Systems 1*, D. S. Touretzky, ed.pp. 186-194 (Morgan Kaufman, San Mateo).

[45] R. Linsker (1989b). How to generate ordered maps by maximizing the mutual information between input and output signals, *Neural Comp.* **1**, 402-411.

[46] J. N. Lythgoe (1979). *The ecology of vision* (Oxford Univesrity Press, Oxford).

[47] K. T. Mullen (1985). The contrast sensitivity of human color vision to red-green and blue-yellow chromatic gratings, *J. Physiol.* **359**, 381-400.

[48] F. L. van Ness and M. A. Bouman (1967). spatial modulation transfer in the human eye, *J. Opt. Soc. Am.* **57**, 401-406.

[49] A. Papoulis (1984). *Probability, random variables, and stochastic processes* (McGraw-Hill, New York).

[50] B. A. Pearlmutter and G. E. Hinton (1986). G-maximization: an unsupervised learning procedure for discovering regularities, in *Proceedings of the AIP conference 151, Neural networks for computing*, J. S. Denker, ed., 333-338 (AIP, New York).

[51] F. Pratt (1942). *Secret and Urgent* (Blue Ribbon Books, New York).

[52] A. N. Redlich (1991). Redundancy reduction as a strategy for unsupervised learning. Inst. for Advanced Study report no. IASSNS-HEP-91/87.

[53] W. F. Schreiber (1956). The measurement of third order probability distributions of television signals, *I. R. E. Trans. Inf. Theory* **IT-2**, 94-105.

[54] C.E. Shannon (1951). Prediction and entropy of printed English. *Bell System Techn. J.* **30**, 50-64.

[55] C.E. Shannon and W. Weaver (1949). *The Mathematical Theory of Communication* (The University of Illinois Press, Urbana).

[56] R. Shapley and C. Enroth-Cugell (1984). Visual adaptation and retinal gain controls, in *Progress in Retinal Research 3*.

[57] S. R. Shaw (1984). Early visual processing in insects, *J. Exp. Biol.* **112**, 225-251.

[58] M. V. Srinivisan, S. B. Laughlin and A. Dubs (1982). Predictive coding: A fresh view of inhibition in the retina, *Proc. R. Soc. London Ser. B* **216**, 427-459.

[59] P. Sterling (1990). Retina, in *The synaptic organization of the brain*, G. Shepherd, ed. (Oxford University Press, New York).

[60] G. Sziklai (1956). Some studies in the speed of visual perception, *I. R. E. Trans. Inf. Theory* **IT-2**, 125-128.

[61] A. M. Uttley (1979). *Information Transmission in the Nervous System* (Academic Press, London).

[62] D. C. Van Essen and C. H. Anderson (1988). Informtion processing strategies and pathways in the primate retina and visual cortex, in *Introduction to Neural and Electronic Networks* (Academic Press, Orlando).

[63] D. C. Van Essen, B. Olshausen, C. H. Anderson, J. L. Gallant, (1991). Pattern recognition, attention, and information bottlenecks in the primate visual system, *Proc. SPIE: Visual Information Processing: From Neurons to Chips*.

[64] D. Warland, M. Landolfa, J. P. Miller, and W. Bialek (1991). Reading between the spikes in the cercal filiform hair receptors of the cricket, in *Analysis and Modeling of Neural Systems*, F. Eeckman, ed., pp. 327-333 (Kluwer Academic).

[65] S. Watanabe (1981). Pattern recognition as a quest for minimum entropy, *Pattern recognition* **13**, 381-387.

[66] S. Watanabe (1985). *Pattern recognition: Human and Mechanical* (Wiley, New York).

Time-Frequency Transforms and Images of Targets in the Sonar of Bats

James A. Simmons

Introduction

A striking characteristic of perception is the *unity* of spatial images. The segregated nature of neural representations of different stimulus features is well documented (in vision, for example, distance, shape, and texture), but the process that brings these features together to form whole images is poorly understood. Images appear to us as having spatial dimensions which define the location and shape of objects along common scales of distance and direction, yet the distance to an object and the details of its shape often are encoded quite differently by the brain. How do the details of shape and texture within each object come to reside in their proper locations within whole visual scenes?

The intricacy of neural representations of stimulus features and the subtlety of the psychological correlates of feature representation in most sensory modalities (e.g., vision, olfaction) make it difficult to develop satisfactory experiments that directly address hypotheses concerning image fusion. In a comparative, neuroethological approach to this vexing problem, the sonar of bats offers a simple, well-defined example of the separateness of the brain's representation of two features of targets from echoes, and of the fusion of these representations into unified images having a common psychological dimension. However, to a reader not familiar with auditory processing in general, or with the even more specialized topic of echolocation, the simplicity and elegance of the bat's target-imaging scheme may not be immediately evident. Instead, the reader is likely to be more impressed, not to say overwhelmed, by the unfamiliar and seemingly arcane details of echolocation signals, neural coding of echoes, and psychophysical techniques for studying the bat's sonar. However, the effort needed to master these details will be repaid by insight into one of the clearest examples yet available about the elusive and mysterious process by which the brain's parallel representations of different features of stimuli emerge into perception itself.

Echolocation: Target Range and Target Shape

Echolocating bats perceive objects with information conveyed in sonar echoes [4, 7, 15]. For example, the big brown bat, *Eptesicus fuscus* [11], emits frequency-modulated (FM) sonar signals to detect, track, identify, and intercept flying insects [7, 22]. This species of bat uses frequencies from 25 to 100 kHz in its sonar sounds, arranged in two prominent descending FM sweeps from about 50 to 25 kHz and from about 100 to 50 kHz. These two sweeps are, respectively, the first and the second harmonics of the sounds. *Eptesicus* emits its echolocation sounds at rates from 5 to over 100 sounds/second, with durations from roughly 10 to 15 ms down to about 0.5 ms [22]. Using echolocation, *Eptesicus* and other "FM" bats can detect insect-sized targets at distances up to about 5 m (echo delay of about 30 ms; [10]) and can locate and discriminate among targets during the approach, prior to capture [8, 16, 31, 32, 33].

The bat's auditory system serves as the sonar receiver, representing echoes by the activity of populations of neurons selective to different combinations of acoustic parameters [18, 27]. Both anatomical and physiological segregation of responses to different echo parameters occurs in the bat's auditory system, and the presence of parallel auditory subsystems having hierarchical organization is well documented. The three principal neural dimensions of auditory representation are *echo frequency,* which is encoded by "place" originating at the cochlea and manifested as frequency tuning of neurons; *echo amplitude,* which is encoded by neurons having different dynamic ranges and manifested both as amplitude tuning and as the number of action potentials per stimulus; and *echo delay,* which is encoded through neural computations that produce delay-selective responses and manifested as delay or range tuning.

The two principal characteristics of an echo for image-forming purposes are its *spectrum* for target shape [9, 13, 19, 25, 22] and its *delay* for target distance, or range [20]. While the spectrum of echoes is displayed along an auditory frequency axis made up of numerous neurons tuned to different frequencies, the delay of echoes is displayed along an auditory time axis made up of numerous neurons tuned to different delays. Physiological experiments confirm the separate nature of these two representations [5], but behavioral experiments demonstrate the fusion of the information they represent onto a single psychological scale of distance [23, 22]. The contributions to the bat's images from auditory frequency and delay representations coexist side-by-side, but they ultimately are combined in the images themselves to create a synthetic or computed dimension of delay.

Eptesicus gathers information about target shape from the spectrum of echoes, but it perceives "shape" in terms of the *range to different parts of the target* [26]. For this to occur, frequency information in the echo spectrum is converted into estimates of the time structure of the echo and thus the range structure of the target. The experiments summarized here critically identify this conversion process as consisting of parallel time-domain and frequency-to-time-domain transforms whose converging outputs create the common delay or range axis of perceived images. The unity of the bat's perceptions appears to arise from properties of the transforms themselves, in spite of the separateness of the initial auditory time representation of echo delay and frequency representation of the echo spectrum.

Experimental Measurement of Sonar Images in Bats

The experiments for studying sonar image-formation by *Eptesicus* use a two-choice discrimination procedure with jittering echoes that yields a graph describing the image of the target to the bat [12, 14, 21, 23]. An electronic target simulator picks up the bat's sonar transmissions and returns them as "echoes" at predetermined delays that correspond to desired target distances (Fig. 7.1). In the basic jitter procedure, bats are trained to discriminate between echoes a_1 and a_2 (Fig. 7.1A) that alternate or jitter in their delay from one sonar emission to the next, and echoes that arrive at a stationary delay, b. During each trial, the bat sits on an elevated Y-shaped platform, broadcasting sonar sounds with durations between 1.5 and 5 ms at rates of 10 to 30 sounds/second to determine whether the echoes from the left or from the right are jittering. The bat's response (rewarded with food) is to move to the left or right towards the loudspeaker that returns the jittering echoes and not to move forward (punished by "time-out") towards the stationary echoes.

Echo delays are typically about 3 ms (corresponding to a target range of about 50 cm), with a_1 and a_2 differing in delay by amounts up to 50 μs-the value of the jitter Δt (Fig. 7.1A) intended for a particular experimental trial. Typically, the bat makes its response during a trial within about one second (arrow in Fig. 7.1A), after scanning the left and right channels of the target simulator with its sonar sounds to determine which side of the apparatus returns the echoes that jitter. Between 40 and 60 trials

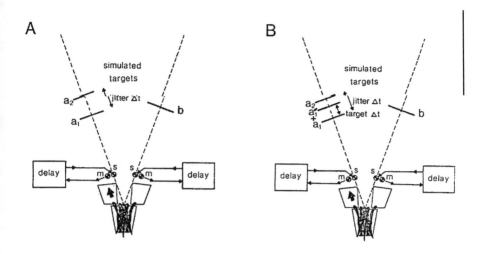

Figure 7.1: Diagrams of two-choice jittering-echo discrimination experiments. Bat's sounds are picked up by microphones (m), delayed, and returned from loudspeakers (s) as echoes. In basic experiment simulating individual single-glint targets (A), echoes a_1 and a_2 alternate in delay from one emitted sound to the next, while b is stationary. In experiment simulating two-glint target (B), echoes a_1 and $a_{1'}$ are combined and alternate as a pair with a_2. In all experiments, size of jitter interval (jitter Δt) is changed in 5 μs steps, using a_2 as probe to trace image of either a_1 or $a_1 + a_{1'}$. Size of the interval between echo components a_1 and $a_{1'}$ (target Δt) is set to 10, 20, or 30 μs to simulate two-part target. As bat scans left or right, only one channel is activated at a time, and only one echo is delivered for each transmission (a_1, $a_1 + a_{1'}$, a_2, or b), requiring bat to remember images for each echo

are conducted at each stimulus condition (a particular value for Δt; for example, from about -5 μs to +50 μs in 5 μs steps).

The critical feature of the jitter procedure is that the stimulus echoes are presented separately, with only one stimulus being delivered (a_1, a_2, or b in Fig. 7.1A) on each sonar emission. Switches built into the target simulator determine whether the bat is aiming its head at the left or the right microphone; they activate only that particular channel of the simulator for any given sonar emission. The delay that occurs for the stimulus echo of any particular sonar emission is selected to be a_1, a_2, or b according to the trial-by-trial schedule of left-right alternations of the positive and negative stimuli. If the channel the bat aims towards is scheduled to deliver jittering echoes, the delay value is changed between sonar emissions to produce the required jitter. Because the jittering echoes appear one-at-a-time, in alternation, the bat must remember the delay of each echo and then compare it with the delay of next echo to detect whether jitter is present. The introduction of this brief memory stage justifies use of the term "image" to describe the results that are obtained.

The image of a simple, single-point target

The purpose of the jitter discrimination experiments is to measure images resulting from the bat's memory for echo delay, so the stimulus echoes have to be presented one-at-a-time. The other chief concern is that the stimulus echoes (a_1, a_2, or b) not overlap with the bat's transmitted sound or with extraneous echoes from the apparatus or the walls of the room for the bat to perform the task. If such overlap occurs, spectral cues might be generated which could substitute for direct perception of echo delay in determining the bat's performance. The overall delay of the stimulus echoes is about 3 ms, while the duration of the sonar sounds used in the task varies from 1.5 to 5 ms in different bats [23]. Several of the bats performed the task without ever using sonar sounds longer than 2 ms, so overlap of transmitted and received sounds did not occur in these cases. Several other bats used sounds as long as 4 to 5 ms in the task, but their performance was the same as the performance of the bats using shorter sounds. Hence, overlap of echoes with emissions could not have been a requirement for the bats to achieve the observed performance, and spectral cues generated by this overlap could not have played a necessary role in shaping the results.

The other source of unintended spectral cues is overlap of stimulus echoes (generated electronically by the simulator) with echoes from the apparatus, including the front faces of the microphones and loudspeakers, the platform, and the walls, ceiling, and floor of the room. There was a time interval of about 2 ms that separated the stimulus echoes from any extraneous echoes; transmitted sounds shorter than 2 ms failed to produce stimulus echoes that overlapped with any echoes from the apparatus that could be measured at the bat's observing position on the Y-shaped platform. Hence, these artifactual spectral cues, too, can be ruled out as a necessary condition for the bats to perform the jitter task.

The bat's image of a single-point target (represented by echo a_1) is "measured" by varying the jitter interval (jitter Δt) between the two jittering echoes, a_1 and a_2, and plotting the percentage of errors achieved by the bat as a function of the amount of jitter up to about 50 μs. Moving echo a_2 to different delays relative to a_1 lets a_2 serve as a probe for tracing the shape of the bat's remembered delay-image of a_1. A further refinement of this tracing procedure is to change the phase of echo a_1 by 180° and then repeat the entire experiment. A combined graph can then be made for both 0° and 180° phase-shifts of echo a_1 as a function of the amount of jitter. The curve for the 180° phase-shift is inverted and then placed beneath the curve for the 0° phase-shift, with a common horizontal delay axis (Fig. 7.2A). The extreme points of the two curves are then connected (see [23] for details). The resulting compound error curve traces the complete image of the single-point target, a_1 [23].

Figure 7.2A shows representative data for one *Eptesicus*. The objective delay of echo a_1 (about 3 ms) is at zero on the horizontal axis (which shows the position of a_2 relative to a_1 in microseconds), and the bat performs at chance levels (about 50% errors) when a_2 is aligned at the same delay as a_1 because in this condition the echoes truly do not jitter. Here, the bat cannot distinguish between a_1/a_2 and b (see Fig. 7.1A) because both sets of echoes are stationary. However, as a_2 departs from a_1 in delay, the bat's performance in Fig. 7.2A improves, with fewer errors in the two-choice task. The presence of the jitter becomes more obvious to the bat, making better performance possible. Note that the bat's performance does not improve monotonically as the difference between a_2 and a_1 becomes greater. In addition to making many errors at a delay difference of zero,

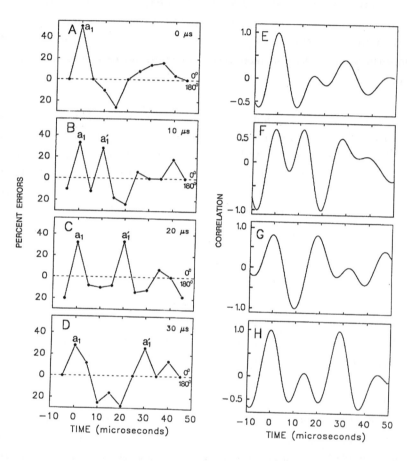

Figure 7.2: Compound error curves based on performance of *Eptesicus* (vertical axis) in jitter discrimination experiment at different values for jitter interval (horizontal axis) with a_1 alone (at $0°$ or $180°$ phase) alternating with a_2 (curve labeled 0 μs; A), or $a_1 + a_{1'}$ (at $0°$ or $180°$ phase) alternating with a_2 (curves labeled 10, 20, or 30 μs; B-D). Images traced by compound curves resemble crosscorrelation functions for a_1 or $a_1 + a_{1'}$ for 0, 10, 20, and 30 μs echo-delay separations (E-H). For details on compound curves, see Ref. [23].

the bat also makes substantial errors for jitters in the region of 25 to 35μs when echo a_1 and echo a_2 are in the same (0°) phase, and substantial errors at 10 to 15μs when echo a_1 is out of phase 180° with a_2.

In effect, the compound curve in Fig. 7.2A traces a kind of silhouette of the distribution of the remembered image of a_1 around its central point at zero, and this distribution has a complex shape of its own. The image has its main peak at the delay of a_1 and also has a somewhat lower side-peak about 30-35 μs away from the main peak. When the phase of a_1 is changed from 0° to 180°, the location of the main error peak moves over to about 10 to 15 μs. The bat evidently perceives the delay of the phase-shifted echo a_1 to be displaced by 10 to 15 μs away from its objective delay at zero, indicating that it uses information about the phase of the echo as part of the process of estimating echo delay.

Amplitude-latency trading: Temporal nature of delay coding

The next experimental manipulation demonstrates that the timing of neural discharges must be the parameter of auditory responses that conveys information about the delay of echoes up through the auditory pathways to the cortex. The procedure is to change the amplitude of either echo a_1 or echo a_2 in the jitter experiment (Fig. 7.1A) and observe whether the apparent delay of that echo shifts to a new position along the time axis of the image portrayed by the performance curve (Fig. 7.2A). The rationale for this experiment is that, as echo amplitude decreases, neural response latency increases in a phenomenon called "amplitude-latency trading" [18]. In simple terms, a quiet sound takes longer to drive a neuron across its threshold then does a loud sound. If the latency of neural responses changes, so should the perceived time-of-occurrence of the stimulus. In *Eptesicus*, auditory evoked-potential response-latency trades with amplitude at about -13 to -18 μs/dB [26], and the perceived delay of echoes should shift by this amount.

Figure 7.3 illustrates the results of the jitter experiment when the amplitude of one or the other jittering echo is changed to induce amplitude-latency trading [23]. When either echo a_1 is decreased by 1 dB or echo a_2 is increased by 1 dB, the error peak marking perceived alignment of a_2 with a_1 shifts to longer delays by about 17 μs (Fig. 7.3, -1 dB a_1 and +1 dB a_2 curves), demonstrating that the information required to judge delay probably is conveyed by the timing of discharges evoked by echoes. Besides indicating the probable temporal character of auditory coding for delay, this effect is the key to dissociating images into their component parts and identifying the nature of their underlying auditory representations [23].

The image of a complex, two-point target

By adding an additional delay component to echo a_1 in the jitter experiment (adding $a_{1'}$ to a_1 in Fig. 7.1B), the image of a complex target containing two reflecting points can be plotted, too. The distance from the bat to each point in the target is different, with the range difference between the two points being simulated by the delay difference between a_1 and $a_{1'}$. Here, the second part of the echo, $a_{1'}$, appears at delay separations from a_1 of 10, 20, or 30 μs (target Δt in Fig. 7.1B). (In Fig. 7.2A, the value of target Δt is 0 μs.) These delay differences correspond to distances of 1.7, 3.4, or 5.2 mm between the nearer and farther parts of the two-point target being simulated electronically. Now, whenever echo a_1 is delivered to the bat in the jittering condition, echo $a_{1'}$ is always coupled with it. Echo a_2 still is used by itself as a probe to plot the image of a_1, except that now a_1 really consists of two delays rather than just one delay. Similarly, echo b still is a stationary stimulus. The jitter interval (jitter Δt in Fig. 7.1B) still varies in 5 μs steps to trace the shape of the image, and jitter values are chosen to bracket the delay values for both a_1 and $a_{1'}$. That is a_2 appears at delays spanning the regions around both a_1 and $a_{1'}$ - for example from -5 to +45 μs.

The sonar signals of *Eptesicus* in jitter discrimination experiments range between 1.5 and 5 ms for different individual bats. These emitted sounds

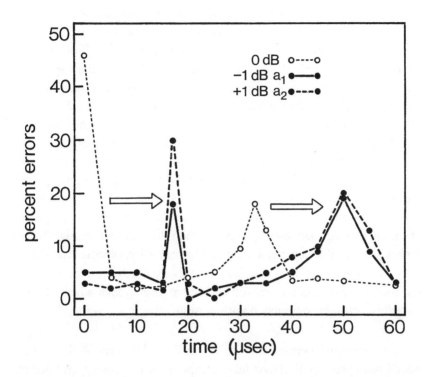

Figure 7.3: Amplitude-latency trading of images: Performance of *Eptesicus* in basic jitter experiment (Fig. 7.1A) with a_1 and a_2 at same amplitude (0 db, dotted line); with a_1 reduced by 1 dB (-1 dB a_1, solid line); and with a_2 increased by 1 dB (+1 dB a_2, dashed line). Whole curve shifts right by about 17 μs for amplitude change of either echo a_1 or a_2, matching physiological amplitude-latency trading and thus indicating neural response-times probably encode delay.

are much longer than the interval between the two echo components, a_1 and $a_{1'}$, so the overall echo reaching the bat's ears from the complex target actually consists of overlapping components instead of two discrete echoes. Thus, there really is only one complex echo, not two simple echoes. The intention of presenting echoes a_1 and $a_{1'}$ together is to deliberately introduce spectral cues by the mechanism of interference (see below). The critical question is whether the bat now can always identify the jittering stimuli from the presence of two overlapping echoes on half of the sonar transmissions towards the jittering channel, or whether the bat instead registers the presence of the second component $(a_{1'})$ of the two-part echo $(a_1 + a_{1'})$ in its remembered image, even though $a_{1'}$ is only 10, 20, or 30 μs later than a_1 itself, and the echo components are mixed together. Thus, the introduction of the second echo component, $a_{1'}$, renders this condition a control experiment to test for the possible role of artifactual spectral cues as well as an experimental condition for the case of a two-point target. The spectral-artifact hypothesis would predict that the bat's performance should remain good (few errors) at all values of jitter because the spectral effects of overlap between a_1 and $a_{1'}$ should be salient enough to cue the bat's discrimination anyway. The alternative hypothesis, that the bat really is perceiving the time-delays of individual echoes, would predict that the bat's performance should be poor at jitter values which place a_2 (the probe echo) in conjunction with a_1 or $a_{\prime}$.

The compound performance curves in Fig. 7.2B-D reveal that the bat indeed does perceive that two delay components are present, and that the bat's performance is not good at all values of jitter. The locations of both a_1 and $a_{1'}$ are labeled on the graphs, with the objective delay of a_1 still at zero on the horizontal axis. The bat makes substantial errors both at zero and at 10 μs (Fig. 7.2B), 20 μs (Fig. 7.2C), and 30 μs (Fig. 7.2D), respectively. By comparing the curves for delay separations of a_1 and $a_{1'}$ from 10 to 30 μs, the movement of the second major peak in the performance curve corresponding to the delay of $a_{1'}$ is immediately evident. This second peak is the component of the image for $a_{1'}$ (Fig. 7.2B-D). Both parts of the target directly appear in the image as error peaks, and the perceived delays of their echoes are closely aligned with the objective delays. Not only does *Eptesicus* perceive the delay of a_1, but it also perceives the delay of $a_{1'}$ as an estimate of time which it remembers. The spectral artifact hypothesis, incidentally, can be rejected as an alternative explanation of the jitter results shown in Fig. 7.2. First, it cannot account for amplitude-latency trading (Fig. 7.3), and second, it does not explain how the bat remembers the delay of the second echo component as a value of *time*.

Dual Time and Frequency Representation of Echoes

The full significance of the fact that the bat perceives both the first and the second echo component of a complex echo becomes apparent only when a critical constraint imposed by the auditory system is considered. Although this is an oversimplification, the short time separations of 10 to 30 μs between echoes in Fig. 7.2B-D would seem to be considerably less than the refractory properties of auditory neurons for discharging to a second echo ($a_{1'}$) so closely following the first (a_1). How does the bat determine that the second echo is present if the first echo has already discharged all of the neurons that might be ready to respond at the time the overall echo is received? The answer to this question turns out to be an unusually well-defined example of the relation between auditory place and temporal codes for creating auditory perceptions.

The bat initially encodes FM sonar emissions and echoes as spectrogram-like arrays of neural discharges marking the time-of-occurrence and amplitude of different frequencies in the FM sweeps [18, 27]. The frequency axis of these auditory "spectrograms" is created by the frequency tuning of parallel neural channels originating at different places along the cochlea. The time axis is created by synchronization of neural discharges within each frequency channel to the passage of the FM sweep of the sonar emission or the echo through that channels particular tuned frequency. The spectrogram thus is represented by the distribution of discharges in time across neurons tuned to different frequencies and different amplitudes at each frequency.

Encoding of echo delay

Each neural frequency channel responds to the emission and the echo separately by encoding the time-of-occurrence of its excitatory frequency in each of these signals [18]. First, the pattern of frequencies over time in the FM sweeps is reproduced in the pattern of neural discharges spread across neurons tuned to different frequencies in the sweeps (see Fig. 10 in [24]. Following separate registration of the transmitted sound and the returning echo, the time that elapses between responses to the emission and to the echo is extracted across different frequencies to represent echo delay. As a result, many higher-level auditory neurons respond selectively to particular echo-delay values [5, 6, 17, 27, 28, 29, 30, 34]. So-called "delay-tuned" neurons with a span of best delays from less that 1 ms to about 25 to 30 ms form a substrate for displaying target range in the auditory cortex. The image of the point-target (a_1 in Fig. 7.2A) may result from integration of delay estimates across the population of delay-tuned neurons that are activated by the stimuli in the jitter experiment. The complication for understanding the complex-target images shown in Fig. 7.2B-D is that delay-tuned neurons register the overall delay of the whole compound echo ($a_1 + a_1$), not the individual echo components a_1 and ($a_{1'}$) by themselves. This limitation originates at peripheral stages of auditory coding of sounds.

The filtering process by which the bat's cochlea separates the frequencies of FM sweeps into different auditory frequency channels is the basis for registering the time-of-occurrence of different frequencies in the emitted sound and then in the returning echo. However, this very capability for encoding echo delay also ensures that the second of two closely-spaced echo components will not evoke neural discharges of its own. The sharpness of tuning of the cochlear filters limits their capacity to respond rapidly enough to pick out both echo a_1 and echo $a_{1'}$ as distinct signals by setting the integration-time of the filters to be longer than the delay interval between the echo components. In *Eptesicus*, echo reception takes place with an integration-time of about 300 to 350 μs [24]. Fig. 7.4 shows this integration-time window measured in *Eptesicus* during

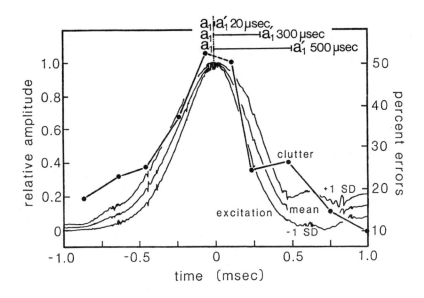

Figure 7.4: Integration-time window for echo detection by *Eptesicus,* measured behaviorally for discrimination of one-component echoes from two-component echoes with various time separations of the two components (bold solid line and data-points labeled "clutter;" right vertical axis for percent errors), or estimated from physiological data on neural tuning (thin solid curves labeled "excitation,' which show mean ±1 standard deviation; left vertical axis for relative amplitude). Approximate half-width of the curves is 300 to 350 μs. From Ref. [24].

a clutter interference experiment in which the bat has to detect test echoes at various delays in the region of the test echoes. Sounds that arrive at intervals shorter than this integration-time will be smeared together into single sound by the cochlea's filters.

The delay separation of echo components a_1 and $a_{1'}$ in the two-glint stimuli (Fig. 7.1B) is only 10 to 30 μs (Fig. 7.2B-D), which is much shorter than the bat's integration-time for echo reception. The implication of the relatively long integration-time of 300-350 μs for echo reception is that the pair of echoes making up the two-glint stimuli in the jitter experiment would have been received as a single, compound echo whose spectrum

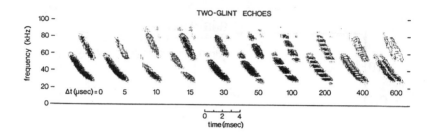

Figure 7.5: Temporal and spectral cues in echoes: Spectrograms showing echoes of 2.5-ms *Eptesicus* sonar sound reflected by complex, two-point target with time separations (Δt) between echo components (a_1 and $a_{1'}$ in Fig. 7.1B) of 0 to 600 μs. Incident sound is similar to spectrogram for $\Delta t = 0$. Spectrograms have 300-400 μs integration-time and show apportioning of information about size of Δt into spectral and temporal cues according to Δt in relation to integration-time.

provides the only clue as to the presence of the second glint, $a_{1'}$.

Encoding of echo spectrum

Due to the relatively long integration-time of the bat's auditory filters, two echo components only 10 to 30 μs apart will be mixed together in the signal coming out of each filter. Consequently, the neural discharges which mark successive frequencies in the FM sweep of the echo will only register a single sweep for the whole, compound echo, not two distinct sweeps for the echo components that make it up. When two echoes mix together within the integration-time of the sonar receiver, spectral interference occurs [1, 3, 24]. The overlapping echo components generate interference peaks and notches in the overall echo spectrum that substitute for direct registration of the echo time separation itself by neural discharges.

Figure 7.5 illustrates interference patterns in echoes reflected by a complex target with various echo-delay separations. The spectrograms show spectral notches and peaks in echoes of an *Eptesicus* sonar sound with a 2.5 -ms duration reflected by a two-point target with time separations ($\triangle t$) between echo components (corresponding to a_1 and $a_{1'}$ in Fig. 7.1B) of 0 to 600 μs. The actual incident sound has a spectrogram similar to the spectrogram for the echo at $\triangle t = 0$. These spectrograms were made with an integration-time of 300-400 μs to approximate auditory spectrograms encoded by *Eptesicus* (width of spectrograms in Fig. 7.5 corresponds to width of window in Fig. 7.4). The frequency spacing of spectral notches is regular and depends upon the size of the echo time separation. Not only does the bat's auditory system encode the arrival-time of the returning compound echo by the timing of each frequency in the FM sweep, it also encodes the pattern of amplitude across frequency in the echo. Within the auditory system, the echo spectrum is reproduced in the pattern of neural discharges to each frequency-discharges that are spread across the population of frequency-tuned neurons rather than the population of delay-tuned neurons [18]. The fundamental dimension of auditory coding begins with the place representation of different frequencies in the cochlea and is carried up through the auditory nervous system as a tonotopic "map" of frequencies across each auditory center in the brain-stem, midbrain, and cortex. This tonotopic axis is the substrate for representing the spectrum of echoes. The spatial distribution of different amounts of neural activity across the tonotopic axis must supply the bat with a display of the locations of spectral peaks and notches at different frequencies.

In Fig. 7.5, as the separation of the overlapping echo components increases, the frequency spacing of the spectral notches decreases and more notches are visible in the spectrograms. Besides the presence of the spectral notches, there is no indication that the echoes actually contain two discrete components-their presence is concealed within the integration-time as long as their time separation is smaller than 300-400 μs. To perceive the magnitude of the time separation between echo components (10, 20, or 30 μs; Fig. 7.2B-D), the bat must process the spectrum of the echo rather than its arrival-time. However, when the echo time separation approaches or exceeds the integration-time, the spectrograms "pull apart" to reveal two distinct sets of FM sweeps at slightly different times (at $\triangle t = 400$ to 600 μs in Fig. 7.5). For such long echo-delay separations, the timing of each echo can be encoded directly by the timing of neural discharges. Thus, these spectrograms show the apportionment of information about the size of Δt into spectral and temporal cues according to Δt in relation to integration-time.

Experimental Dissociation of Images

To demonstrate the reality of dual time and spectral processing of information about echo delays, the amplitude-latency trading effect can be used to distinguish between those image elements derived from the timing of frequencies in echoes and those derived from the amplitude at certain frequencies [26]. The rationale for dissociating image elements is that the representation of echo $a_{1'}$ is necessarily different for different time separations of a_1 and $a_{1'}$ (see Fig. 7.1B), as illustrated by the spectrograms of two-glint echoes in Fig. 5. The integration-time for echo reception is about 300-350i μs in *Eptesicus*, and the spectrograms were made with the same integration-time to illustrate the partitioning of information about $a_{1'}$ relative to a_1 into spectral and temporal cues.

For short time separations up to 200 μs (Δt in Fig. 7.5), peaks and notches in the spectrum provide all the information available about the presence and timing of $a_{1'}$. Echo a_1 arrives first and evokes neural discharges that register its perceived delay, which undergoes amplitude-latency trading by about -15 μs/dB over a wide amplitude range [26]. Because spectral notches rather than separately-evoked neural discharges encode echo $a_{1'}$ (see Fig. 10 in [24]), $a_{1'}$ should *not* undergo amplitude-latency trading independent of a_1. That is, changing the amplitude of echo $a_{1'}$ ought to produce no change in the perceived delay of the second echo component. Fig. 7.6A-B shows the results of a jitter experiment to test this proposition. For a 20 μs time separation of a_1 and $a_{1'}$, the perceived delay of $a_{1'}$ indeed does not trade with amplitude. In contrast, both a_1 and $a_{1'}$ shift together by -15 μs/dB if a_1 alone is changed in amplitude (not shown in Fig. 7.6). This tandem trading occurs because the spectral cues for $a_{1'}$ are referred to the absolute delay of a_1 in the course of being expressed along the time axis in the image. If a_1 and $a_{1'}$ are closer together than the receiver's integration-time (Fig. 7.4), there is only a single, spectrally-complex echo as far as the receiver is concerned, and it is not surprising that the two image components trade as a unit.

For long time separations of 400 μs and more, the spectrograms of a_1 and $a_{1'}$ pull apart and appear as separate FM sweeps (Fig. 7.5). Each echo evokes its own set of neural discharges (Fig. 10 in [24]), and $a_{1'}$ should undergo amplitude-latency trading on its own. Fig. 7.6F-G shows the results of a jitter experiment with a relatively long echo-delay separation of 500 μs. Now, the perceived delay of $a_{1'}$ indeed does trade with amplitude by about -15 μs/dB independent of a_1 when the delay separation is as long as 500 μs. A separate volley of neural discharges with its own amplitude-latency trading capacity must represent the delay of the second echo component.

For intermediate time separations of 200 to 400 μs (Fig. 7.4), both spectral notches and partially separated FM sweeps are visible in the spectrograms (Fig. 7.5). The spectral notches should contribute an estimate for the delay of $a_{1'}$ relative to a_1, and so should neural discharges evoked by the partially distinct sweeps in the waveform of $a_{1'}$ (Fig. 10 in [24]). However, only that contribution to the image of $a_{1'}$ originating in discharges evoked by $a_{1'}$ itself should undergo amplitude-latency trading when the amplitude of $a_{1'}$ is changed. The spectral notches caused by $a_{1'}$ would be referred to the delay of a_1 and should only trade when a_1, not $a_{1'}$, is changed in amplitude. Thus, changing the amplitude of echo $a_{1'}$ should split its image into two parts-one part locked to a_1, and another part shifting in perceived delay according to amplitude-latency trading for $a_{1'}$. Fig. 7.6C-E shows the results of a critical experiment using the jitter technique to determine whether such image-splitting occurs. The image of $a_{1'}$ indeed does dissociate into two parts for an intermediate time separation of 300 μs. One part trades at -15 μs/dB and presumably represents direct determination of the delay of $a_{1'}$ from its own neural

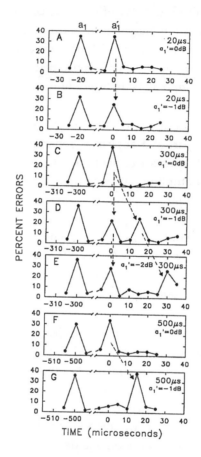

Bat Sonar

Figure 7.6: Dissociation of image into time and spectral components: Representative performance of *Eptesicus* as function of jitter interval in two-glint jitter discrimination experiment with short (20 μs; A-B), intermediate (300 μs; C-E), and long (500 μs; F-G) separations of a_1 and $a_{1'}$ in relation to 300-350 μs integration-time for echo reception. Change in amplitude of $a_{1'}$ does not shift latency of the image for short separation (B), entirely shifts image for long separation (G), and splits image for intermediate separation (D-E). Splitting of $a_{1'}$ image demonstrates independence of parallel temporal and spectral representations of $a_{1'}$, with spectral representation attached to image of a_1 for latency trading.

discharges, while the other part remains associated with a_1 (it trades with changes in a_1 amplitude) and presumably represents an indirect estimate of delay derived from the spectrum of $a_1 + a_{1'}$.

The integration-time of 300-350 μs defines a perceptual boundary between the use of auditory temporal and spectral codes for representing the separation of a_1 and $a_{1'}$. The predicted splitting of the error peak for $a_{1'}$ at time separations near this boundary (Fig. 7.6C-E) demonstrates the existence of the corresponding separate temporal and spectral paths to the eventual image of $a_{1'}$. What is the nature of the operations that take place in each signal-processing path?

Time-frequency Transforms and Image Formation

Eptesicus can distinguish between two targets that reflect echoes having interference notches at different frequencies [13, 25, 22]. However, Figs. 7.2B-D and 6 reveal that the bat perceives the shape of the echo spectrum in terms of an estimate of the delay separation of echo components a_1 and $a_{1'}$ required to create the observed pattern of peaks and notches across frequencies. It does not merely perceive the fact that the echo spectra are different. The physiological reality of separate delay and spectral codes is well-demonstrated, but the bat evidently merges the information carried by each representation into a unitary image having a single psychological scale-that of range. The image of $a_{1'}$ in Fig. 7.2B-D expresses wholly in the time domain information that initially can only have been encoded across frequencies from the interference spectrum of a_1 and $a_{1'}$ mixed together. What is the nature of the process that converts the echo spectrum into a delay estimate?

The time and frequency domains are reciprocally related through transforms [2]. A transform converts a function from one metric domain to another. For example, Fourier and related transforms make time-and frequency-domain descriptions of signals interchangeable. The practical significance of transforms for applied mathematics in general, and for neuroscience in particular, is that a transform of a signal may prove to be more amenable to numerical representation and fast evaluation than the original signal. That is, displays or descriptions of signals for different purposes often can be obtained more conveniently by first transforming the data into a domain where the mathematical function to be evaluated is simpler than the function that would have to be used in the original domain. For the brain, "simpler" may mean "more readily encoded" by large numbers of parallel neurons whose interconnections do the work of the required evaluation of functions. After calculations are done, the result then can be transformed back into the original domain of the signal itself if those dimensions are intrinsically more desirable than the dimensions of the transform.

In sonar, target range is naturally associated with echo delay, and, although aspects of delay computation could well take place in the frequency domain, it seems intuitively more convenient to eventually express the result in entirely in units of time rather than wholly or partially in units of frequency. Then the target's image really would have a dimension characteristic of objects-distance-rather than dimensions characteristic of of sounds-frequency and time. Furthermore, a single point in range is a single delay estimate across all frequencies. This same delay estimate expressed in the frequency domain involves the numerical values of amplitudes at all frequencies being encoded in the transform to specify the placement of spectral peaks and notches that are equivalent to the single echo delay.

Suppose that the desired goal of spatial imaging is to show each discrete element of an object or scene at its corresponding location in space. Elements of objects are single points in space, whereas transforms typically spread the magnitude of each point in a signal across many or all points in the emergent domain. Disentangling each element of an object from a display of all the points in the transform seems cumbersome unless the transform is operated in reverse to get back to the more natural-and spatial-domain. To recover the desired uniqueness of parts of objects occupying individual points in space, the display ultimately ought to be in spatial units. The bat's brain manifestly achieves virtually real-time display of sonar targets using large numbers of parallel neural computational elements. The creation of an auditory frequency dimension as part of the representation of emissions and echoes must surely render the imaging computations more tractable for neuron-like elements, but the bat evidently goes to the trouble of transforming information from frequency back to time coordinates when it assembles the images it perceives. The resulting images have the natural dimension of range. It is in this context that the bat's sonar addresses a general problem in perception-conversion of a distributed, multidimensional neural code initially created for the convenience of the brain back into images that correspond to real space.

Sonar Images as the Product of Transforms

Figure 7.2E-H shows the crosscorrelation functions for the echoes delivered to the bat (a_1 and $a_{1'}$) during the jitter discrimination experiments described above. The time separation of the two echo components (0 to 30 μs) is directly readable from the crosscorrelation functions themselves, using the locations of the two major peaks in the function as estimates of the delay of the two echo components. These functions display all of the information that can be extracted from the echoes about the distance to the targets, including the range separation of the two reflecting elements responsible for the separate echo components. The crosscorrelation function is a time-domain transform of the transmitted signal and the echoes taken as a unit. The same information can be displayed in the frequency domain, however, as the cross-spectrum of the echoes with the transmission, which is an alternative transform of the emitted sound and its echoes. Not only are the crosscorrelation function and cross-spectrum capable of being obtained from the waveform of the emitted sound and the echoes, but they can be obtained from each other.

The curves summarizing the performance of *Eptesicus* in Fig. 7.2A-D resemble the corresponding crosscorrelation functions in Fig. 7.2E-H, suggesting that the bat perceives the target as a time-domain transform of the echoes in relation to the transmission. This is true for a single-point target (Fig. 7.2A) and for two-point targets with echo-delay separations up to 30 μs (Fig. 7.2B-D). The image of the single-point target (a_1) is obtained from the timing of neural discharges to the echo in relation to the discharges to the transmission. This can be demonstrated by changing the amplitude of the echo and determining that the bat perceives the echo delay to shift according to the amplitude-latency trading ratio measured for neural responses in *Eptesicus* (Fig. 7.3). Evidently the integration of delay estimates across a population of delay-tuned neurons in the bat's brain can achieve the equivalent of crosscorrelation. Here, the role of the frequency domain is to parcel out the frequencies in the FM sweeps of the sonar transmission and echoes so that a series of delay estimates can be made in parallel across different frequencies. The neural computations required to produce the image of a_1 take place on the timing of neural discharges within each frequency channel. However, the image of the second part of the two-point target ($a_{1'}$) cannot come from this same integrative process because separate neural discharges do not occur to mark the arrival-time of the second echo component.

The constraint imposed by the integration-time of the bat's ear (300 to 350 μs; Fig. 7.4) ensures that the bat's estimate of the delay of $a_{1'}$ can only come from its auditory representation of the interference spectrum of the two echo components mixed together-that is, from the frequencies of spectral peaks and notches. The neural representation of this interference spectrum is displayed across the tonotopic axis of auditory centers in the brain, not by delay-tuned neurons but by frequency-tuned neurons. Somehow the bat converts the pattern of neural responses across the tonotopic axis into a new estimate of delay for $a_{1'}$ that is then associated with the delay already estimated for a_1 to produce an image of the entire complex target along the dimension of delay or range. The exact nature of the frequency scale of this tonotopic axis must be a crucial element in the transformation of the spectral representation into an estimate of delay separation. In *Eptesicus* , this tonotopic scale is approximately hyperbolic with frequency, or linear with time [23]. It is thus possible that the anatomical location of the neural activity encoding spectral peaks and notches serves the bat as a neural map of the delay separation of the echo components.

In Figure 7.2B-D, the image of a two-part target as a whole resembles the sum of separate crosscorrelation functions for each of the two parts of the complex echo. As the time separation of a_1 and $a_{1'}$ changes from 0 to 30 μs, the separate crosscorrelation functions for a_1 and $a_{1'}$ can be seen to interfere by reinforcing and cancelling one another. This interference between functions is a critical observation: For the images of a_1 and $a_{1'}$ to interfere with each other at different delay separations, they must be expressed in the same metric domain. That is, the numbers that make up these demonstrably separate physiological representations nevertheless must interact arithmetically with each other at the level of perception. This is a very strong statement about the nature of perceived images-that there is a kind of calculus which regulates the interactions between parts of images that emerge from very different representations. This calculus specifically affects the binding of different stimulus features, represented in different ways by the brain, into whole images of objects. What kind of mechanism is the substrate for this calculus? Is it sufficient that the two representations-of the delay and the spectrum of echoes-coexist at the same moment in time for the resulting perception to merge their contents into a single psychological dimension? Or, does a higher-level neural network detect the simultaneity of the delay and spectral representations and create new delay-tuned neurons to encode the delay of the second echo component relative to the first? Does the hyperbolic tonotopic scale in *Eptesicus* make delay-tuned neurons for $a_{1'}$ by default, without the need for a higher-level process at all?

The bat's sonar demonstrates that transforms play a large role in forming images, and that they are not limited to the peripheral coding of stimuli but are actively carried out at higher levels of the brain as well. The example of feature-binding for target range and shape in the case of echolocation by *Eptesicus* is unusually clear-cut in the sense that the binding process can be identified mathematically as a frequency-to-time transform (for $a_{1'}$) operating in tandem with a time-to-time transform (for a_1). Ongoing physiological experiments in this species of bat offer the hope of identifying specific neural circuits that perform the binding operation.

Acknowledgments

The research summarized in this article been supported by the Office of Naval Research (Grant No. N00014-89-J-3055), by the National Institute of Mental Health (Grant No. MH-00521), by the System Development Foundation (Grant No. 57), and by prior grants from the National Science Foundation.

References

[1] R. A. Altes (1984). Texture analysis with spectrograms. *IEEE Trans. Sonics Ultrasonics*, **SU-31**, 407-417.

[2] R. N. Bracewell (1990). Numerical Transforms. *Science*, **248**, 697-704.

[3] K. J. Beuter (1980). A new concept of echo evaluation in the auditory system of bats. In *Animal Sonar Systems*, R.-G. Busnel and J. F. Fish. eds.,pp. 747-761 (Plenum, New York).

[4] R.-G. Busnel, and J. F. Fish (1980). *Animal Sonar Systems* (Plenum, New York).

[5] S.P. Dear, J. Fritz, T. Haresign, M. Ferragamo, and J. A. Simmons (1992). Tonotopic and functional organization in the auditory cortex of the big brown bat, *Eptesicus fucus*, *J. Neurophysiol.* in press.

[6] A. S Feng, J. A. Simmons, and S. A. Kick, (1978). Echo detection and target-ranging neurons in the auditory system of the bat Eptesicus fuscus. *Science*, **202**, 645-648.

[7] D. R. Griffin (1958). *Listening in the Dark* (Yale University Press, New Haven).

[8] D. R. Griffin, J.H. Friend, and F.A. Webster (1965). Target discrimination by the echolocation of bats. *J. Exp. Zool.*, **158**, 155-168.

[9] J. Habersetzer, and B. Vogler (1983). Discrimination of surface-structured targets by the echolocating bat *Myotis myotis* during flight. *J. Comp. Physiol.*, **152**, 275-282.

[10] S. A. Kick (1982). Target detection by the echolocating bat, *Eptesicus fuscus*, *J. Comp. Physiol.*, **145**, 431-435.

[11] A. Kurta, and R. H. Baker (1990). *Eptesicus fuscus. Mammalian Species*, **356**, 1-10.

[12] D. Menne, I. Kaipf, I. Wagner, J. Ostwald, and H.-U. Schnitzler (1989). Range estimation by echolocation in the bat Eptesicus fuscus: Trading of phase versus time cues, *J. Acoust. Soc. Am.*, **85**, 2642-2650.

[13] J. Mogdan and H.U. Schnitzler (1990). Range resolution and the possible use of spectral information in the echolocating bat, *Eptesicus fuscus*, *J. Acoust. Soc. Am.*, **88**, 754-757.

[14] C. F. Moss and H. U. Schnitzler (1989). Accuracy of target ranging in echolocating bats: Acoustic information processing, *J. Comp. Physiol. A*, **165**, 383-393.

[15] P. Nachtigall and P. W. B. Moore (1988). *Animal Sonar: Processes and Performance* (Plenum, New York).

[16] G. Neuweiler (1990). Auditory adaptations for prey capture in echolocating bats. *Physiol. Rev.*, **70**, 615-641.

[17] W. E. O'Neill and N. Suga, (1982). Neural encoding of target range and its representation in the auditory cortex of the mustached bat. *J. Neurosci.*, **47**, 225-255.

[18] G. D. Pollak and J. H. Casseday (1989). *The Neural Basis of Echolocation in Bats* (Springer, New York).

[19] S. Schmidt (1988). Evidence for a spectral basis of texture perception in bat sonar, *Nature*, **331,** 617-619.

[20] J. A. Simmons (1973). The resolution of target range by echolocating bats, *J. Acoust. Soc. Am.*, **54,** 157-172.

[21] J. A. Simmons (1979). Perception of echo phase information in bat sonar, *Science*, **204,** 1336-1338.

[22] J. A. Simmons (1989). A view of the world through the bat's ear: The formation of acoustic images in echolocation, *Cognition*, **33,** 155-199.

[23] J. A. Simmons, M. Ferragamo, C. F. Moss, S. B. Stevenson, and R. A. Altes (1990). Discrimination of jittered sonar echoes by the echolocating bat, Eptesicus fuscus: The shape of target images in echolocation, *J. Comp. Physiol. A*, **167,** 589-616.

[24] J. A. Simmons, E. G. Freedman, S. B. Stevenson, L. Chen and T. J. Wohlgenant, T.J. (1989). Clutter interference and the integration time of echoes in the echolocating bat, *Eptesicus fuscus*, *J. Acoust. Soc. Am.*, **86,** 1318-1332.

[25] J. A. Simmons, W. A. Lavender, B. A. Lavender, D. A. Doroshow, S.W. Kiefer, R. Livingston, A.C. Scallet and D. E. Crowley (1974). Target structure and echo spectral discrimination by echolocating bats, *Science*, **186,** 1130-1132.

[26] J. A. Simmons, C. F. Moss and M. Ferragamo (1990). Convergence of temporal and spectral information into acoustic images of complex sonar targets perceived by the echolocating bat, *Eptesicus fuscus*, *J. Comp. Physiol. A*, **166,** 449-470.

[27] N. Suga, (1988). Auditory neuroethology and speech processing: Complex-sound processing by combination-sensitive neurons. In *Auditory function*, G.M. Edelman, W.E. Gall and W.M. Cowan eds., pp. 679-720 (Wiley, New York).

[28] N. Suga and J. Horikawa (1986). Multiple time axes for representation of echo delays in the auditory cortex of the mustached bat, *J. Neurophysiol.*, **55**, 776-806.

[29] N. Suga and W. E. O'Neill (1979). Neural axis representing target range in the auditory cortex of the mustached bat, *Science*, **206**, 351-353.

[30] W. E. Sullivan (1982). Neural representation of target distance in auditory cortex of the echolocating bat *Myotis lucifigus, J. Neurophysiol.*, **48**, 1011-1032.

[31] F. A. Webster and O.G. Brazier (1965). Experimental studies on target detection, evaluation and interception by echolocating bats. T.D.R. AMRL-TR-65-172, Aerospace Medical Division, USAF Systems Command.

[32] F. A. Webster and O.G. Brazier (1968). Experimental studies on echolocation mechanisms in bats. T.D.R. AMRL-TR-67-192, Aerospace Medical Division, USAF Systems Command.

[33] F. A. Webster and N. I. Durlach (1963). Echolocation systems of the bat. Report No. 41-G-3, MIT Lincoln Lab., Lexington, MA.

[34] D. Wong and S. L. Shannon (1988). Functional zones in the auditory cortex of the echolocating bat *Myotis lucifigus, Brain Res.*, **453**, 349-352.

Optimal Signal Processing in the Nervous System

William Bialek

A Physicist's View of the Sensory Systems

An organism uses its senses to gather information about the environment. This information - the presence and trajectories of mates and predators, the location of food, the obstacles to movement, ... - is carried to the organism by a variety of physical signals. The task of the sensory system is to detect and estimate these signals, ultimately extracting information of interest to the organism. The theme of these lectures is that the sensory systems must be an understandable part of the brain because we, as physicists, solve the same sorts of problems when we do experiments. A toad using his eyes to hunt bugs in the dead of night is not so different from an astrophysicist tracking a dim object across the sky.

We begin with the detector elements themselves, the sensory receptor cells. The ability of the organism to extract information about the environment is limited in a most fundamental way by the noise performance of these cells. There are physical sources of noise which are irreducible in detectors of given dimensions --- the random arrival of photons, Brownian motion in mechanical sensors, diffusive fluctuations in chemical sensors, and so on. It is a remarkable fact that nature has constructed exquisite detectors which are limited by these fundamental physical considerations. I have always been especially impressed by the thermosensors of certain cave beetles that live in France. Single cells in the antennae of these creatures give reliable responses when the temperature of the surrounding air drifts by less than 10^{-3} K over several seconds, and if one estimates the thermodynamic temperature fluctuations in the detector elements this performance is essentially equal to the physical limit. Reference [17] provides a review of the evidence for the approach to fundamental noise limits in sensory receptor cells, including the case of the cave beetle. For me much of the motivation for thinking about the neural processing of sensory signals is to understand how the brain makes use of these wonderful devices.

Given receptor cells with certain noise characteristics, we need to arrange these cells so as to provide the most effective sampling of the environment. As an example, in building a vision system we must decide how to divide the visible spectrum among our receptor cells, how to distribute the optical axes of the detectors (should we sample homogeneously, or build a "fovea"?), whether the detectors should be fixed to the body or mobile, whether to explicitly gather depth information through the use of multiple views, ... Again there are fundamental physical considerations which enter the discussion, such as the effect of diffraction in setting the photoreceptor aperture, sum rules which constrain the receptor spectral sensitivities, and noise limitations on temporal sampling with moving detectors. This is one area where I think most people agree that physically motivated "design principles" are relevant to real organisms, even if the comparison between theory and experiment is still fuzzy.

Many organisms need to work in conditions where the signal-to-noise ratios are especially poor. In several cases the receptor cell is followed by a well-defined stage of analog pre-processing, presumably to help "clean up" the signals before transmission to higher processing stages. The general problem of signal processing at low signal-to-noise can be given a formulation which borrows heavily from our understanding of functional integrals in quantum mechanics and statistical mechanics, and this approach leads to successful model-independent predictions for the structure of the pre-processor.

Animals receive information from the world in the form of continuous functions of time --- sound pressure at the eardrum, light intensity at each point in the visual field, the concentration of various substances in the air, ... At a very early stage in processing, however, these continuous signals are converted into discrete sequences of identical pulses, called action potentials or spikes. In each sensory neuron the spikes are all identical; information is carried only by the arrival times. The dynamics by which spikes are generated and propagated over relatively long distances from cell to cell are understood in molecular detail [48, 91]. From a physical point of view, however, this stage of processing constitutes an *encoding* of the incoming analog signals, and it is natural to ask whether there are any principles which govern the structure of this code. Is the code is structured so that the spike train provides the maximum possible information about the sensory world, or so that this information is represented in a maximally efficient manner? We will see that a new theoretical approach to the coding problem motivates a series of experiments which give clear answers to these questions. These results in turn focus attention on a new set of theoretical questions which we have only just begun to attack.

Once signals have been sent to the central processor, the problem of signal processing becomes more specialized. In favorable cases one can identify the computations of interest to particular creatures, and in the simpler creatures it is even possible to point to specific cells which carry out these computations. For all of these computations there is a limit to reliability imposed by noise in the sensory receptors themselves, and we shall see that in several instances the performance of the organism approaches these fundamental limits. To understand how this is possible we need a theory of "optimal estimation," in which one tries to compute some non-trivial functional making maximal use of the noisy input data. This theory can be cast into a form familiar once again from statistical mechanics, and we will then use standard methods to extract the mathematical structure of the optimal estimator and compare it with the structure and dynamics of real neural circuits.

These lectures are essentially a guided tour through the sensory systems of different organisms, with stops chosen to illustrate the issues summarized above. I hope that you can discern the outlines of a unified physical perspective on the structure and function of these systems. The ideas are drawn from work done in collaboration with L. Kruglyak, M. Landolfa, E. R. Lewis, J. P. Miller, K. Moortgat, W. G. Owen, F. Rieke, D. Ruderman, R. R. de Ruyter van Steveninck, D. Warland, W. Yamada, and A. Zee. With luck, some of the spirit of fun which has driven this work comes through in the written description. I have also tried to follow the more informal colloquial style of the original lectures, rather than yielding to the usual strictures of scientific publications.

Receptor Sampling

Given a collection of receptor cells, our first problem in designing a sensory system is to arrange these cells in such a way as to provide an effective sampling of the world. To make our design problem well-defined, we take some properties of the receptor as given. In most cases it makes sense to assume that the temporal impulse response and output noise characteristics of the cell are fixed. Other characteristics such as the spatial aperture of a photoreceptor can be modified by accessory structures.

Compound eye design

The best studied example of the sampling problem is provided by the insect compound eye, which is even discussed in Feynman's undergraduate lectures [37]; for a general review of optics in the visual system see Ref. [52]. Feynman considers a bee's eye of radius R, in which individual facets of the compound eye have a size d; the bee thus divides the world into pixels of angular dimension $\delta\phi \sim d/R$ radians. By making the facets smaller one can improve the angular resolution, but this is only true until geometrical optics breaks down and diffraction takes over. Then if we have a physical aperture of linear dimension d, the effective aperture is of angular subtense $\delta\phi \sim \lambda/d$, where λ is a typical wavelength of the light being used to form the image. To optimize the angular resolution one must seek a compromise between these different regimes, which predicts that $d \sim \sqrt{\lambda R}$. Feynman notes that this is prediction is correct within ten percent for the particular species of bee he was looking at.

Feynman's discussion has antecedents which can traced back into the last century [60]. By 1952 Barlow had surveyed some twenty-seven different insect eyes, finding that the square-root relation between facet size and eye size holds, to a good approximation, over a factor of 25 variation in R [5]; see also [51]. On the other hand, there are many insects which apparently undersample the world by as much as a factor of three, and compound eyes of creatures dwelling in murky waters can have facets as much as 100 times as large as expected from the diffraction limit. Perhaps physicists are just too simple-minded, and biology is governed by the historical accidents of evolution rather than by any clean theoretical principles. A more intriguing possibility is that the optimization of angular resolution fails to capture the essential physics of the problem.

When it's dark an insect which divides the world into tiny pixels suffers the problem that each pixel captures only a few photons. If the aperture of each photoreceptor were increased the angular resolution would be degraded but the intensity resolution within each pixel would be improved. Similarly, if the receptor cells have limited time resolution and the insect flies at high speed, large numbers of tiny pixels would carry redundant information because of motion blur. These observations suggest that, at least under some conditions, maximizing the angular resolution would be a mistake. In 1977 Snyder, Stavenga, and Laughlin [98, 99] suggested that the tradeoff between angular resolution and signal-to-noise could be quantified in terms of an information theoretic optimization principle, specifically that the eye should be designed to have the largest possible information capacity for representing images. This approach resulted in formal versions of the intuitive (and correct!) prediction that insects flying slowly in bright light should reach the diffraction limit, while quick night-flyers should open up their apertures. In the application of these results to the well-studied case of *Musca domestica*, the relatively high angular velocities of the bug play a crucial role in the correct prediction of undersampling.

I think one should view these results as semi-quantitative. Fully quantitative predictions require a more careful treatment of time-resolution in the receptor and careful consideration of the statistics of natural images. One would also like to think about the continuous transmission of information in time as the scene is varying. Nonetheless the work of Snyder *et al.* is crucial, for it teaches us that the diversity of functions found in different organisms can emerge from a single, physically motivated theoretical principle.

Information capacity and eye movements

We can understand the basic tradeoffs in the work of Snyder *et al.* by considering a related problem. Our visual system gives us a precise image of a small (roughly 2 square degrees) region - the fovea - and a much coarser sampling of the surround. We move our eyes so that this high resolution region scans the scene. If we scan slowly, then at each location we can use a long integration time to build up a high signal-to-noise ratio. But slow scanning means that we do not visit many independent regions of the scene per unit time. What's the optimal strategy?

Let us try a crude model. Suppose that every τ seconds we can look at a completely independent piece of the visual world. By integrating for τ seconds we achieve some signal-to-noise ratio $SNR(\tau)$. If the signals and noise are both Gaussian, then from Shannon [95] we know that we gain an amount of information $I = \log_2[1 + SNR(\tau)]$ bits about each piece of the world; for an introduction to information theory see Atick's contribution to these proceedings. Since we see a new, independent part of the scene each τ seconds, the rate at which we gain information

$$R_{info} = \frac{1}{\tau} \log_2[1 + SNR(\tau)]. \qquad (8.1)$$

If have a signal $s(t)$ and a spectral density of noise given by $N(\Omega)$, the signal-to-noise ratio is given by

$$SNR = \int \frac{d\Omega}{2\pi} \frac{1}{N(\Omega)} \left| \int dt\, e^{+i\Omega t} s(t) \right|^2. \qquad (8.2)$$

If $s(t)$ is non-zero only for some *short* time τ then it easy to show that $SNR(\tau \to 0) \propto \tau^2$. At long times --- long compared to the correlation time of the noise --- the signal-to-noise ratio will grow in proportion to the integration time, except in pathological cases like $1/f$ noise. Thus SNR grows quadratically at short times, slowing to linear growth at long times. Unless there is something very strange about the spectral density of the noise, this slowing will be monotonic, and we can say that

$$2 < \frac{d \ln SNR(\tau)}{d \ln \tau} < 1. \qquad (8.3)$$

The condition for maximizing R_{info} can be rewritten as

$$\ln[1 + SNR(\tau_*)] = \frac{SNR(\tau_*)}{1 + SNR(\tau_*)} \frac{d \ln SNR(\tau)}{d \ln \tau}\bigg|_{\tau = \tau_*}, \qquad (8.4)$$

where τ_* is the optimum sampling time. Then from Eq. (8.3) we have $\ln[1 + SNR(\tau_*)] < 2$, or $SNR(\tau_*) < 6.4$.

This may seem counterintuitive: To gain the most information about the world one should operate at relatively low signal-to-noise ratios! A more optimistic view is that the possibility of gaining truly novel information about a different part of the world wins out over the accumulation of detail in any small region. How does this crude calculation compare with experiment? Under normal conditions we move our eyes roughly three times per second, and the movements themselves occupy ~ 40 msec. When confronted with a novel image we tend to hold our eyes fixed for no more than ~ 150 msec, then flick to a new feature; this time scale defines the boundary of "pre-attentive" vision. From preliminary data on the noise characteristics of the receptor cells in the primate fovea [92] one can make rough estimates of $SNR \sim 4$ if one adopts a $\sim 100 - 250$ msec integration time in moderate daylight. Certainly we are in the right ballpark, though much remains to be done.

The cochlear compromise

The auditory system also faces an interesting sampling problem. Given a collection of transducer cells --- the hair cells --- the ear divides the range of interesting frequencies among these cells, in some cases using the mechanical tuning properties of accessory structures to insure that each cell receives a limited frequency band. In mammals, for example, this tuning is evident in the basilar membrane, which supports a "travelling wave" along its length; tuning is achieved by the fact that the local mechanical properties of the membrane vary with distance along the membrane [116]. This picture of a propagating wave in a medium with spatially varying properties leads to an interesting design problem first pointed out by Zweig et al. [117]: If the mechanical properties vary too slowly, the system cannot focus different frequency bands into different hair cells; if mechanical properties vary too rapidly one will build a mirror for the travelling wave. The natural compromise is to operate almost at the threshold for wave reflection, which is mathematically the point at which the WKBJ approximation begins to break down. Analysis of experiments on basilar membrane motion suggest that the mammalian ear indeed operates near this point. On the other hand, de Boer and MacKay have suggested that one can build completely reflectionless systems which achieve arbitrarily precise focusing, so there is no need for compromise [23]. There are, however, subtle mathematical points in comparing these different points of view on the generation of reflections. I think there is an important open problem here.

Other sampling problems

There are a number of other interesting problems in the sampling of the sensory world. Many sensory systems are constructed in layers, and at each layer there is some sort of "map" of the sensory world [50]. In the visual cortex of mammals, for example, the mapped parameters include not only location in the visual field (as on the retina itself) but also ocular dominance (right vs. left eye), local spatial frequency and orientation. In the barn owl auditory system one finds that after several stages of processing there is a layer of cells which form a map of sound source location, which is computed from the differences in intensity and phase for sounds arriving at the two ears. These maps constitute "resamplings" of the world, and several authors have considered the possibility that these sampling strategies are in some sense optimal, maximizing efficiency [90], smoothness [34], or information transmission [106], or alternatively minimizing the cost of connecting regions of the map which contain related data [68]. Most of these ideas are pointed at understanding the key features of the maps in cat and monkey visual cortex, and of course these are very complex systems in which to attempt a direct comparison of theory and experiment. The case of the cricket cercal mechanoreceptor --- which forms a map of wind direction using a handful of neurons --- thus may provide an important test case [106].

Another sampling problem is color vision: Why have color vision at all, and why just three types of receptors? Clearly the ability to see colors adds a richness to the world which is lost to animals which see only in grey scales. There is information contained in the spatial variation of reflectance spectra across a scene, but neighboring photoreceptors look in different directions, so that in a color vision system some combinations of location and spectrum will be missed. In addition there are physical constraints (sum rules, for example) on the possible forms of the photopigment absorption spectra. This problem remains to be given a fully quantitative formulation.

Some bits and pieces should be noted. First, Barlow [9] pointed out that because the pigment absorption spectra are broad, the visual system smoothes the reflectance spectra of objects in the scene, perhaps to the point where three samples are sufficient to reconstruct the spectra by the usual Shannon-Nyquist arguments [26]. Maloney [61] studied the statistics of real reflectance spectra, and found that they are quite smooth even without filtering through the pigments. Recent experiments seem to get more directly at the "frequency" response of color vision in humans [24], but it is not clear that these different approaches are converging.

One can make progress by looking at more specialized animals, for example the bees (and their relatives) which use color vision to discriminate among different flowers. Chittka and Menzel have sampled the spectral reflectances of a large number of flower species, and asked where the absorption maxima of three receptor pigments should be located to optimize discriminability among the different species, or alternatively to maximize the variance at the output of each receptor type [32]. The results are quite impressive, with the observed absorption maxima of 40 species of *Hymenoptera* forming tight clusters around the predicted optima. This is probably an example of co-evolution toward the optimum, since the flowers also "have an interest" in being uniquely identifiable by their pollinators. Finally one should remember that not all animals are trichromatic or achromatic. There seem to be shrimps with ten different receptors, at least along one stripe of the compound eye [62]!

Analog Pre-processing

Receptor cells are inevitably noisy. In vision, noise must be a serious problem for any organism which is active in the darkest times of night. Here we explore the evidence that organisms can perform near optimal processing of visual signals at low photon flux, and discuss the general strategies for signal processing in this low SNR limit. It turns out that no matter what feature of the world the organism might want to estimate, the first step in processing the receptor outputs is always the same. This universal first stage of "matched filtering" is one example of analog pre-processing, which seems to occur in almost all visual systems from insects to primates.

Optimal processing of single photon signals

One of the great and classic pieces of work in biophysics concerns the ability of the visual system to count single photons. In the 19th century a number of physicists and physiologists measured the minimum energy required for the visibility of a brief flash in a dark room. Apparently [25] in 1916 H. A. Lorentz discussed these experiments with his biological colleague Zwaadermaker and calculated that one could see roughly one hundred photons incident on the cornea. There is considerable scattering and loss within the eye itself, so that at most a few tens of photons can be expected to reach the retina. Lorentz wondered if the real limit to vision might be set by the quantum nature of light itself.

One would like a more dramatic signature of single photon counting, and the basic idea seems to have occurred to two groups at the same time, around 1940. Suppose that we are looking at dim light flashes from a conventional light source. The intensity I of the flash determines the mean number of photons $\langle n \rangle = \alpha I$, while the actual number of photons n in a single flash is a random variable with a Poisson distribution, $P(n) = e^{-\langle n \rangle} \langle n \rangle^n / n!$. If we say "I saw it" when at least K photons arrive then the probability of seeing is

$$P_{\text{see}}(I) = \exp(-\alpha I) \sum_{n=K}^{\infty} \frac{(\alpha I)^n}{n!}. \tag{8.5}$$

There are two key points. First, the response of humans to dim light flashes is predicted to be probabilistic --- there is a *probability* of seeing --- but this reflects the stochastic arrival of photons at the retina rather than some biological variability. Second, the function $P_{\text{see}}(I)$ is diagnostic of the threshold photon count K. If we look at P_{see} *vs.* $\log I$ the unknown parameter α corresponds to a simple translation, while K determines the shape of the curve.

Hecht, Shlaer and Pirenne [45] measured $P_{\text{see}}(I)$ and found excellent fits to Eq. (8.5) with $K = 5 - 7$. It would thus seem that humans can "see" as few as five photons, and under the conditions of these experiments these photons are distributed over many photoreceptors. Single photon arrivals at individual photoreceptors must thus generate signals which contribute to the detection process. At about the same time deVries [107] realized that the random arrival of photons would set a limit to intensity discrimination and other supra-threshold tasks, and soon Rose [83] suggested that various aspects of human vision could be measured on an absolute scale and compared with artificial devices.

In an independent series of experiments van der Velden also measured $P_{\text{see}}(I)$ and found that Eq. (8.5) was a good fit with $K = 2$ [108]. It appears that deVries played a key role in helping van der Velden to analyze these data; I am grateful to J. Kuiper for telling me about the history of this period. The fit to $K = 2$ led (briefly) to much speculation about "coincidence detection" and related mechanisms. Barlow resolved the conflict between the two P_{see} measurements by proposing that there exists a level of noise in the visual system, so that even in the dark there is some probability of registering a non-zero photocount at the output of the receptor [7]. In this case an observer could avoid falsely reporting that he "sees" in the dark by choosing a high threshold K; on the other hand this means that some real flashes will be missed. Conversely one could be unconcerned about false positive responses and then operate at a low value of K. Indeed the two measurements of P_{see} were taken on observers with very different false positive rates; when observers are specifically asked to adopt different false positive rates the threshold photocount K varies as expected from Barlow's dark noise hypothesis [104, 105]. These arguments are part of a more general realization that most apparent "thresholds" for sensation or perception represent adjustable criteria which are set to achieve a desired reliability in the presence of noise [43, 55].

In an absolutely remarkable experiment Sakitt [89] generalized the P_{see} measurement by asking observers to rate their perception of the intensity of dim flashes using integers from 0 to 6. The mean rating varied linearly with the flash intensity, but more importantly the probability of giving a rating larger than K obeyed an equation of the same form as (8.5) for each K, with all seven curves being fit by the same α. The observer is behaving exactly as expected if the rating number is the number of photons counted!

To finish out the story of photon counting, the first recordings of single-photon responses from photoreceptor cells were reported in experiments on the horseshoe crab Limulus [40]. It is much more difficult to characterize single-photon responses in the vertebrate retina because the photoreceptors are electrically coupled; each cell produces a reliable photocurrent in response to single photons, but the resulting voltage response is distributed among dozens of cells. In the late 1970s Baylor, Lamb and Yau developed a technique in which they draw individual receptor cells into a micropipette so that all the current flowing across the cell membrane flows up the interior of the pipette. The key is to make a very tight seal between the cell membrane and the walls of the pipette, a problem which had been solved by Neher and Sakmann in their development of the "patch clamp" method [91]. In 1979-80 Baylor and co-workers reported detailed results on the photocurrents and noise in toad rod cells [14, 15]. Single photons trigger a highly reproducible photocurrent pulse of roughly one picoamp in magnitude and one second in duration. Similar experiments have now been done in a variety of species, including the macaque monkeys whose visual systems are so close to our own [16]. Nearly seventy years after Lorentz' original questions to his biologist friend, these experiments provided clean and direct evidence of single photon sensitivity.

Careful analysis of Sakitt's data yields an accurate estimate of the dark noise, that is of the rate of (subjective) photon counting in the dark. Recordings from single rod cells reveal spontaneous photon-like events which almost certainly reflect the spontaneous isomerization of the visual pigment rhodopsin. If we convert Sakitt's estimate into a rate per rod cell, this rate is in excellent agreement with the observed spontaneous event rate in monkey rods. This strongly suggests that the reliability of night vision is set by noise in the photoreceptor array itself, not by noise or inefficiencies in the subsequent neural processing.

The comparison of psychophysical and receptor noise levels has been beautifully confirmed in measurements on the behavioral and neural responses of a frog, where it possible to vary the temperature over a range corresponding to more than order of magnitude in receptor cell dark noise levels. The experiments of Aho *et al.* [2] demonstrate that the "output" (behavior or retinal ganglion cell) noise level follows the temperature dependence measured for the "input" (receptor cell) noise; if one extrapolates to 37°C the data match those for monkeys and humans. Taken together these observations strongly suggest that under dark-adapted conditions --- where the signal-to-noise ratio for vision is by definition low --- organisms can make optimal use of the photoreceptor signals, extracting all the information relevant to their behavioral tasks and adding little if any noise in the process [12, 18, 33]. What is the mathematical structure an optimal processor?

Universal pre-processing at low photon flux

A collection of photoreceptors receives photons at a time-varying rate $R(t)$. This is the rate of a random Poisson process, with the photons themselves arriving at times t_μ. These arrivals are signaled by current pulses of stereotyped waveform $I_0(t - t_\mu)$, but these pulses are embedded in a background of noise $\delta I(t)$. Let us assume that the current noise is Gaussian with some spectral density $N(\Omega)$, although much of the formal development can be generalized. The receptor current is

$$I(t) = \sum_\mu I_0(t - t_\mu) + \delta I(t). \tag{8.6}$$

The simplest problem of computational vision is just to estimate the light intensity --- the time-varying photon arrival rate $R(t)$ --- from the current trace. We are especially interested in the limit where photons arrive only rarely (it's dark outside), so the signal-to-noise ratio is low.

Fred Rieke, Geoff Owen and I [19, 76, 78] have approached this problem of optimal estimation in two ways. First, in the spirit of Wiener's original discussion [114], we try to find a filter $f(\tau)$ such that if we pass the current through this filter the result is the best possible estimator of $R(t)$ in the least-square sense. Specifically, we write our estimate

$$R_{est}(t) = \int d\tau f(\tau) I(t - \tau) \tag{8.7}$$

and choose the filter $f(\tau)$ which minimizes

$$\chi^2 = \int dt |R(t) - R_{est}(t)|^2. \tag{8.8}$$

The result, at low SNR, is that

$$\tilde{f}(\Omega) = \int d\tau e^{+i\Omega\tau} f(\tau) = S_R(\omega)\frac{\tilde{I}_0^*(\Omega)}{N(\Omega)}. \tag{8.9}$$

This optimal filter for the reconstruction of the light intensity has two pieces. First there is a term $\tilde{I}_0^*(\Omega)/N(\Omega)$ determined entirely by the signal and noise characteristics of the receptor cell. It is a "matched filter" which insures that the output amplitude at each frequency reflects the signal-to-noise ratio at that frequency. The second term, $S_R(\Omega)$, is the power spectrum of the fluctuations in photon arrival rate, and incorporates our *a priori* knowledge about the signals impinging upon the receptor cell. Note that the matched filter does not have any such dependence on the ensemble of stimuli.

The decomposition of the reconstruction filter into a matched filter and an ensemble-specific filter can be seen more generally from a probabilistic point of view. Everything that we know about the signal by virtue of observing the receptor current is contained in the conditional probability distribution $P[R(t)|I(t)]$. Any estimate of what is happening in the outside world is controlled by the structure of this distribution. We think of the possible signals $R(t)$ as living in a high dimensional space, for example the space of all the Fourier coefficients needed to specify the function. Before we make any measurements on the receptor current, we know only that $R(t)$ is chosen from some *a priori* distribution $P[R(t)]$, which we imagine as a cloud of points in this high dimensional space. Once we observe the current $I(t)$ this cloud of possible waveforms is shifted and possibly narrowed around our best estimate of the real signal. The problem is analogous to what happens to a collection of molecules as they diffuse. At first we have a cloud in random positions, but if we apply an external force, e.g. in electrophoresis or centrifugation, the cloud takes a definite mean trajectory and narrows around the position of minimum energy. To understand the problem of estimating the signal in a noisy background we have to understand this effective "force" on the distribution of possible waveforms.

To calculate the distribution $P[R(t)|I(t)]$ we use Bayes' theorem,

$$P[R(t)|I(t)] = \frac{P[I(t)|R(t)]P[R(t)]}{P[I(t)]}, \tag{8.10}$$

where $P[R(t)]$ is the *a priori* distribution defining the probability of different signals occurring in a given natural or experimental stimulus ensemble. We then introduce the photon arrival times t_μ, since

$$P[I(t)|R(t)] = \int Dt_\mu P[I(t)|\{t_\mu\}]P[\{t_\mu\}|R(t)], \tag{8.11}$$

where $\int Dt_\mu$ is shorthand for integration over all photon arrival times $t_1, t_2, \ldots, t_N$ and a sum over all photocounts N. The distribution of arrival times $P[\{t_\mu\}|R(t)]$ is the standard Poisson expression

$$P[\{t_\mu\}|R(t)] = \frac{1}{N!} \exp\left\{ -\int d\tau R(\tau) \right\} R(t_1)R(t_2)R(t_3)\cdots\cdots R(t_N).$$

$$\tag{8.12}$$

Finally, if the current noise δI is Gaussian, we can write

$$
\begin{aligned}
P[I(t)|\{t_\mu\}] &= Z^{-1}\exp\left\{ -\frac{1}{2}\int dt \int dt' \left[I(t) - \sum_\mu I_0(t - t_\mu) \right] \right. \\
&\quad \times \left. \int \frac{d\omega}{2\pi}\frac{e^{-i\omega(t-t')}}{N(\omega)} \left[I(t') - \sum_\nu I_0(t' - t_\nu) \right] \right\},
\end{aligned}
\tag{8.13}
$$

with normalization Z and the current noise power spectrum $N(\omega)$.

We are interested in the case of low photon flux, where the signal-to-noise ratio is of necessity low. Hence the driven component of the current variations is small compared to the typical current fluctuation, and we can use this approximation to expand Eq. (8.13),

$$
P[I(t)|\{t_\mu\}] \approx Z^{-1} \exp \left\{ -\frac{1}{2} \int dt \int dt' I(t) \int \frac{d\omega}{2\pi} \frac{e^{-i\omega(t-t')}}{N(\omega)} I(t') \right.
$$
$$
\left. + \int dt \int dt' I(t) \int \frac{d\omega}{2\pi} \frac{e^{-i\omega(t-t')}}{N(\omega)} \sum_\mu I_0(t' - t_\mu) \right\}.
$$

$$(8.14)$$

With this approximation it is straightforward to substitute into Eq. (8.11) and perform the integrals over the t_μ and sum over photocounts:

$$
P[R(t)|I(t)] \approx P[R(t)] \exp \left\{ -\int dt R(t)[1 - e^{y(t)}] \right\}, \tag{8.15}
$$

$$
y(t) = \int d\tau F(\tau) I(t - \tau) \qquad F(\tau) = \int \frac{d\Omega}{2\pi} e^{-i\Omega\tau} \frac{\tilde{I}_0^*(\Omega)}{N(\Omega)}. \tag{8.16}
$$

To carry our "external force" analogy to completion, we think of the distribution for $R(t)$ as being like the Feynman path integral weighting factor for the trajectory of a physical system where R is the coordinate [36]. Then the Hamiltonian for the system in isolation is related to the *a priori* distribution of signals, $H_0[R] \propto -\ln P[R(t)]$, and our observations on the receptor current contribute an external time dependent force $1 - \exp y(t)$.

In the low SNR limit we see explicitly that the effective force which determines our estimate of the stimulus waveform is controlled by the output of the matched filter. Thus *the optimal estimator of signals in the low photon flux limit always involves the receptor currents passed through the matched filter.* There is no additional dependence on the currents and the structure of the matched filter does not depend on what problem we are trying to solve. In particular, if we try to estimate some non-linear functional of the time-dependent light intensity --- for example in the problem of motion detection, as discussed below --- the optimal estimator is a non-linear functional of the receptor currents, but one can decompose this non-linear functional to isolate a term which involves linear filtering of the receptor currents through the matched filter $F(\tau)$. This situation is schematized in Fig. 8.1.

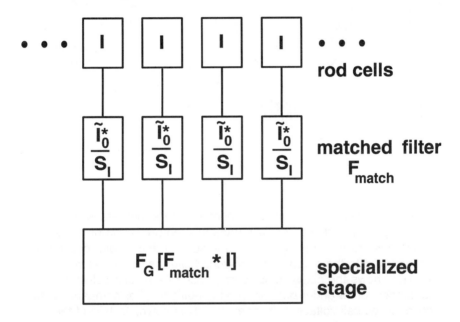

Figure 8.1: A schematic view of signal processing at low photon flux. Matched filters serve as universal pre-processors, with characteristics determined only by the signal and noise spectra of the photodetector. Specific tasks and the statistics of the input signals affect only the subsequent, generally non-linear stages of processing. Figure from Ref. [76]; note that the rod current noise spectrum, N in the text, is written here as S_I.

The matched filter is the *universal* first stage in optimal signal processing at low photon flux, and it would make sense if one could place this filter as early as possible in the system. In the vertebrate retina all photoreceptor signals pass through the bipolar cells, and this is anatomically the first stage of signal processing. If the first mathematical stage of optimal signal processing can be identified with the first anatomical stage, then we have

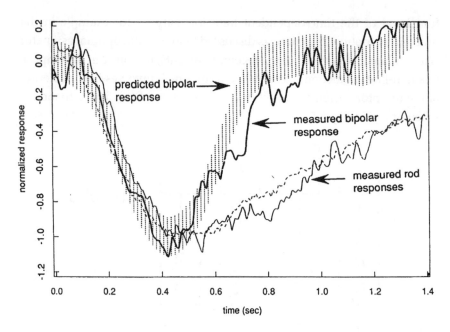

Figure 8.2: Bipolar cell impulse response and the first stage of optimal processing. Hashed lines show parameter-free prediction of the bipolar cell response to a dim flash at $t = 0$, assuming that bipolar cell voltage is proportional to the output of the matched filter as described in the text. Solid line shows the actual impulse response of a bipolar cell in the tiger salamander retina under dark-adapted conditions. For comparison, two records of rod cell voltage responses are also shown. Details of the data analysis are given in Refs. [76,78]

a parameter-free theory for the dynamics of the bipolar cells! Specifically if the voltage response of the bipolar cell is proportional to the output $y(t)$ of the matched filter, then the bipolar cell impulse response is determined entirely by the signal and noise characteristics of the photoreceptor. This is not a model for the bipolar cell; in fact there is nothing in this prediction which incorporates any knowledge the mechanisms in real bipolar cells. Optimal processing provides us with a theory of what the nervous system *should* compute rather than a highly parameterized model of how it computes.

In Fig. 8.2 we see the agreement between theory and experiment, in the tiger salamander retina. As far as we know this is the first case in which it has been possible to give a truly predictive theory for the computationally significant dynamics of a neuron.

Saturation and matched coding

In the analysis of pre-processing we have isolated a stage of processing which is consists of a (matched) linear filter. No cell is linear in response to all inputs; all sensory cells saturate, but signals in the natural world do not become arbitrarily large, so that one does not really need an infinite dynamic range. The interesting question is whether there is some optimal strategy for mapping the limited range of natural signals onto the limited dynamic range of the cell. Laughlin [54] argued on information theoretic grounds that one would like to map equally probable ranges of the input signal into equal ranges of the output variable, or more precisely into equally discriminable ranges. If this is correct then a normalized input/ouput relation should be proportional to the cumulative probability distribution for the signal, which means that the form of the cell's saturation behavior would be determined entirely by the statistics of the natural input signals to which the system is matched. This idea and its (successful) experimental test are reviewed by Atick in his contribution to these proceedings. I want to draw attention to several issues raised by this work: What is the correct generalization to time-dependent signals? Do the saturation characteristics adapt to different scene statistics, or is there some universal form for the contrast distribution in natural scenes which can be hard-wired into the system? Are the cell outputs "read" with constant voltage resolution, or is signal-dependent noise important? These are interesting open questions which can be addressed with existing experimental techniques.

Efficient Coding

Almost all sensory systems encode incoming data in trains of identical action potentials streaming down an array of primary sensory neurons. The sensory data is usually most naturally described in an analog language --- continuous functions of time such as sound pressure or light intensity, chemical concentrations, and so on. The sensory neuron performs a discretization of these signals, in which the only available output information is the occurrence time of each individual spike. While a great deal is known about the phenomenology of this encoding in different sensory systems, relatively little is understood about the structure of the code from an information theoretic point of view [71].

Feature detection and redundancy reduction

Various authors have suggested that the limited dynamic range of spiking neurons forces the system to throw away information at the first possible stage, signalling only the occurrence of particular features which may be of special interest to the animal. The best-known statement of this approach is contained in the work of Lettvin *et al.* on "What the frog's eye tells the frog's brain" [56], although the basic idea of feature detection appears much earlier in Barlow's work [6]. By the time of the Lettvin paper, and of Hubel and Wiesel's discovery of feature selectivity in mammalian cortex, Barlow had gone on to suggest a very different point of view [8]. Concerned about the fact that photoreceptors vastly outnumber retinal ganglion cells in the cat, he suggested that the array of ganglion cells must provide a more economical encoding of the visual world than would be given just by transmitting a stream of receptor voltage signals. This is possible only because signals in the real world have a high degree of redundancy --- intensities and color at nearby points are highly correlated.

In some way the transformations which the nervous system performs before the generation of spikes must serve to reduce redundancy, and at least qualitatively it is clear that lateral inhibition and light adaptation both serve such a function.

Redundancy reduction cannot be the whole story, since at low signal-to-noise ratios redundancy has the positive effect of reinforcing the real signal against the incoherent noise background. Atick and Redlich formulated the problem of minimizing redundancy while holding fixed the magnitude of the information transfer, which allows for interpolation between the low- and high-SNR regimes [3]. If one identifies this variation in SNR with the variation in background light intensity, it is possible to account not only for the observation of lateral inhibition but also for the form of the ganglion cell receptive field as a function of adaptation level. Atick, Li and Redlich have extended this approach to discuss the encoding of color, and have been able to show that very general information theoretic considerations lead to interactions between chromatic and spatial processing which depend on the extent of overlap between the spectral sensitivities of the different receptors [4]. These predictions are in agreement with observations on ganglion cells in primates and goldfish, which represent very different limits of the theory. These ideas are described in detail by Atick in his contribution to these proceedings, and closely related arguments were offered by Srinivasan et al. [100]

The work of Atick et al. serves to illustrate the general structure of any information theoretic approach to the prediction of neural coding strategies. One starts with the hypothesis that some quantity --- perhaps a measure of coding efficiency, such as redundancy --- is being optimized, subject to certain constraints. To relate these information-theoretic notions back to real neurons one needs a model of the coding process as well as knowledge of the signal and noise characteristics of the receptor cells which feed the encoder neurons. Finally to discuss information transfer one needs statistical data about the source of signals, that is one must (in general) know the probability distribution from which natural signals are being drawn.

Maximizing information transmission

As a different example of these ideas, Linsker has explored the consequences of assuming that signal transfer from one layer of neurons to
the next serves to maximize information transmission subject to some
constraints of architecture [58]. Once again one can predict receptive
fields qualitatively like those found at different layers of the visual system,
and perhaps more importantly it's possible to show that relatively simple
local "learning rules" can lead to the optimization of information transfer,
without any global oracle to measure the actual transfer rates.

Dan Ruderman, Tony Zee and I have studied a version of the maximum
information transmission problem in the low SNR limit, and we find a
choice of the constraints which results in a mapping from this optimization
problem onto the variational method in quantum mechanics, so that the
optimal receptive fields are bound state wave functions in some effective
potential determined by the statistics of the input images [21]. We assume
that images are defined by a scalar field $\phi(x)$ on a two dimensional surface
with coordinates x. This image is sampled by an array of cells whose
outputs Y_n are given by

$$Y_n = \int d^2x\, F(x - x_n)\phi(x) + \eta_n, \tag{8.17}$$

where the cell is located at site x_n, its spatial transfer function or receptive
field is defined by F, and η is an independent noise source at each
sampling point. We will assume for simplicity that the noise source is
Gaussian, with $\langle \eta^2 \rangle = \sigma^2$. Our task is to find the receptive field F which
maximizes the information provided about ϕ by the set of outputs $\{Y_n\}$.
This transinformation is given by Shannon's expression,

$$I = \frac{1}{\ln 2} \int D\phi \int d^N Y\, P[\{Y_n\}|\phi(x)] P[\phi(x)] \ln\left(\frac{P[\{Y_n\}|\phi(x)]}{P[\{Y_n\}]}\right), \tag{8.18}$$

where $P[\phi(x)]$ is the *a priori* distribution of contrast patterns. The conditional distribution $P[\{Y_n\}|\phi(x)]$ is determined by Eq. (8.17), and the integrated distribution

$$P[\{Y_n\}] = \int D\phi \, P[\{Y_n\}|\phi(x)] P[\phi(x)]. \tag{8.19}$$

The problem of computing $\ln P[\{Y_n\}]$ and then averaging over the $\{Y_n\}$ is very similar to the problem of performing quenched averages in disordered systems. Indeed one can view the information transfer as the difference between the entropies of coupled ϕ and Y fields calculated in the annealed and quenched limits. As a result it is convenient to calculate the information transfer using the replica method, and we find a natural expansion in $1/\sigma$, or equivalently in the signal-to-noise ratio:

$$I = \frac{1}{2\sigma^2} \sum_\mu \langle F_\mu^2 \rangle \; - \; \frac{1}{4\sigma^4} \sum_{\mu\nu} \langle F_\mu F_\nu \rangle^2 - \frac{1}{6\sigma^6} \sum_{\mu\nu\lambda} \langle F_\mu F_\nu F_\lambda \rangle^2$$
$$+ \; \frac{1}{6\sigma^6} \sum_{\mu\nu\lambda} \langle F_\mu F_\nu \rangle \langle F_\nu F_\lambda \rangle \langle F_\lambda F_\mu \rangle + ..., \tag{8.20}$$

where $F_\mu = \int d^2x \, F(x - x_\mu)\phi(x)$. For most of our discussion we will look at low signal-to-noise ratios, keeping only the first term of the expansion.

It will be convenient to work in Fourier (momentum) space, where the relevant quantities are the power spectrum of the signal,

$$S(k) = \int d^2y \exp(-ik \cdot y) \langle \phi(x+y)\phi(x) \rangle, \tag{8.21}$$

and $\tilde{F}(k) = \int d^2x \exp(-ik \cdot x) F(x)$, the spatial frequency representation of the receptive field. In terms of these quantities the low SNR limit is

$$I \approx \frac{N}{2 \ln 2\sigma^2} \int \frac{d^2k}{(2\pi)^2} |\tilde{F}(k)|^2 S(k). \tag{8.22}$$

To make our definition of the noise level σ meaningful we must constrain the total "gain" of the filters F. One simple approach is to normalize the functions F in the usual L^2 sense,

$$\int d^2x F^2(x) = \int \frac{d^2k}{(2\pi)^2} |\tilde{F}(k)|^2 = 1. \tag{8.23}$$

If we imagine driving the system with spatially white images, this condition fixes the total signal power passing through the filter.

Even with normalization, optimization of information capacity is still not well-posed. To avoid pathologies we must constrain the scale of variations in k–space. This makes sense biologically since cells in the visual system typically have rather local interconnections. We implement this constraint by introducing a penalty proportional to the mean square spatial extent of the receptive field,

$$\int d^2x \, x^2 F^2(x) = \int \frac{d^2k}{(2\pi)^2} |\nabla_k \tilde{F}(k)|^2. \tag{8.24}$$

With all the constraints we find that, at low signal to noise ratio, our optimization problem becomes that of minimizing the functional

$$C[\tilde{F}] = (1/2)\alpha \int \frac{d^2k}{(2\pi)^2} |\nabla_k \tilde{F}(k)|^2 \quad - \quad \frac{1}{2\ln 2\sigma^2} \int \frac{d^2k}{(2\pi)^2} |\tilde{F}(k)|^2 S(k)$$

$$- \Lambda \int \frac{d^2k}{(2\pi)^2} |\tilde{F}(k)|^2, \tag{8.25}$$

where Λ is a Lagrange multiplier and α measures the strength of the locality constraint. Optimal filters are solutions of the variational equation,

$$-\frac{\alpha}{2} \nabla_k^2 \tilde{F}(k) - \frac{1}{2\ln 2\sigma^2} S(k)\tilde{F}(k) = \Lambda \tilde{F}(k). \tag{8.26}$$

We recognize this as the Schrodinger equation for a particle moving in k-space, in which the mass $M = \hbar^2/\alpha$, the potential $V(k) = -S(k)/2\ln 2\sigma^2$, and Λ is the energy eigenvalue. Since we are interested in normalizable F, we are restricted to bound states, and the optimal filter is just the bound state wave function.

There are in general several optimal filters, corresponding to the different bound states. Thus each sampling point should be served by a set of filters rather than just one. Indeed, throughout the visual system one finds a given region of the visual field being sampled by many cells with different spatial frequency and orientation selectivities.

If the signal spectra $S(k)$ are isotropic, so that features appear at all orientations across the visual field, all of the bound states of the corresponding Schrodinger equation are eigenstates of angular momentum. There are many cells in the early visual pathways with circularly symmetric receptive fields, corresponding to $m = 0$. But other cells have receptive fields with a single optimal orientation, not the multiple optima expected if the filters F correspond to angular momentum eigenstates. One would like to combine different angular momentum eigenfunctions to generate filters which respond selectively to a small range of orientations. In general, however, the different angular momenta are associated with different energy eigenvalues and hence it is impossible to form linear combinations which are still solutions of the variational problem.

We *can* construct receptive fields which are localized in orientation if there is some extra symmetry or accidental degeneracy which allows the existence of equal-energy states with different angular momenta. Recently Field [38] has measured the power spectra of several natural scenes. As one might expect from discussions of "fractal" landscapes, these spectra are scale invariant, with $S(k) \approx A/|k|^2$. It is easy to see that the corresponding quantum mechanics problem is a bit sick --- the energy is not bounded from below. In the present context, however, this apparent sickness is a saving grace. The equivalent Schrodinger equation is

$$-\frac{\alpha}{2}\nabla_k^2 \tilde{F}(k) - \frac{A}{2\ln 2\sigma^2|k|^2}\tilde{F}(k) = \Lambda\tilde{F}(k). \tag{8.27}$$

If we take $q = (\sqrt{2|\Lambda|/\alpha})k$, then for bound states ($\Lambda < 0$) we find

$$\nabla_q^2 \tilde{F}(q) + \frac{B}{|q|^2}\tilde{F}(q) = \tilde{F}(q), \tag{8.28}$$

with $B = A/\alpha \ln 2\sigma^2$. Thus we see that the energy Λ can be scaled away; there is no quantization condition. These sorts of quantum mechanics problems were discussed many years ago by Case [31], and more recently by Carreau *et al.* [30] from a path-integral point of view. In the quantum case one can impose the additional requirement that different solutions are orthogonal, which restores the quantization condition, but this has no analog in the information theoretic problem; I am grateful to M. Carreau for bringing these references to my attention during the lectures. We are free to choose any value of Λ, but for each such value there are several angular momentum states. Since they correspond to the same energy, superpositions of these states are also solutions of the original variational problem. The scale invariance of natural images is the symmetry we need in order to form localized receptive fields.

The details of the solutions to Eq. (8.28) are not so important here, since our formulation of the problem is still rather a cartoon of the real visual system. Let us look at the main qualitative points. First, for a given value of B, which measures the signal-to-noise ratio, there exists a finite set of angular momentum states; these states can then be superposed to give receptive fields with localized angular sensitivity. In fact *all* linear combinations of m−states are solutions to the variational problem at low signal to noise ratio, so the precise form of orientation tuning is not determined. If we examine the higher terms in our expansion [Eq. (8.20)] we find terms which will select different linear combinations of the m−states and hence determine the precise orientation selectivity. These terms involve multi-point correlation functions of the image. At the lowest SNR, corresponding to the first term in our expansion, we are sensitive only to the two-point function (power spectrum) of the signal ensemble, which carries no information about angular correlations. A truly predictive theory of orientation tuning must thus rest on measurements of angular correlations in natural images; as far as we know such measurements have not been reported.

Second, at very small B there are only $m = 0$ solutions. This is the limit of very low SNR or equivalently very strong locality constraints (large α). The circularly symmetric receptive fields that one finds in this limit are center-surround in structure --- there is a positive lobe in the center of the field and a wider and shallower negative lobe surrounding this center; the surround becomes more prominent as the signal-to-noise ratio is increased. These predictions are in qualitative accord with what one sees in the mammalian retina, where, as noted above, the signal-to-noise ratio in individual cells is indeed quite low and the connections are very local --- the receptive field center of a foveal ganglion cell may consist of just a few photoreceptors. As one proceeds to the cortex single cells collect inputs from many lower-order cells, enhancing the SNR, and the range of connections is increased; correspondingly orientation selectivity becomes possible. In lower vertebrates the range of connections within the retina is longer as well; orientation selectivity is thus possible at the level of the ganglion cell.

Third, if we examine the $m = 0$ solutions in more detail we find that the center and surround are just the beginning. The receptive field has further concentric structure which is again of opposite sign to the surround; because the amplitude at any single point is very small, this can be difficult to measure. Recently measurements have been reported on retinal ganglion cells of the cat using square patch stimuli with a wide range of sizes [57]. As the patch becomes larger the response of the cell first increases then decreases, corresponding to the center and surround, respectively, but then one sees a gradual recovery of response to much larger area stimuli. This "disinhibitory region" is in qualitative accord with the theoretical prediction.

Finally, it is interesting that, at low SNR, there is no preferred value for the length scale. Thus the optimal system may choose to sample images at many different scales and at different scales in different regions of the image. The experimental variability in spatial frequency tuning from cell to cell may thus not represent biological sloppiness but rather the fact that any peak spatial frequency constitutes an optimal filter in the sense defined here.

Interlude: Coding and decoding in spike trains

Despite the theoretical efforts described above, our understanding of neural coding from an information theoretic point of view remains rather poor. If one looks carefully at this work, all of the "neurons" are cartoon devices which take analog inputs and deliver analog outputs. One can mutter kind words about firing rates, but nonetheless one has not come to grips with the fundamental fact about sensory neurons, namely that they spike. A corollary of this is that there is essentially no experiment in the literature which gives a convincing measure of the rate at which the spike train of a primary sensory neuron provides information about a time-varying, reasonably natural stimulus.

The traditional approach to neural coding characterizes the *encoding* process: For an arbitrary stimulus waveform $s(\tau)$, what can we predict about the spike train? This process is completely specified by the conditional probability distribution $P[\{t_i\}|s(\tau)]$ of the spike arrival times $\{t_i\}$ conditional on the stimulus $s(\tau)$. In practice one seldom characterizes the distribution in its entirety. Most experiments focus on the lowest moment

$$r[t; s(\tau)] = \sum_N \int dt_1 dt_2 \cdots dt_N P[\{t_i\}|s(\tau)] \sum_{i=1}^{N} \delta(t - t_i), \qquad (8.29)$$

which is the firing rate as function of time given the stimulus $s(\tau)$.

The classic experiments of Adrian and others established that, for static stimuli, the resulting constant firing rate provides a measure of stimulus strength [1, 73]. This concept is easily extended to any stimulus waveform which is characterized by constant parameters, such as a single frequency or amplitude for a sine wave. Much of the effort in studying the encoding of sensory signals in the nervous system thus reduces to probing the relation between these stimulus parameters and the resulting firing rate. Generalizations to time-varying firing rates, especially in response to periodic signals, have also been explored.

These measurements of average neural responses to a each of a limited set of stimuli lead to the definition of receptive fields, temporal filter characteristics, *etc.*. Sophisticated versions of this approach --- such as the white noise and reverse correlation methods [35] --- lead to the development of models which specify the firing rate of the neuron as a function of time given some arbitrary stimulus.

The problem of neural coding is often phrased with reference to these classical measurements on the firing rate. One asks if the spike train of a single neuron contains more information than is captured in a description of rate *vs.* time, or conversely if the pattern of firing rates across an array of neurons provides a complete description of the code. We believe that this emphasis on firing rate measures is misplaced, at least for some signal processing problems. Specifically one should realize that, like the stimulus waveform itself, the rate is a continuous function of time. But the organism only has access to the discrete sequence of spikes, not to this continuous function. To say that information is coded in firing rates is of no help unless one explains how the organism could estimate these continuously varying rates from observations of its own spike trains.

Real-time signal processing with neural spike trains must involve some sort of interpolation between the spikes that allows the organism to estimate a continuous function of time. One way of formulating this problem is to ask how one can go from the spike train to the time-varying firing rate, and then invert the stimulus-response relation. We see no reason to prefer this indirect approach to the more direct decoding problem: How can we go from the spike train to the stimulus waveform itself?

The decoding problem has a general probabilistic formulation, but now the appropriate probability distribution is $P[s(\tau)|\{t_i\}]$, that is the distribution of stimulus waveforms conditional on observation of the spike train. From this distribution we can estimate the waveform using maximum likelihood, finding that function of time $s_{est}(\tau)$ which maximizes $P[s(\tau)|\{t_i\}]$. This maximum likelihood prescription defines a decoding algorithm in which one takes the spike train $\{t_i\}$, forms the function $P[s(\tau)|\{t_i\}]$, and then optimizes $s(\tau)$. The accuracy of the estimate is related to the behavior of the distribution in the neighborhood of the maximum likelihood waveform, essentially the (generalized) variance of the distribution.

The two conditional probabilities describing encoding and decoding are related through Bayes' theorem,

$$P[s(\tau)|\{t_i\}]P[\{t_i\}] = P[\{t_i\}|s(\tau)]P[s(\tau)].$$

In practice the encoding distribution $P[\{t_i\}|s(\tau)]$ is difficult to measure with a completeness and accuracy which would allow true decoding. One approach to solving this problem is thus to design experiments which directly estimate $P[s(\tau)|\{t_i\}]$. For a single cell in the fly visual system it was possible to make fairly good estimates of this distribution, at least for spike trains $\{t_i\}$ restricted to short sequences. The structure of these distributions suggested that one really could *decode* the spike train, providing a real-time estimate of the unknown waveform $s(t)$ solely from observations of the spike arrival times. Preliminary attempts at such reconstructions led us to wonder whether this idea of decoding could be put on a firmer theoretical basis. The experiments to measure $P[s(t)|\{t_i\}]$ were done by de Ruyter van Steveninck, and are described in his thesis [84] and in Ref. [85].

We would like to explore a a simple model for the statistics of neural spike trains as they encode a continuously varying signal $s(t)$. The simplest model is to assume that the probability of spike generation depends on $s(t)$ but not on the history of the spike train itself. Then the spikes form an inhomogeneous Poisson process, where the probability of observing N

spikes at times t_i on the interval $(0, T)$ given the signal $s(t)$ is

$$P[\{t_i\}|v(t)] = \frac{1}{N!} \exp\left[-\int_0^T d\tau r(\tau)\right] r(t_1)r(t_2)\cdots r(t_N), \qquad (8.30)$$

with $r(t)$ the rate function determined by $s(t)$. Evidence for the near-Poisson character of neural firing has been found in the mammalian auditory nerve [47, 94], and in retinal ganglion cells firing has been described as a Poisson process driven by the Poissonian arrival of photons at the retina and slightly modified by "dead time" [103].

How does the rate depend upon $s(t)$? In the Poisson regime a neuron can be thought of as firing each time the internal noise currents drive the cell voltage across some threshold. The probability that this occurs in any unit time is exponentially small if the threshold is high. External signals bias the mean current and are thus expected to change the rate in an exponential fashion, and this can be verified in simulations of realistic models for the dynamics of spike generation. For general time-varying signals we still expect an exponential relation but the signal will be filtered. Thus $r(t) = \lambda \exp[\int d\tau f(\tau)s(t-\tau)]$.

The "exponential-Poisson" model has been applied previously to the study of neural coding and to the interpretation of statistical experiments on neural firing [47]. The approach described here was developed some years ago in collaboration with Tony Zee [22]. Our basic strategy is to note that everything we know about the signal $v(t)$ by virtue of having observed the spikes $\{t_i\}$ is contained in the distribution $P[v(t)|\{t_i\}]$, as determined by Bayes' theorem

$$P[v(t)|\{t_i\}] = \frac{P[\{t_i\}|v(t)]P[v(t)]}{P[\{t_i\}]}. \qquad (8.31)$$

To evaluate this distribution we need the *a priori* distribution for the signal, $P[v(t)]$. This is determined by the characteristics of the natural (or experimental) stimulus ensemble, suitably filtered by the response of the pre-synaptic cells. For simplicity we choose $P[v(t)]$ to be that of stationary Gaussian noise,

$$P[v(t)] = Z^{-1} \exp\left[-(1/2)\int \frac{d\omega}{2\pi}\frac{|\tilde{v}(\omega)|^2}{S(\omega)}\right],\tag{8.32}$$

with $\tilde{v}(\omega) = \int dt e^{+i\omega t}v(t)$ and $S(\omega)$ the power spectrum.

After much discussion, then, we have identified our problem: Our knowledge of the signal $s(t)$ as derived from $\{t_i\}$ is summarized by

$$P[s(t)|\{t_i\}] \propto$$
$$\exp\left[-\lambda\int_0^T dt e^{f*s(t)} + \sum_{i=1}^N f*s(t_i) - \frac{1}{2}\int\frac{d\omega}{2\pi}\frac{|\tilde{s}(\omega)|^2}{S(\omega)}\right],\tag{8.33}$$

where $f*s(t) = \int d\tau f(\tau)s(t-\tau)$. It will be convenient to think of this distribution in terms of statistical mechanics or quantum mechanics, where $P[s(t)|\{t_i\}] \sim \exp\{-A[s(t);\{t_i\}]\}$ defines the effective action A for a particle in one dimension to take a trajectory $s(t)$ while being "kicked" at the times t_i. If we want to estimate the signal waveform $s(t)$ given the spike train $\{t_i\}$, the best estimator in the least-squares sense is the conditional mean,

$$s_{\text{est}}(t) = \int Ds P[s(\tau)|\{t_i\}]s(t),\tag{8.34}$$

which is just the quantum or thermal expectation value. Thus estimating the signal from the spike train is just the problem of computing expectation values for a physical system subject to a time-dependent force composed of impulses at the spike arrival times. The result is therefore of the form

$$s_{\text{est}}(t) = \sum_i K_1(t-t_i) + \frac{1}{2}\sum_{ij} K_2(t-t_i, t-t_j) + ...,\tag{8.35}$$

where the K_n's are as usual related to the correlation functions of the theory in which the force is set equal to zero.

The expansion in Eq. (8.35) isn't very useful unless we have some reason to think that it will be dominated by the first few terms. We know that neurons can be highly non-linear devices, so that if tried to expand the firing rate in powers of the stimulus, low-order terms would *not* be sufficient under natural conditions. But our expansion is very different --- rather than describing the input/ouput relation of the neuron, we are trying to describe a hypothetical 'black box' which takes the spike train as input and returns an estimate of the stimulus waveform $s(t)$. To get a feeling for the difference between the estimation problem and the more conventional input/output analysis, consider a simple detector whose average output x is proportional to the input s; in fact we can divide out the proportionality constant and write

$$x = s + \eta, \tag{8.36}$$

where η is the noise. The essence of the problem can be understood without worrying about the time-dependence, so in Eq. (8.36) the various quantities are just real numbers rather than functions of time. Characterizing the input/output relation of this system is trivial --- on average, the output is equal to the input. Is the estimation problem equally trivial?

Everything that we know about the signal s by virtue of observing x is contained in the distribution

$$P(s|x) \;=\; \frac{P(x|s)P(s)}{P(x)} = \frac{1}{Z(x)} P(\eta = x - s)P(s)$$

$$\propto \; \exp\left[-H(x - s) - \frac{1}{2\langle s^2 \rangle}s^2\right], \tag{8.37}$$

where for simplicity we assume that the signal is chosen from a Gaussian distribution and we write the distribution of the noise in terms of some effective Hamiltonian $H(\eta)$ such that thermal fluctuations in this ensemble simulate the real noise distribution. We are interested in calculating the best estimator in the least-square sense, $s_{\text{est}} = \int ds s P(s|x)$. It is easy to see that if the noise is Gaussian, so that $H(\eta) = \eta^2/2\langle\eta^2\rangle$, then $s_{\text{est}} = K_1 x$. The 'kernel' $K_1 = SNR/(SNR + 1)$, where the signal-to-noise ratio $SNR = \langle s^2\rangle/\langle\eta^2\rangle$. This is reasonable result: to "decode" the output of a linear detector we use a linear system whose gain saturates at high SNR; at low SNR we scale down the output of the detector since we know that most of what are seeing must be noise.

When the noise is not Gaussian, the Hamiltonian H is no longer a simple quadratic function, and the calculation of the best estimate is essentially the same as calculating the response of some *non-linear* system to an external force $F \propto x$. This means that the best estimator will not be a linear function of the detector output, even thought the detector itself is linear and the signals are chosen from a Gaussian distribution --- non-Gaussian noise is sufficient to destroy linear decodability! Of course if the SNR is very high, the distribution $P(x|s)$ approaches a delta function and hence its exact shape (the distribution of the noise) is irrelevant, but we know that biological detectors seldom operate in this limit.

This simple example shows us that the possibility of decoding the spike train with a linear filter such as K_1 in Eq. (8.5) (or a near-linear filter if we include a small K_2) really has nothing to do with the conventional notion of linearity in the input/ouput or stimulus/firing rate behavior of the neuron. It is even possible that the cell is linear by the conventional measures but because the noise in the cell's response is non-Gaussian linear decoding won't work. What then is required for the expansion to be dominated by its first term? The key is to think about the correlation time τ_c of the signal (as seen through the filter f) in relation to the typical inter-spike intervals. The occurrence of a single spike tells us something about the signal within a time window of roughly $\pm\tau_c$ around the spike itself. If the next spike occurs after an interval $\tau >>> \tau_c$, then this spike gives us independent information about the signal and the contributions of the two spikes to our estimate of the stimulus waveform must just add: Our expansion of the optimal estimate in Eq. (8.35) is really an expansion in $\langle r\rangle\tau_c$, and this can be verified by detailed perturbation theory calculations.

Linear decodability thus defines a regime of neural dynamics in which each significant variation in the signal (on time scale τ_c) triggers of order one spike or less. This is almost the opposite picture from that suggested in rate coding models, where information is carried only in windows of time which contain several spikes, enough to form a reasonable estimate of the firing rate over the window. Is there any evidence concerning the value of $\langle r \rangle \tau_c$? For auditory neurons, it is clear that modulations in biologically significant sounds (speech, bat echolocation, frog calls, cricket chirps, ...) occur on short time scales, $5 - 20$ msec, so that sensory neurons generate at most a few spikes during the correlation time of the input signal. In the fly visual system, as we discuss below, movements across the visual field result in the generation of a compensating torque within 30 msec, during which time the movement-sensitive neurons generate just a few spikes each. In the mammalian visual cortex, pre-attentively discriminable textures produce an average of 1 to 3 spikes per cell within the 50--100 msec behavioral decision time [49], while optimally chosen moving gratings produce modulations of less than 3 spikes per 100 msec (see, for example, Ref. [75]). These scattered observations strongly suggest that real neurons are at least close to the regime of linear decodability.

Returning to the model of encoding as a Poisson process, one can prove that the rate at which the spike train provides information about the stimulus is bounded by

$$R_{\text{info}} \leq \left\langle r(t) \log_2 \left[\frac{r(t)}{\langle r \rangle} \right] \right\rangle. \tag{8.38}$$

The key to the proof is that the Possion process has the maximum entropy of any point process with the same mean rate, and the inequality is saturated precisely in the limit $\langle r \rangle \tau_c \to 0$ which would guarantee linear decoding! This provides us with a further hint that linear decoding may make sense for real neurons if the code has been 'designed' to maximize information transmission.

Before moving to experimental tests of these ideas, it is good to look back at the literature and see some connections with previous work. Our approach to the characterization of the neural code is nearly opposite the conventional one, where one seeks a model of the neuron's response to known external signals. Ideas much closer to our point of view were first advocated by Fitzhugh [39], who used individual spike trains to make choices among a limited set of input stimuli. Barlow and Levick [13] carried out a similar analysis in experiments on detection and discrimination of intensity in the cat retina. This work focused, however, on forced-choice discrimination among a small number of possible signals.

Johannesma, Gielen and Hesselmans developed a probabilistic formalism to address the more general problem of estimating an unknown, time-varying signal [42, 46], arriving independently at many of the ideas in Refs. [20, 22, 85]. In his early applications of white noise methods to the auditory system, de Boer (see the review in [35]) emphasized the interpretation of the reverse correlation function --- the mean stimulus which triggers a spike --- as the "feature" of the stimulus waveform which is signalled by the occurrence of a spike. It was suggested [42] that one could estimate the stimulus waveform simply by adding up these features. This raises the problem that as the spikes come more frequently the reverse correlation functions centered on successive spikes overlap and can provide conflicting estimates of the stimulus. The correct resolution of these conflicts requires that we attach measures of confidence to the different estimates, and this is accomplished by measuring the relevant probability distributions, as described above.

Information transmission in sensory neurons

Until recently there were no direct, model independent measures of the rate at which the spike train $\{t_i\}$ in a sensory neuron provides information about continuously varying sensory signals $s(t)$. Estimates ranged over several orders of magnitude; for extreme examples see Refs. [59, 64, 101]. The information which the spike train provides about the stimulus is given by

$$I[\{t_i\} \to s(\tau)] = \int Dt_i \int Ds P[s(\tau); \{t_i\}] \log_2 \left(\frac{P[s(\tau); \{t_i\}]}{P[s(\tau)]P[\{t_i\}]} \right),$$

(8.39)

where $\int Dt_i$ is shorthand for integration over all arrival times $t_1, t_2, ..., t_N$ and summation over all spike counts N on the time interval $0 < t < T$, and $\int Ds$ as usual denotes an integration over the space of functions $s(0 < t < T)$. Information is measured in bits, hence $\log_2$. $P[s(\tau)]$ is the *a priori* distribution from which the signal is drawn in a given experimental or natural situation. We start with the simplest case where this distribution is Gaussian and hence completely characterized by the signal power spectrum (two-point function) $S(\Omega)$. We can rewrite the transinformation in the form

$$I[\{t_i\} \to s(\tau)] = - \int Ds P[s(\tau)] \log_2 P[s(\tau)]$$
$$- \int Dt_i P[\{t_i\}] \left[-\int Ds P[s(\tau)|\{t_i\}] \log_2 P[s(\tau)|\{t_i\}] \right].$$

(8.40)

The second term is just the entropy of the distribution $P[s(\tau)|\{t_i\}]$, that is the entropy of the signals with fixed spike train, averaged over the distribution of spike trains $P[\{t_i\}]$. But the entropy of a distribution is always less than that of Gaussian distribution having the same mean and variance. This leads us immediately to to a lower bound on the transmitted information,

$$I \geq \frac{1}{2\ln 2} \int Dt_i P[\{t_i\}] \mathrm{Tr} \ln[S\hat{N}^{-1}(\{t_i\})],$$

(8.41)

where $\hat{N}(\{t_i\})$ is the covariance of the fluctuations in $s(t)$ around the conditional mean,

$$\langle s(t)\rangle_{\{t_i\}} = \int DsP[s(\tau)|\{t_i\}]s(t). \tag{8.42}$$

Finally since $\langle \ln x\rangle \leq \ln\langle x\rangle$, the rate of information transmission

$$R_{\text{info}} = \lim_{T\to\infty} I[\{t_i\} \to s(\tau)]/T \geq \frac{1}{2\ln 2}\int \frac{d\Omega}{2\pi}\ln\left[\frac{S(\Omega)}{\bar{N}(\Omega)}\right], \tag{8.43}$$

where $\bar{N}(\Omega)$ is the power spectrum of fluctuations around the conditional mean averaged over spike trains. Note that although $\hat{N}(\{t_i\})$ describes non-stationary fluctuations, the average over spike trains restores time-translation invariance and allows the definition of the power spectrum. If we construct some arbitrary estimator which takes as input the spike train $\{t_i\}$ and returns some estimate of $s(t)$, then the power spectrum of errors in this estimate $N_{\text{est}}(\Omega)$ will always be greater than or equal to $\bar{N}(\Omega)$, since the conditional mean is also the optimal least-square estimator. Hence

$$R_{\text{info}} \geq \frac{1}{2\ln 2}\int \frac{d\Omega}{2\pi}\ln\left[\frac{S(\Omega)}{N_{\text{est}}(\Omega)}\right]. \tag{8.44}$$

Clearly if we choose a bad estimator this bound will be far below the true information rate. The ratio of the estimator-based bound to the true information rate give us the fraction of the available information about $s(t)$ which is captured by the estimated waveform.

We thus arrive at a simple experimental strategy: Try to construct a box which takes as input the spike train $\{t_i\}$ and delivers as output an estimate $s_{\text{est}}(t)$ of the unknown, continuous stimulus chosen from the ensemble $P[s(\tau)]$. If we can parameterize this box, choose the parameters so as to minimize the mean-square deviation (χ^2) between the estimate and the true signal. Finally, the power spectrum of the errors will provide a lower bound on the information rate through Eq. (8.44). But the simple model discussed above provides us with a form for the optimal estimator, namely Eq. (8.35). In the model the detailed structure of the kernels can be calculated, but it is probably too much to ask of the model for these to be quantitatively applicable to real cells. Instead we choose the kernels to minimize χ^2 using a subset of our data, and then test the quality of the estimate on a distinct subset.

This approach was first tested in experiments on the identified neuron H1 in the fly visual system. The cell responds to rigid horizontal movements across the visual field, and produces a signal involved in the stabilization of straight flight. In this case the linear term K_1 is sufficient to give extremely precise estimates of the angular velocity waveform, and the addition of K_2 does not significantly reduce χ^2. Over a limited bandwidth we can make a stronger statement, as described below: The noise level in the estimate approaches the limits imposed by noise in the photodetector array, so it would be impossible to improve the estimate no matter how many terms in the series one wishes to include.

The success of linear decoding now extends to sensory neurons in a wide variety of systems: A simple mechanical sensor in the cricket [112, 113], a vibratory sensor in the frog inner ear [76, 81], the two acoustic sensors of the frog inner ear [77], and (in preliminary experiments) the optic nerve of the salamander [65]. In each case "success" is first defined perturbatively to mean that inclusion of K_2 does not substantially reduce the errors in the estimated signal waveforms; in the one case where small but real improvements do occur, it was checked that K_3 does not produce statistically significant changes. In all these cases the errors in the reconstruction are Gaussian to a good approximation, so we expect that our bound in Eq. (8.44) will be quite tight. The cricket provides the highest information rates, but the information per spike is comparable in the frog.

Finally, although the absolute noise levels in the reconstructed waveforms can be quite small --- sufficient in the cricket to give nanometer precision in estimating the displacement of the sensory hair --- the signal-to-noise ratio is moderate, of order one or less in many cases. If the Gaussian signals are imbedded in a background of Gaussian noise, then from Shannon we know that the information rate is $R_{info} = \Delta f \log_2(1 + SNR)$, where Δf is the bandwidth. Evidently widening the bandwidth provides a greater increase in information than a comparable increase in signal-to-noise. In the cricket, an information transmission rate of ~ 300 bits per second arises (roughly) from an SNR of unity over a 300 Hz bandwidth. Let us look at these results in detail.

Crickets have two "prongs" sticking out of their backsides. These structures (two cerci; one is a cercus) are covered with sensory hairs which can be deflected by air motion. Each hair grows out of a neuron which sends its output end (axon) up into the cricket's body. When the hair is bent, the neuron changes its spike activity, roughly increasing the firing rate in response to motion one direction and decreasing it in response to motion in the opposite direction; the mechanics of the hair attachment confine the motion to one axis, and different orientations are represented by different hairs along the cercus. The different orientations provide information about the direction of the source of the air motion, and this ultimately goes into the construction of the sensory map discussed above. In the cockroach this system mediates an escape reaction so that the air motion which results from an approaching object causes the bug to scurry away from the object; this is why cockroaches are hard to stomp. In the cricket, alas, it is less clear how the cercal system functions in behavior, but they are a favorite system for many different aspects of neurobiology. I learned about crickets from John Miller while I was in Berkeley, and two of our students, David Warland and Mike Landolfa, did the key experiments which led to direct measures of information transmission and coding efficiency in this system; all of the analysis methods were developed by David together with Fred Rieke [76, 79, 112].

The sensory hairs provide us with a pure example of the coding problem: Given the sequence of spikes in the sensory neuron, how much information do we have about the trajectory of the hair? The experiment is to grab hold of the hair, deliver random displacements $s(t)$, and record the resulting spike arrival times. We then use a long stretch of such recordings to optimize the filter $K_1(t)$ with which we try to decode the spike train [as in Eq. (8.35)], and test our decoding algorithm on some new stretch of data which did not enter the optimization procedure; this insures that we are not somehow "fitting" to details of a particular stimulus waveform, but rather learning to decode signals chosen from some ensemble. In these initial experiments the ensemble was Gaussian noise with a flat spectrum from 25 to 525 Hz. An example of the stimulus, spike train and reconstruction is shown in Fig. 8.3. The reconstructed waveform clearly interpolates between the spikes, and in some places gives a close match to details of the stimulus on very short time scales. This tells us that the bandwidth of the system is large. On the other hand, the typical errors are comparable to the stimulus itself, so the overall signal to noise ratio is about unity. Qualitatively, the estimate of the signal at any instant of time is imprecise, so that we obtain of order one bit of information (the signal is positive or negative, for example), but because of the large bandwidth this estimate is updated very often, so the number of bits per second is large.

The quantitative analysis of the reconstructions proceeds by separating our estimate of the stimulus into a 'signal' which is deterministically related to the stimulus (but could be systematically biased) and an 'effective noise' which is uncorrelated with the stimulus. As a first attempt at separating systematic and random errors we write, for each frequency,

$$\tilde{s}_{\text{est}}(\omega) = g(\omega)\left[\tilde{s}(\omega) + \tilde{n}_{\text{eff}}(\omega)\right], \tag{8.45}$$

where $g(\omega)$ is a frequency dependent gain introduced to correct for systematic errors and $\tilde{n}_{\text{eff}}(\omega)$ is the random noise, defined as an effective input noise level. To calculate g and n_{eff} we divide the experiment into segments, and Fourier transform the stimulus and reconstruction in each segment. The result is a collection of Fourier components $\{\tilde{s}^{(n)}(\omega), \tilde{s}_{\text{est}}^{(n)}(\omega)\}$, where n denotes segment number. For each frequency ω we make a plot of stimulus Fourier components $\tilde{s}^{(n)}(\omega)$ against reconstruction Fourier components $\tilde{s}_{\text{est}}^{(n)}(\omega)$; each segment contributes one point to this plot. The

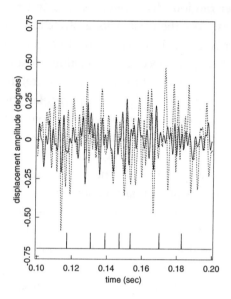

Figure 8.3: Signal (dashed line) and reconstruction (solid line) in a single mechanosensor afferent from the cricket cercal system. As described in the text, the stimulus consists of broad band Gaussian random motions of the sensory hair, and spikes (shown at bottom) are recorded with an electrode inside the cell's axon. The reconstructions are accomplished by linear filtering of the spike train. Note that the spike sequence is extremely sparse, yet the reconstruction succeeds in capturing some of the high-frequency details in the waveform.

gain $g(\omega)$ is the slope of the best linear fit through these points, and the effective noise $\tilde{n}_{\text{eff}}(\omega)$ measures the scatter, along the stimulus axis, about this best fit line.

It is remarkable that the effective noise is *extremely* Gaussian (Fig. 8.4), which means that our bound on the information transmission rate in terms of the power spectrum of the reconstruction "noise" [Eq. (8.44)] should be fairly accurate. Our extraction of $\tilde{n}_{\text{eff}}(\omega)$ corresponds to the conventional practice of referring noise back to the input, which is slightly different than just computing the power spectrum of the error $s - s_{\text{est}}$. This effective input noise spectrum is shown together with the signal spectrum in Fig. 8.5; we

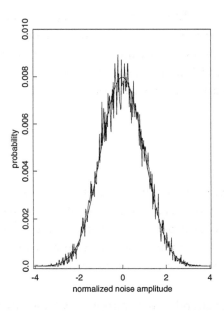

Figure 8.4: The distribution of the effective noise for reconstruction of hair displacement in the cricket cercus. The distribution is formed as a normalized histogram, using samples of the noise at each frequency; to merge data from different frequencies we normalize by the standard deviation of that frequency component, which is proportional to the square-root of the power spectrum. With this normalization, Gaussian noise of any spectrum will generate a unit-variance Gaussian distribution, which is also shown. The agreement is excellent, strongly supporting the approximation of the reconstruction errors by a Gaussian noise source.

see that $SNR = 1$ over a bandwidth of $\sim 300\,\mathrm{Hz}$, as promised. Finally doing the integral, the information rate in this experiment is 294±6 bits/s. This corresponds to 3.2±0.07 bits per spike.

In 1952 MacKay and McCulloch pointed out that a system which keeps track of spike arrival times or inter-spike intervals could in principle convey several bits per spike, much more than crude rate codes [59]. We believe that our work is the first to confirm this prediction. Could the cell in fact transmit more information? There is an absolute upper bound to the transmitted information set by the entropy of the spike sequences themselves: The entropy measures (roughly) the number of distinguishable sequences given some timing precision, and the information rate measures the number of distinguishable stimulus waveforms; clearly one cannot distinguish more waveforms than spike trains in any coding scheme. Thus we arrive at a measure of coding efficiency,

$$\epsilon = R_{\text{info}}/S, \tag{8.46}$$

where S is the entropy per unit time of the spike train. To assess the approach of the cell to this fundamental limit, we estimated the spike train entropy from the same experiments where we found the high information rates. In practice, we cannot get an exact number for the entropy, so we need to be careful to bound the entropy from above.

One simple upper bound to the entropy is obtained by assuming that all the spikes are independent, occurring at some rate r. Then we divide time into bins of width $\Delta\tau$, the probability of a spike (a '1') is $p = r\Delta\tau$ and the probability of no spike (a '0') is $q = 1 - r\Delta\tau$. The entropy per unit time of these binary strings is then

$$S^{(0)} = -\frac{1}{\Delta\tau}[p\log_2 p + q\log_2 q] \tag{8.47}$$

$$\approx \frac{r}{\ln 2}\left[\ln\left(\frac{1}{r\Delta\tau}\right) + 1\right], \tag{8.48}$$

where in the approximation we make use of $p \ll 1$. More accurate bounds can be obtained by assuming that inter-spike intervals are independent but consistent with the measured distribution rather than just the first moment $1/r$, and so on. Clearly the entropy depends on the precision $\Delta\tau$ with which the spikes can be timed; so does the quality of our reconstructions. Since we don't know the timing precision of the nervous system we study the whole function $\epsilon(\Delta\tau)$.

In both the cricket and the bullfrog, information rates are within a factor of two of the spike train entropy, corresponding to a coding efficiency in excess of 50%, as shown in Fig. 8.6. This implies that although the

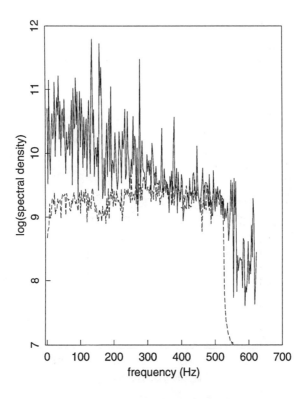

Figure 8.5: Signal and effective noise spectra for the cricket experiments of Fig. 8.3. The noise spectrum is jagged because we use small frequency bins, thus limiting the number of independent samples that can contribute to each point; these errors lead to a slight underestimation of the information rate. Note that the noise level at low frequencies is very high, so that information is concentrated in the 200-500 Hz region. Across this bandwidth the SNR is roughly unity, as explained in the text.

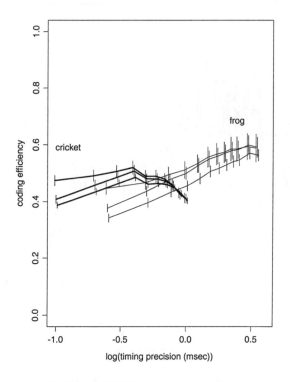

Figure 8.6: Coding efficiency as a function of timing precision in the cricket cercus and the frog sacculus The cricket encodes higher frequencies at higher firing rates and hence reaches maximal efficiency at higher time resolution. Progressively higher efficiency curves correspond to tighter bounds on the spike train entropy, using the succession of approximations described in the text.

spike sequences are highly variable, most of this variability is in fact being used to convey information about the stimulus, a possibility emphasized by Bullock in his reviews of coding and reliability in the nervous system [27, 28]. We have some hints from preliminary experiments that these cells can adapt so as to maintain high coding efficiencies under a wide range of conditions.

Toward a theory of optimal coding in spiking neurons

Having seen that the representation of sensory signals in neural spike trains can be highly efficient and that large amounts of information can be transmitted by single sensory neurons it is natural to return to our starting point and ask if some optimization principles for the code itself can be formulated. As a first try at the theory of optimal coding in spiking neurons, consider the following model problem. A spiking neuron receives its signal $s(t)$ in a background of Gaussian noise $\eta(t)$. The combination of signal and noise is filtered,

$$y(t) = \int dt' F(t - t')[s(t') + \eta(t')]. \tag{8.49}$$

and a spike is generated each time t_i the signal y crosses a threshold θ with positive slope. The optimal coding problem is to choose the filter $F(t)$ and the threshold θ so as to maximize the information which the spike arrival times $\{t_i\}$ provide about the stimulus $s(t)$.

In the limit of small signal-to-noise ratio the problem of optimal coding simplifies tremendously, and it is a straightforward matter to relate the information transfer rate to the filter and the threshold. Optimizing this rate, however, is not a well-posed problem, because the information can always be increased by making filters which have longer lasting transient responses. This would mean, however, that the information takes a very long time to find its way into the spike train. Presumably there is a

penalty for such long delays --- it is not much help to be very certain
that a tiger was lunging at us one minute ago. The exact structure of
the optimal filter depends on the structure of this penalty, but there is
a simple choice which leads to a remarkable result: The optimal filters
are solutions of a Schrodinger equation, very much analogous to the
situation found by Ruderman, Zee and I in the problem of optimizing
spatial receptive fields. As in the spatial case, this means that there will
be multiple solutions, corresponding to multiple cell types. The idea that
a very general, information theoretic optimization principle leads to the
prediction of multiple cell types is extremely important, since it suggests
that some of the obvious diversity of the nervous system can be tamed
and understood from a single coherent theoretical point of view.

In addition to results on the form of the filter, optimization of the threshold
leads to the conclusion that, for example, retinal ganglion cells should
adapt not only to the background light intensity but also to the variance
of the contrast as seen through its spatio-temporal receptive field. There is
preliminary evidence for this in the tiger salamander retina [65], we have
seen similar effects in the cricket cercal system, and one could argue that
for low frequency auditory neurons the well-known adaptation to sound
intensity (which is just the pressure variance) is the same phenomenon.
If the threshold adapts to maintain optimal information transmission,
then within this class of models it also turns out that linear decoding
should be a good approximation. This suggests that our two fundamental
observations on the structure of the neural code --- the success of simple
decoding algorithms and the very high information rates --- may be tied
together by a principle of optimal coding.

Specialized Computations

Different organisms are interested in different aspects of their sensory
environment. To meet these diverse needs, a variety of specialized

computations are carried out. Obviously it is essential that the results of these computations --- the neural computations which regulate behavior --- be reliable. No calculation which begins with real sensory data can be perfect, however, since this data itself is inevitably noisy. Thus there is a fundamental limit to the reliability of animal behavior. In this section we examine the evidence that these limits to the reliability of neural computation are actually reached, and discuss some implications.

Nanosecond precision in bat echolocation

Perhaps the most remarkable example of reliability in neural computation is provided by the target range estimation process if echolocating bats, specifically the species *E. fuscus* described by Simmons in these proceedings. A long series of experiments by Simmons and co-workers has demonstrated that this bat can discriminate echo delay differences as small as 10 nanoseconds [96, 97].

Is it really so hard to detect ten nanosecond differences in time delay? We imagine that the bat hears a known pulse shape $p_0(t - \tau)$ in a background of noise $\eta(t)$. How reliably can the bat detect the difference between a reference value of τ and a slightly different value $\tau + \delta\tau$? If the noise is white with spectral density N_0, then the probability of observing a particular sound pressure waveform $p(t)$ is

$$P[p(t)|\tau] \propto \exp\left[-\frac{1}{2N_0} \int dt |p(t) - p_0(t - \tau)|^2\right] \qquad (8.50)$$

if the echo delay has the value τ. Similarly,

$$P[p(t)|\tau - \delta\tau] \propto \exp\left[-\frac{1}{2N_0} \int dt |p(t) - p_0(t - \tau - \delta\tau)|^2\right]. \qquad (8.51)$$

Now if we (or the bat) observe a particular waveform $p(t)$, we can decide whether the delay was τ or $\tau + \delta\tau$ by computing the logarithm of the relative likelihood,

$$
\begin{aligned}
\lambda[p(t)] &= \ln\left(\frac{P[p(t)|\tau]}{P[p(t)|\tau + \delta\tau]}\right) \\
&= \frac{1}{N_0}\int dt\, p(t)[p_0(t - \tau) - p_0(t - \tau - \delta\tau)] \\
&\approx \frac{\delta\tau}{N_0}\int dt\, p(t)\frac{dp_0(t - \tau)}{dt},
\end{aligned} \tag{8.52}
$$

where we have specialized to the case of small $\delta\tau$.

It is easy to see that $\lambda[p(t)]$ is a Gaussian random variable, with slightly different mean values in response to the two different delays. The signal to noise ratio is

$$
\begin{aligned}
SNR &= \frac{1}{\langle(\delta\lambda)^2\rangle}\left[\langle\lambda[p(t)]\rangle_{\tau+\delta\tau} - \langle\lambda[p(t)]\rangle_\tau\right]^2 \\
&= (\delta\tau)^2\frac{1}{N_0}\int dt\left[\frac{dp_0(t - \tau)}{dt}\right]^2.
\end{aligned} \tag{8.53}
$$

Thus the reliability of discrimination is determined by the typical time derivatives of the echo waveform and by the background noise level. The performance of the bat is essentially equal to that predicted from this optimal signal-to-noise ratio [97]. Indeed, if one is not careful to count how many pulses the bat is using, it may even seem that the animal surpasses this limit, which is of course impossible.

This is a clear example where the animal makes optimal use of the available signal to achieve its impressive behavioral performance. To reach this limit the bat must carry out certain very a specific computation, Eq. (8.52). Although we know the mathematical structure of these computations, we do not know how these computations are realized in the bat's brain.

Aspects of human vision

Humans are apparently very good at recognizing symmetry and regularity in images. Indeed, as scientists we must take care not to see too much regularity in our data. Are we really good at this task, or are we fooling ourselves? Barlow and co-workers [10] generated random dot patterns which had a statistical tendency toward bilateral symmetry --- some of the randomly placed dots had near-symmetric partners across a reflection axis, while others did not --- and asked observers to discriminate between these patterns and a set of random dot patterns with no correlations across the symmetry axis. In such experiments the observer gets just one look at one randomly chosen pattern, so that, by chance, one of the nominally asymmetric patterns may appear nearly symmetric, and conversely. In the real experiment this confusion was enhanced by blurring the images so that individual dots were not visible and by forcing the observer to make a decision in a time so short that scrutiny (scanning the image with eye movements) was impossible.

If we convert the percent correct discrimination performance into effective signal-to-noise ratios for the recognition of symmetry, human observers achieve signal-to-noise ratios within a factor of two of the limits imposed by randomness in the stimulus itself. This conclusion is robust to many changes in the details of the symmetry recognition task --- one can, within limits, change the degree of blurring, randomize the symmetry axis, and changes the rules of dot placement while still maintaining near optimal performance. These results have been confirmed and extended by Tapiovaara [102]. The approach of human observers to the theoretical limit is not restricted to this one task; similar results are obtained for the recognition of selected patterns in noisy backgrounds, the detection of variations in density for random dot patterns, and so on [11, 29].

Movement estimation in the fly visual system

To assess the approach of neural computation to some theoretical limits of reliability or noise level we must have a system in which one can make measurements at more than one stage in the processing chain. In the case of flies one can study visually-guided flight behavior both by measuring flight paths during natural behaviors [53, 109, 110, 111] and by examining the torques produced by tethered flies in response to controlled movements of patterns across the visual field [72, 74]. One can also achieve very stable recordings from the photoreceptor cells of the compound eye, allowing the measurements of signal and noise characteristics at the input to the computation of motion. Finally, the output of the movement computation is carried by a handful of identified cells in the lobula complex [44].

The concept of an identified neuron is truly remarkable, and provides a concrete antidote to the prejudices of some physicists that the nervous system is a hopelessly messy place. To take the specific case of the motion-sensitive cell H1 in the fly visual system, one can search in a region of the fly's brain for a cell with certain fairly crude response characteristics --- it is excited by motion in one direction, inhibited by motion in the opposite direction, and nearly unaffected by motion in the perpendicular direction; it integrates over most of the visual field, rather than being selective for objects of a given size. The cell defined by these criteria is essentially unique. Quantitative experiments on different flies of the same species show that the detailed response characteristics are reproducible at (at least) the ten percent level, and if one fills the cell with a dye and looks under the microscope the picture is almost identical in every fly of that species. If any given experiment leaves you wanting to know more about the system, you can go back the next day to a new fly with complete confidence that you are studying exactly the same cell. These remarks should not be taken to imply that everything about the system is rigidly defined or "hard-wired;" in fact we know that the response of H1 adapts in remarkable ways to the recent visual experience of the fly, and this will be important in thinking about the possible optimality of the movement estimate.

As a personal note, I should probably have learned about identified neurons when I was a student (yes, I actually took a neurobiology course once). Maybe I did, but the significance of the idea for the quantitative reproducibility of experiments never really sank in. When I was a postdoc in the Netherlands I had the good fortune to have the office next door to Rob de Ruyter van Steveninck, who was doing his thesis work studying

H1. When he told me about this system --- and in particular the fact that he could make reliable recordings from H1 for periods of up to five days --- I was an immediate convert to the community of insect lovers. Clearly the fly provides a system in which we can be as quantitative as we wish, since we essentially never run out of data; only now that I have collaborated with groups working on many different systems do I realize how special this situation really is. Returning to our theme, Rob, Fred Rieke, David Warland and I have carried out a series of experiments and analyses which characterize the precision of the movement computation which is coded in the spike train of H1 [20, 80, 84, 85, 86, 87].

Following the approach described above, we have decoded the spike train of H1 to recover estimates of the trajectory of random patterns moving across the visual field; Fig. 8.7 shows an example of one segment in these reconstructions. These results provide graphic evidence that the spike train of H1 contains enough information to infer --- in real time, without averaging --- details of the stimulus waveform on times comparable to the typical inter-spike interval, as suggested by the behavioral reaction times [53], and the structure of the reconstruction algorithm confirms that the optimal estimate of the waveform at one instant of time is controlled by the timing of at most a handful of spikes. Comparing these estimated waveforms with the real trajectories we found that the effective noise level in our reconstruction corresponds to an angular displacement noise of $\sim 10^{-4}$ (degrees)^2/Hz. This implies that, with a behaviorally relevant integration time of ~ 30 msec, one can use the output of H1 to estimate trajectories to a precision of ~ 0.1 degree, roughly one order of magnitude smaller than the spacing between photoreceptors in the lattice of the compound eye and a factor of five smaller than the width of the diffraction pattern around the photoreceptor aperture. This is well within the regime corresponding to "hyperacuity" in human vision [115], and we have confirmed these results in experiments where we ask directly about the discriminability of step displacements using single examples of the H1 spike train.

Hyperacuity is a remarkable phenomenon, but it is not fundamentally

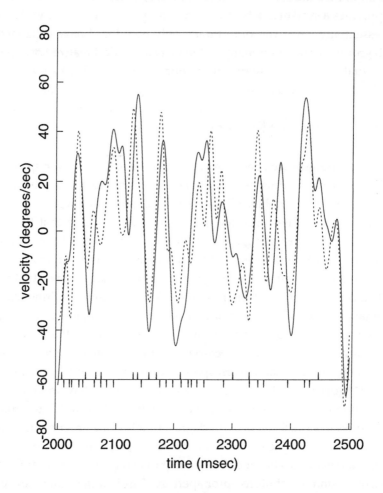

Figure 8.7: Time-dependent angular velocity signals (dashed line) are reconstructed (solid line) by filtering the spike train of H1 (bottom). Positive spikes represent the response of H1 to the stimulus as shown, negative represent the response to the negative of the stimulus as would be seen by the H1 cell on the other side of the head during rigid rotation of the fly.

mysterious. Microscopists have known for a century that one can resolve displacements far below the nominal "diffraction limit" set by the Rayleigh criterion provided that one has sufficient signal-to-noise ratio. In the fly de Ruyter van Steveninck [84] has recorded from the photoreceptor cells under conditions identical to those used for the H1 experiments, characterizing the signal and noise behavior of these cells so that we can actually calculate the limits to hyperacuity on the assumption that the fly is the optimal processor of the voltage signals in the array of photoreceptors. The key point is that in response to a sudden movement the fly can generate a torque within 30 ms. But for small amplitude signals, the photoreceptors themselves need nearly 10 ms to reach their peak response. Even with very large signals --- which result in an almost instantaneously noticeable photoreceptor voltage --- there is a delay of roughly 15 ms before H1 will respond. For small displacements, then, we expect that there is a delay of almost 25 ms before H1 can report the full signal corresponding to a displaced pattern But this leaves only 5 ms during which the fly can be "looking" at the full signal as it tries to estimate the appropriate compensating torque. This short time is essentially equal to the correlation time in the photoreceptor voltage noise (~ 8 ms), so rather than looking at time-averaged signals the fly must base its behavioral decisions on a single snapshot of the voltage array.

Based on de Ruyter's experiments we can simulate what the fly "sees" when patterns are displaced by small fractions of the photoreceptor spacing. Figure 8.8 shows an example of these simulations, carried out by N. Socci; grey levels are proportional to the voltage excursion in each cell. I think this gives us graphic evidence that *on the time scales of relevance to fly behavior the signals represented in the photoreceptor array are very noisy*, and that as a result the hyperacuity task is hard. In this section I would like to show how we can make these observations precise, developing the theory of optimal step-discrimination in some detail, and then go on to discuss the structure of the optimal real-time estimator in relation to the known architecture of the fly's visual system. The other half of the story --- the limiting effective noise level in the estimation task --- is described in Refs. [18, 80].

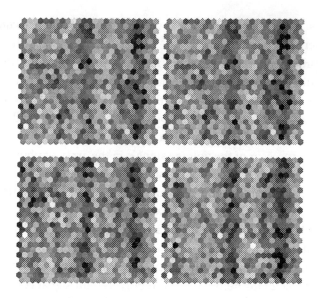

Figure 8.8: Snapshots of the fly photoreceptor voltage array in response to small displacements. Left panel, displacement of a random bar pattern by 0.24° (from top to bottom). Right panel, displacement of the same pattern by 0.36°. Simulations are based on signal and noise characteristics of the receptor cells measured under conditions identical to those used in the experiments on H1. As described in Refs. [84,86,87], the arrival times of the first few spikes following the step are sufficient to discriminate between these different displacements with a reliability of better than 75%.

The compound eye of *Calliphora* samples the world quite coarsely, with the photoreceptors forming a hexagonal lattice with a horizontal spacing of 1.35°. Viewed through the photoreceptor aperture, the experimental stimuli have relatively low contrast. Under these conditions the photoreceptor voltage responds linearly to changes in contrast; we describe the linear photoreceptor response in terms of spatial and temporal transfer functions. In addition to the deterministic average response, the photoreceptors generate voltage noise. Direct measurements of the voltage noise have been made under identical experimental conditions as the experiments on H1 [84]; the noise is Gaussian and independently in different photoreceptors.

From these observations we write the voltage in the nth photoreceptor as

$$V_n(t) = \int d\phi M(\phi - \phi_n) \int d\tau T(\tau) C(\phi, t - \tau) + \delta V_n(t), \qquad (8.54)$$

where the receptor occupies angular position ϕ_n in the lattice, $T(\tau)$ is the temporal response to a short flash of uniform contrast with an amplitude of 1 contrast unit, and $M(\phi)$ is the normalized angular sensitivity profile, which is determined largely by the optics of compound eye. The statistics of the noise are given by the power spectrum $N(\Omega)$, where

$$\langle \delta V_n(t) \delta V_m(t') \rangle = \delta_{nm} \int \frac{d\Omega}{2\pi} e^{-i\Omega(t-t')} N(\Omega). \qquad (8.55)$$

Because our patterns are vertically uniform, we can consider a line of receptors; different rows of photoreceptors in the compound eye report the same movement signals with independent noise sources. The angular sensitivity profile $M(\phi)$ is described as a normalized Gaussian,

$$M(\phi) = \frac{1}{\sqrt{2\pi}\Delta\phi} \exp\left[-\frac{\phi^2}{2(\Delta\phi)^2}\right], \qquad (8.56)$$

with $\Delta\phi = 0.51°$. The temporal impulse response function $T(\tau)$ and the noise spectral density $N(\Omega)$ are from the numerical data of [84].

We use patterns composed of random light and dark bars with width $\phi_B = 0.029°$, much smaller than the photoreceptor aperture $\Delta\phi$. For these patterns the contrast signal to receptor n,

$$C_n(t) = \int d\phi M(\phi - \phi_n) C(\phi - \theta(t)), \qquad (8.57)$$

consists of the sum of a large ($\Delta\phi/\phi_B \sim 60$) number of independent random variables, the intensities of the individual bars. By the central limit theorem, $C_n(t)$ is well-approximated as a Gaussian random variable. The mean contrasts, $\langle C_n(t) \rangle$, are zero since dark ($C = -1$) and light ($C = +1$) bars are equally likely.

We are interested in small displacements $\theta(t)$, in which case we can expand the contrast $C(\phi - \theta(t))$ in a Taylor series:

$$
\begin{aligned}
C_n(t) &= \int d\phi M(\phi - \phi_n) C(\phi - \theta(t)) \\
&\approx \int d\phi M(\phi - \phi_n) C(\phi) - \theta(t) \int d\phi M(\phi - \phi_n) C'(\phi),
\end{aligned}
$$
(8.58)

where $C'(\phi) = dC(\phi)/d\phi$. If the pattern is stationary at $\theta = 0$ before a stepwise displacement, the contrast signal prior to the step is

$$
C_n^- = \int d\phi M(\phi - \phi_n) C(\phi).
$$
(8.59)

After a step of magnitude θ_0 the signal is

$$
C_n^+ = \int d\phi M(\phi - \phi_n) C(\phi) - \theta_0 \int d\phi M(\phi - \phi_n) C'(\phi).
$$
(8.60)

The step discrimination task is to decide between two angular steps given the photoreceptor voltages as described above. The steady state voltages in each receptor are given by

$$
V_n^\pm = \tilde{T}(\Omega = 0) C_n^\pm + \delta V_n,
$$
(8.61)

where the noise δV_n is averaged over the integration time of the movement estimate, resulting in some variance σ_V^2. The steady-state transfer function

$$
\tilde{T}(\Omega = 0) = \int d\tau T(\tau) \approx 3\,\mathrm{mV/unit\ contrast}.
$$
(8.62)

Since the contrast signals and noise voltages are Gaussian random variables, the receptor voltages $V_n^{\pm}$ are also Gaussian. The covariances of the steady-state voltages are given by

$$
\begin{aligned}
\langle V_n^- V_m^- \rangle &= |\tilde{T}(0)|^2 \int d\phi d\phi' M(\phi - \phi_n) M(\phi' - \phi_m) \langle C(\phi) C(\phi') \rangle \\
&\quad + \langle \delta V_n \delta V_m \rangle \\
&= |\tilde{T}(0)|^2 \frac{\phi_B}{\Delta\phi} \frac{1}{2\pi^{1/2}} \exp\left[-\frac{(\phi_n - \phi_m)^2}{4(\Delta\phi)^2} \right] + \delta_{nm}\sigma_V^2, \\
\langle V_n^+ V_m^+ \rangle &= \langle V_n^- V_m^- \rangle, \\
\langle V_n^- V_m^+ \rangle &= |\tilde{T}(0)|^2 \frac{\phi_B}{\Delta\phi} \frac{1}{2\pi^{1/2}} \left[1 + \frac{\theta_0(\phi_n - \phi_m)}{2(\Delta\phi)^2} \right] \\
&\quad \times \exp\left[-\frac{(\phi_n - \phi_m)^2}{4(\Delta\phi)^2} \right].
\end{aligned}
\tag{8.63}
$$

We assume that noises before and after the step are uncorrelated, and we neglect terms of second order in the small displacement θ_0.

All of the information available about step discrimination from the photoreceptor array is contained in the conditional probability distribution $P[\theta|\vec{V}]$, where we have formed a vector $\vec{V}$ from the set $\{V_n^-, V_n^+\}$. Using Bayes' Theorem, we write this distribution as

$$
P[\theta|\vec{V}] = \left(P[\vec{V}] \right)^{-1} P[\vec{V}|\theta] P[\theta].
\tag{8.64}
$$

From the discussion above we can write

$$
P[\vec{V}|\theta] = Z^{-1} \exp\left[-\frac{1}{2} \vec{V}^T \cdot C^{-1}(\theta) \cdot \vec{V} \right],
\tag{8.65}
$$

where $C(\theta)$ is the step-dependent co-variance matrix, N is the total number of photoreceptors, and Z is the normalization constant. Our task is to choose between θ_0 and θ_1 from observation of $\vec{V}$. To maximize the probability of a correct decision [43] we define a decision variable λ as

$$\lambda(\vec{V}) = \ln \left(\frac{P[\theta_0|\vec{V}]}{P[\theta_1|\vec{V}]} \right). \tag{8.66}$$

If $\lambda(\vec{V}) > 0$, we guess that the step was of magnitude θ_0, and if $\lambda(\vec{V}) < 0$ we guess θ_1.

In the limit that we are stimulating a large number of receptor cells $\lambda(\vec{V})$ is the sum of a large number of random variables, and by the central limit theorem the distribution of λ given some step size will be well approximated by a Gaussian. The discrimination problem is then exactly the problem of detecting a signal in a Gaussian noise background, where the signal is the difference in the means of $P[\lambda|\theta_0]$ and $P[\lambda|\theta_1]$, and the noise is the variance of the distribution. After some calculation we find that the signal-to-noise is

$$SNR = (1/2)(\theta_0 - \theta_1)^2 \text{Tr} \left[C^{-1}(\theta_0) \frac{\partial C(\theta_0)}{\partial \theta_0} C^{-1}(\theta_0) \frac{\partial C(\theta_0)}{\partial \theta_0} \right]. \tag{8.67}$$

To proceed we make use of the fact that the co-variance matrix matrix $C(\theta)$ is translationally invariant, so it is diagonalized in a spatial Fourier representation; at the end one must sum over all spatial frequencies to effect the trace (Tr). The result is that

$$SNR = \frac{N\phi_0(\theta_0 - \theta_1)^2}{4(\Delta\phi)^4} \int_{-\pi/\phi_0}^{\pi/\phi_0} \frac{dq}{2\pi} \left(\frac{\partial G(q)}{\partial q} \right)^2 \frac{1}{1 + 2G(q)}, \tag{8.68}$$

$$G(q) = S \sum_n \exp \left[inq\phi_0 - (n\phi_0/2\Delta\phi)^2 \right] \tag{8.69}$$

$$S = \frac{1}{2\pi^{1/2}} \frac{|\tilde{T}(0)|^2}{\sigma_V^2} \frac{\phi_B}{\Delta\phi}, \tag{8.70}$$

where we have simplified to the experimentally relevant case where both θ_0 and θ_1 are small compared to the photoreceptor aperture. The quantity S measures the signal-to-noise ratio in a single receptor cell, and $S \approx 0.7$ from the data cited above. Because the parameter $\beta = \exp[-(\phi_0/2\Delta\phi)^2] \approx 0.16$ is small, one can make a simple analytic calculation which is accurate to 10% or better; the result is

$$(SNR)^{1/2} = \frac{|\theta_0 - \theta_1|\phi_0}{(\Delta\phi)^2} \frac{S}{(1+S)^{1/2}} \left(\frac{N}{2}\right)^{1/2} \exp[-(\phi_0/2\Delta\phi)^2]. \tag{8.71}$$

Inserting all the numerical factors, we find that for the optimal processor, $SNR = 1$ is reached for discrimination of $|\theta_0 - \theta_1| = 0.06°$ displacement differences. In the experiments on H1 we find that displacement differences of $0.12°$ lead to values of SNR slightly larger than unity for discriminations based on the first two spikes following the step displacement. Thus we conclude that the observed performance of H1 is within a factor of two of the theoretical limit imposed by noise in the photoreceptors.

What is the structure of the optimal estimator for angular velocity, and how does this theory of optimal estimation relate to known properties of the fly visual system? If the fly is looking at some rigid contrast pattern $C(\phi)$ which moves relative to the fly along some trajectory $\theta(t)$, then each photoreceptor produces a voltage [cf. Eq. (8.54)]

$$V_n(t) = \int d\tau T(\tau) \int d\phi M(\phi - \phi_n) C[\phi - \theta(t - \tau)] + \delta V_n(t). \tag{8.72}$$

This description of the photoreceptor voltage responses determines the conditional probability distribution

$$P[\{V_n(t)|\theta(\tau)\}] = \frac{1}{Z[\theta(\tau)]} \exp\left[-\frac{1}{2}\sum_n \int \frac{d\Omega}{2\pi} \frac{|\tilde{V}_n(\Omega) - \bar{V}_n(\Omega)|^2}{N(\Omega)}\right], \tag{8.73}$$

where

$$\bar{V}_n(\Omega) = \tilde{T}(\Omega) \int dt \exp(+i\Omega t) \int d\phi M(\phi - \phi_n) C[\phi - \theta(t)]. \qquad (8.74)$$

Everything that we know about the trajectory $\theta(\tau)$ by virtue of observing the array of receptor voltages is contained in the distribution

$$P[\theta(\tau)|\{V_n(t)\}] = \frac{P[\{V_n(t)|\theta(\tau)\}]P[\theta(\tau)]}{P[\{V_n(t)\}]}, \qquad (8.75)$$

which combines the response characteristics of the cells with our *a priori* knowledge of the statistics of $\theta(t)$. In the experiments of Refs. [20, 85], for example, the statistics of the trajectories $\theta(t)$ correspond to a simple random walk, or angular diffusion with diffusion constant $D \sim 4 \deg^2/\sec$.

Our task is to understand how one can use all of this information to estimate the angular velocity $\dot{\theta}(t)$ based on the data contained in the $\{V_n(t)\}$. Let us take as the optimal estimator of $\dot{\theta}(t)$ the conditional mean,

$$\langle \dot{\theta}(t) \rangle_{\{V_n(t)\}} = \int [d\theta] \dot{\theta}(t) P[\theta(\tau)|\{V_n(t)\}], \qquad (8.76)$$

which is the average trajectory that gives rise to our observations. Alternatively one could look for the most likely trajectory, but the theory for the average trajectory is more straightforward.

We start by thinking about the low signal-to-noise limit, where the mean receptor voltages $\bar{V}_n(\Omega)$ are small compared to the typical amplitudes of the voltage noise ($\propto \sqrt{N(\Omega)}$). In this limit one can evaluate the functional integral of Eq. (8.76) as a perturbation series in $\bar{V}_n$, and we find that the optimal estimator can be written as

$$\dot{\theta}_{est}(t) = \langle \dot{\theta}(t) \rangle_{\{V_n(\tau)\}}$$

$$= \frac{DS_C}{\pi^{1/2}} \int \frac{d\Omega}{2\pi} \int \frac{d\Omega'}{2\pi} e^{-i(\Omega+\Omega')t} \frac{1}{-i(\Omega+\Omega')}$$

$$\times \sum_{nm} \left[\frac{\tilde{T}^*(\Omega)}{N(\Omega)} g_{nm}(\Omega)\tilde{V}_n(\Omega) \right] \left[\frac{\tilde{T}^*(\Omega')}{N(\Omega')} \tilde{V}_n(\Omega') \right]. \qquad (8.77)$$

where the filters g_{nm} are given by

$$g_{nm}(\Omega) = \int d\tau e^{i\Omega\tau} \frac{\phi_n - \phi_m}{[(\Delta\phi)^2 + D|\tau|]^{3/2}} \exp\left[-\frac{1}{4} \cdot \frac{(\phi_n - \phi_m)^2}{(\Delta\phi)^2 + D|\tau|} \right] ;(8.78)$$

as before D is the angular diffusion constant of the motion we are trying to estimate and $\Delta\phi$ is the width of the photoreceptor aperture, $M(\phi) \propto \exp[-\phi^2/2(\Delta\phi)^2]$.

We see that the optimal estimator at low signal-to-noise ratios has the form of a correlator, multiplying spatially displaced pairs of photoreceptor voltages which have been passed through different temporal filters. This is a generalization of the scheme proposed forty years ago by Hassenstein and Reichardt from their experiments on insect optomotor behavior; for a review of these ideas see Refs. [72, 74]. What we have shown is that this is not just the simplest scheme for motion detection, it is also the optimal scheme for motion estimation under certain conditions. The phenomenological filters of the original discussion now take specific forms related to the characteristics of the photoreceptors and the ensemble of stimuli. There is ample evidence that the cell H1 is at least approximately described by a Reichardt-like model, and one can take this evidence as qualitatively supporting the existence of neural dynamics as predicted from the principle of optimal processing. Several points should be emphasized:

1. In principle, the optimal estimator includes contributions from all pairs of receptors. The peak amplitude (in the time domain) of the filter g_{nm} is proportional to $1/(n-m)^2$, so that nearest neighbors dominate, although contributions from more distant neighbors should be observable. Such higher-neighbor terms have been seen in experiments where one stimulates individual facets of the compound eye [93].

2. Before entering the filters g_{nm} which are dedicated to the task of movement estimation, all photoreceptor signals pass through a filter $\tilde{T}^*(\Omega)/N(\Omega)$. This is in accord with the general arguments described above, where we found that such a filter --- matched to the signal and noise characteristics of the photoreceptor but independent of the stimulus ensemble or the computational task --- is the universal first step in any optimal estimation task at low signal to noise ratio. By analogy with the bipolar cells in the vertebrate retina, we might imagine that this filtering is achieved in the lamina, the first stage of signal processing in the fly.

3. The specialized filters g_{nm} have dynamics which depend on the statistics of the movement signal, in this case just the diffusion constant D. More generally, if small amplitude motions are superposed on a constant velocity --- as in natural flight --- then the dynamics of the g_{nm} depend on this constant velocity. Thus we predict an adaptive generalization of the correlation model, in which not just the overall sensitivity of the cell but also its filter characteristics must adapt to background motions. This is observed in H1 [88], and the fact that this velocity-dependent adaptation occurs locally strongly suggests that the filters g_{nm} reside in a portion of the fly visual system which retains the orderly geometrical relations of the retina itself, perhaps in the medulla.

4. The outputs of the filters g_{nm} are correlated with signals from neighboring cells, then summed. Presumably this integration step resides in H1 itself. Optimal movement estimation thus involves at least three mathematically distinct stages of computation, which we can tentatively identify with three layers of processing circuitry in the lamina, the medulla, and the lobula complex.

5. We have evidence for near-optimal performance under conditions where the total signal to be computed [e.g., in Eq. (8.77)] involves a sum over roughly 2500 elementary movement detectors, and it is this total signal which exhibits a signal-to-noise ratio near unity. But then the movement signal from each isolated pair of photoreceptors is at a signal-to-noise (amplitude) ratio of just $1/\sqrt{2500} \sim 1/50$. Each multiplication step must

therefore be done with about 5 to 6 bits of accuracy ($\log_2 50 = 5.64$) in order that the subsequent averaging among parallel channels be effective in revealing the signal. Thus, although one could argue that H1 achieves its impressive performance as a result of averaging over many inputs, it is clear that non-linear computations must be done *before* this averaging occurs, and this places requirements on the precision of the individual steps in the processing chain.

Finally it should be emphasized that the correlation scheme is *not* optimal at higher signal-to-noise ratio. Although the exact form of the optimal estimator at high SNR is not yet known, it is clear that, if H1 is to maintain optimality, not just the time constants or gain parameters but indeed the qualitative character of the computation must change under different stimulus conditions. In a beautiful series of experiments, Franceschini and co-workers [41] have probed the dynamics of H1 under conditions far removed from those discussed here --- high SNR inputs to just two receptor cells, as opposed to low SNR inputs to thousands of cells. Under these conditions the non-linearities in the pathway to H1 are very different from a simple multiplication or correlation. A crucial challenge for the theory is to see if this sort of adaptive variation in computational strategy can be predicted. One suspects that this may be a more general phenomenon, so that the theory of optimal processing might guide the search for analogous adaptive behavior in other sensory systems.

Conclusions

The idea that biological systems may approach some fundamental limits to sensory performance is very old. There is a tendency, however, to see such performance as an isolated curiosity, perhaps significant only for its amusement value in the physics community. I hope that surveying a large collection of optimal and near-optimal systems helps to counter this view. Indeed I think it is remarkable that one can find a sense in which the computations of visual movement in the fly are the same as the computations of target range in the bat --- both make optimal use of the signal-to-noise available at the sensory input. It is this universality which makes the notion of optimal processing so attractive from a theoretical point of view.

What do we need to establish optimality as a general principle for the early stages of sensory signal processing? One approach is to continue the quantitative comparison of neural performance with optimal performance in the largest possible variety of systems. In particular we would like to move away from the consideration of one cell at a time and think about computation and coding in arrays of cells. As an example, in the visual system the coding efficiency of the retina is really a property of the entire ganglion cell array and not any single cell. Methods are now available to record simultaneously from large numbers of retinal ganglion cells [66, 67], and our decoding algorithms have a natural generalization to the multi-cell case, so I believe that the tools are available to attack these problems.

Ideally we would like to find some dramatic qualitative signatures of optimality. There is a strong hint from the work reviewed here the general phenomenon of sensory adaptation must be much richer than usually suspected if optimal performance is to be maintained over a wide range of conditions --- filters must adapt their dynamics, spiking neurons must adapt their thresholds in relation to the stimulus variance, and even the qualitative character of non-linear operations must shift as a function of signal-to-noise ratio. All of this is much more than just shifting the operating point of the system to compensate for static backgrounds, which is the textbook picture of adaptive responses.

As things stand, one is forced to illustrate the different issues in the design of sensory systems by looking at different pieces of different organisms. Obviously it would be nice to make a complete story straight through one system. This naturally leads to the questions of whether there is a "simplest case." One candidate is the auditory system of noctuid moths, which mediates complex bat-evading flight control using inputs from just two sensory neurons [82]. Another possibility is to take a step back from the nervous system and look at what may be the simplest (and evolutionarily oldest) systems of all, the chemosensors of single-celled organisms; recent progress on this problem is described in Kruglyak's contribution to these proceedings.

The fact that biological systems can function at or near the optimum for processing of signals in noisy backgrounds means that the theory of optimal signal processing should provide us with a quantitative and predictive theory for the dynamics of real neural circuits. We have seen some examples of this theory at work, in the discussion of pre-processing and in the movement estimation problem, and this experience suggests that statistical mechanics provides a natural setting for these problems. Perturbation expansions of the relevant functional integrals are equivalent to expansions in SNR, while saddle-point evaluations of these functional integrals (mean-field, or semi-classical approximations) should give us an approach to the high SNR limit. This deserves to be explored more fully; one clear need is for a statistical mechanics model of the ensemble which generates the signals in the natural world.

In some sense the problem of optimal signal processing is the extreme version of a much older problem, namely understanding how the organism is able to perform reliable computations. Inspired in part by the difficulties of building reliable electronic computers (and apparently also by conversations with Gell-Mann and Brueckner), von Neumann [69] posed the problem of how to perform reliable computations even when the components of the computing engine are highly unreliable. The apparent variability of neural responses has led many people to think that this problem is relevant for the nervous system. Von Neumann's own opinion on this issue is hard to discern; his attempts to compare electronic and natural computers were cut short [70]. Somehow the words 'reliable computation with unreliable components' conjure up an image of a system which barely manages to work in spite of itself. The arguments presented here point more in the direction of a system which has sophisticated adaptive strategies for seeking essentially perfect solutions.

How is all this possible? Darwin tells us that evolution proceeds by opportunistic means; beneficial variations endure, but there is no goal toward which animals evolve. In some global sense none of the systems we have studied is optimal --- you can build better photodetectors with kilovolt power supplies for amplification and a large bath of liquid helium to freeze out the noise. At best we are trying to find constrained optima: Given an array of photodetectors, how do we process their output to provide the most information about objects of interest? It seems plausible that there is a large evolutionary pressure on animals to make better estimates of the locations of predators and prey; it may still be surprising that this pressure is large enough to drive systems to their local optima. Some of the mystery can be taken out by realizing that archaebacteria --- bacteria which are the modern analogs of the most ancient creatures --- can count single photons [63].

Eons ago the seas were populated by creatures with 'primitive' sensory systems. But the data of Ref. [63] suggest that these primitive systems were already capable of performing near the fundamental limits set by the laws of physics; bacteria had discovered photons millions of years before Einstein and Planck. It would seem that biology found a set of near ideal strategies for signal processing almost from the beginning, and what we see in modern organisms is the working out of these strategies for ever more sophisticated tasks. If all these words are correct, then there is something universal behind all the apparent diversity of the sensory systems, and perhaps we physicists have a chance of understanding brains after all.

Acknowledgements

I have already thanked my many collaborators who helped make this work so much fun. In addition, conversations with many people have (hopefully) helped to make the ideas more precise: J. Atick, H. B. Barlow, D. Baylor, S. Block, N. Franceschini, D. Glaser, A. Libchaber, M. Meister, J. Simmons, and G. Zweig.

References

[1] E. D. Adrian (1928). *The Basis of Sensation* (W. W. Norton, New York).

[2] A.-C. Aho, K. Donner, C. Hyden, L. O. Larsen, and T. Reuter (1988). Low retinal noise in animals with low body temperature allows high visual sensitivity, *Nature* **334**, 348-350.

[3] J. J. Atick and A. N. Redlich (1990). Towards a theory of early visual processing, *Neural Comp.* **2**, 308-320.

[4] J. J. Atick, Z. Li, and A. N. Redlich (1992). Understanding retinal color coding from first principles, *Neural Comp.* **4**, 559-572.

[5] H. B. Barlow (1952). The size of ommatida in apposition eyes, *J. Exp. Biol.* **29**, 667-674.

[6] H. B. Barlow (1953). Summation and inhibition in the frog's retina *J. Physiol.* **119**, 69-88.

[7] H. B. Barlow (1956). Retinal noise and absolute threshold, *J. Opt. Soc. Am.* **46**, 634-639.

[8] H. B. Barlow (1961). Possible principles underlying the transformation of sensory messages, in *Sensory Communication*, W. Rosenblith, ed. p. 217-234 (MIT Press, Cambridge MA).

[9] H. B. Barlow (1982). What causes trichromacy? A theoretical analysis using comb-filtered spectra, *Vision Res.* **22,** 635-643.

[10] H. B. Barlow (1980). The absolute efficiency of perceptual decisions, *Philos. Trans. R. Soc. Lond. Ser. B* **290,** 71-82.

[11] H. B. Barlow (1981). Critical limiting factors in the design of the eye and visual cortex, *Proc. R. Soc. Lond. Ser. B* **212,** 1-34.

[12] H. B. Barlow (1988). Thermal limit to seeing, *Nature* **334,** 296.

[13] H. B. Barlow and W. Levick (1969). Three factors limiting the reliable detection of light by the retinal ganglion cells of the cat, *J. Physiol.* **200,** 1-24.

[14] D. A. Baylor, T. D. Lamb, and K.-W. Yau (1979). Responses of retinal rods to single photons, *J. Physiol.* **288,** 613-634.

[15] D. A. Baylor, G. Matthews, and K.-W. Yau (1980). Two components of electrical dark noise in toad retinal rod outer segments, *J. Physiol.* **309,** 591-621.

[16] D. A. Baylor, B. J. Nunn, and J. F. Schnapf (1984). The photocurrent, noise and spectral sensitivity of rods of the monkey *Macaca fascicularis, J. Physiol.* **357,** 575-607.

[17] W. Bialek (1987). Physical limits to sensation and perception, *Ann. Rev. Biophys. Biophys. Chem.* **16,** 455-478.

[18] W. Bialek (1990). Theoretical physics meets experimental neurobiology, in *1989 Lectures in Complex Systems, SFI Studies in the Sciences of Complexity, Lect. Vol. II*, E. Jen, ed., pp. 513-595 (Addison-Wesley, Menlo Park CA).

[19] W. Bialek and W. G. Owen (1990). Temporal filtering in retinal bipolar cells: Elements of an optimal computation?, *Biophys. J.* **58,** 1227-1233.

[20] W. Bialek, F. Rieke, R. R. de Ruyter van Steveninck, and D. Warland (1991). Reading a neural code, *Science* **252**, 1854-1857.

[21] W. Bialek, D. L. Ruderman, and A. Zee (1991). Optimal sampling of natural images: A design principle for the visual system?, in *Advances in Neural Information Processing 3*, R. Lippmann, J. Moody, and D. Touretzky, eds., pp. 363-369 (Morgan Kaufmann, San Mateo CA).

[22] W. Bialek and A. Zee (1990). Coding and computation with neural spike trains, *J. Stat. Phys.* **59**, 103-115.

[23] E. de Boer and R. MacKay (1980). Reflections on reflections, *J. Acoust. Soc. Am.* **67**, 882-890.

[24] V. Bonnardel and F. J. Varela (1991). a frequency view of colour: Measuring the human sensitivity to square-wave spectral power distributions, *Proc. R. Soc. Lond. Ser. B* **245**, 165-171.

[25] M. A. Bouman (1961). History and present status of quantum theory in vision, in *Sensory Communication*, W. Rosenblith, ed., pp. 377-401 (MIT Press, Cambridge MA).

[26] L. Brillouin (1962). *Science and Information Theory* (Academic Press, New York NY).

[27] T. H. Bullock (1970). The reliability of neurons, *J. Gen. Physiol.* **55**, 584-656.

[28] T. H. Bullock (1976). Redundancy and noise in the nervous system: Does the model based on unreliable neurons sell nature sort?, in *Electrobiology of Nerve, Synapse, and Muscle*, ed. J. Reuben, D. Purpura, M. V. L. Bennett, and E. Kandel, pp. 179-185 (Raven Press, New York).

[29] A. E. Burgess, R. F. Wagner, R. J. Jennings, and H. B. Barlow (1981). Efficiency of human visual signal discrimination, *Science* **214**, 93-94.

[30] M. Carreau, E. Farhi, S. Guttman, and P. F. Mende (1990). The functional integral for quantum systems with Hamiltonian unbounded from below, *Ann. Phys.* **204**, 186-207.

[31] K. Case (1950). Singular potentials, *Phys. Rev.* **80**, 797-806.

[32] L. Chittka and R. Menzel (1992). The evolutionary adaptation of flower colours and the insect pollinators' colour vision, *J. Comp. Physiol. A* **171**, 171-181.

[33] K. Donner (1989). The absolute sensitivity of vision: Can a frog become a perfect detector of light induced and dark rod events?, *Phys. Scr.* **39**, 133-140.

[34] R. Durbin and G. Mitchison (1990). A dimension reduction framework for understanding cortical maps, *Nature* **343**, 644-647.

[35] J. J. Eggermont, P. I. M. Johannesma, and A. M. H. J. Aertsen (1983). *Q. Rev. Biophys.*, **16**, 341-414.

[36] R. P. Feynman and A. R. Hibbs (1965). *Path Integrals and Quantum Mechanics*. (McGraw Hill, New York).

[37] R. P. Feynman, R. Leighton, and M. Sands (1963). *The Feynman Lectures on Physics, Volume I*. (Addison-Wesley, Reading MA).

[38] D. Field (1987). Relations between the statistics of natural images and the response properties of cortical cells, *J. Opt. Soc. Am. A* **4**, 2379-2394.

[39] R. Fitzhugh (1958). A statistical analyzer for optic nerve messages, *J. Gen. Physiol.* **41**, 675-692.

[40] M. G. F. Fourtes and S. Yeandle (1964). Probability of occurrence of iscrete potential waves in the eye of *Limulus*, *J. Gen. Physiol.* **47**, 443-463.

[41] N. Franceschini, A. Riehle, and A. le Nestour (1989). Directionally selective motion detection by insect neurons, in *Facets of Vision*, R. C. Hardie and D. G. Stavenga, eds. pp. 360-390 (Springer-Verlag, Berlin).

[42] C. C. A. M. Gielen, G. H. F. M. Hesselmans, and P. I. M. Johannesma, Sensory interpretation of neural activity patterns, *Math. Biosci.* **88**, 15-35 (1988).

[43] D. M. Green and J. A. Swets (1966). *Signal Detection Theory and Psychophysics.* (Wiley, New York).

[44] K. Hausen (1984). The lobular complex of the fly: Structure, function, and significance in behavior, in *Photoreception and vision in invertebrates,* M. Ali, ed., pp. 523-559 (Plenum, New York NY).

[45] S. Hecht, S. Shlaer, and M. H. Pirenne (1942). Energy, quanta, and vision, *J. Gen. Physiol.* **25,** 819-840.

[46] P. I. M. Johannesma (1981). Neural representation of sensory stimuli and sensory interpretation of neural activity, *Advances in Physiological Sciences* **30,** 103-126.

[47] D. H. Johnson (1974). the response of single auditory nerve fibers in the cat to single tones: Synchrony and average discharge rate (Dissertation, Massachusetts Institute of Technology).

[48] B. Katz (1966). *Nerve, Muscle, and Synapse.* (McGraw-Hill, New York).

[49] J. J. Knierem and D. C. van Essen (1992). Neuronal responses to static textures in area V1 of the alert Macaque monkey, *J. Neurophys.* **67,** 961-980.

[50] E. I. Knudsen, S. du Lac, and S. D. Esterly (1987). Computational maps in the brain, *Ann. Rev. Neurosci.* **10,** 41-65.

[51] J. W. Kuiper (1962). The optics of the compound eye, *Symp. Soc. Exp. Biol.* **16,** 58-71.

[52] M. F. Land (1988). The optics of animal eyes, *Contemp. Phys.* **29,** 435-455.

[53] M. F. Land and T. S. Collett (1974). Chasing behavior of houseflies (*Fannia canicularis*): A description and analysis, *J. Comp. Physiol.* **89,** 331-357.

[54] S. B. Laughlin (1981). A simple coding procedure enhances a neuron's information capacity, *Z. Naturforsch.* **36c,** 910-912.

[55] J. L. Lawson and G. E. Uhlenbeck (1950). *Threshold Signals.* (McGraw-Hill, New York).

[56] J. Y. Lettvin, H. R. Maturana, W. S. McCulluch, and W. H. Pitts (1959). What the frog's eye tells the frog's brain, *Proc. I. R. E.* **47**, 1940-1951.

[57] C.-Y. Li, X. Pei, Y.-X. Zhow, and H.-C. von Mitzlaff (1991). Role of the extensive area outside the X-cell receptive field in brightness information transmission, *Vision Res.* **31**, 1529-1540.

[58] R. Linsker (1989). An application of the principle of maximum information preservation to linear systems, in *Advances in Neural Information Processing 1*, D. Touretzky, ed., pp. 186-194 (Morgan Kaufmann, San Mateo CA).

[59] D. MacKay and W. S. McCulloch (1952). The limiting information capacity of a neuronal link, *Bull. Math. Biophys.* **14**, 127-135.

[60] A. Mallock (1894). Insect sight and the defining power of compound eyes, *Proc. R. Soc. Lond. Ser. B* **55**, 85-90.

[61] L. T. Maloney (1986). Evaluation of linear models of surface reflectance with small numbers of parameters, *J. Opt. Soc. Am. A* **3**, 1673-1683.

[62] N. J. Marshall, M. F. Land, C. A. King, and T. W. Cronin (1991). The compound eye of mantis shrimps *(Crustacea, Hoplocarida, Stomatopoda)*. II. Colour pigments in the eyes of stomatopod crustaceans: Polychromatic vision by serial and lateral filtering, *Phil. Trans. R. Soc. Lond. Ser. B* **334**, 57-84.

[63] W. Marwan, P. Hegemann, and D. Oesterhelt (1988). Single photon detection by an archeabacterium, *J. Mol. Biol.* **199**, 663-664.

[64] J. W. McClurkin, L. M. Optican, B. J. Richmond, and T. J. Gawne (1991). Concurrent processing and complexity of temporally encoded messages in visual perception, *Science* **253**, 675-677.

[65] M. Meister (1991). personal communication.

[66] M. Meister, J. Pine, and D. A. Baylor (1992). Recording visual signals in the retina with a multielectrode array, preprint.

[67] M. Meister, R. O. L. Wong, D. A. Baylor, and C. J. Shatz (1991). Synchronous bursts of action potentials in ganglion cells of the developing mammalian retina, *Science* **252**, 939-943.

[68] G. Mitchison (1991). Neuronal branching patterns and the economy of cortical wiring, *Proc. R. Soc. Lond. Ser. B.* **245**, 151-158.

[69] J. von Neumann (1956). Probabilistic logics and the synthesis of reliable organisms from unreliable components, In *Automata Studies*, C. E. Shannon and J. McCarthy eds., pp. 43-98 (Princeton University Press, Princeton NJ).

[70] J. von Neumann (1958). *The Computer and the Brain.* (Yale University Press, New Haven CT).

[71] D. H. Perkel and T. H. Bullock (1968). Neural coding, *Neurosci. Res. Prog. Sum.* **3**, 405-527.

[72] T. Poggio and W. Reichardt (1976). Visual control of orientation behavior in the fly. Part II. Towards the underlying neural interactions, *Q. Rev. Biophys.* **9**, 377-438.

[73] F. Ratliff, ed. (1974). *Studies on Excitation and Inhibition in the Retina* (Rockefeller University Press, New York NY).

[74] W. Reichardt and T. Poggio (1976). Visual control of orientation behavior in the fly. Part I. A quantitative analysis, *Q. Rev. Biophys.* **9**, 311-375.

[75] R. C. Reid, R. E. Sodak, and R. M. Shapley (1991). Directional selectivity and spatiotemporal structure of receptive fields of simple cells in cat striate cortex, *J. Neurophys.* **66**, 505-529.

[76] F. Rieke (1991). *Physical Principles Underlying Sensory Processing and Computation* (Dissertation, University of California at Berkeley).

[77] F. Rieke, D. Bodnar, and W. Bialek (1992). Coding of natural sound stimuli by the bullfrog auditory nerve: Phase, amplitude and information rates, in *Proceedings of the Third International Congress of Neuroethology, Montreal.*

[78] F. Rieke, W. G. Owen, and W. Bialek (1991). Optimal filtering in the salamander retina, in *Advances in Neural Information Processing 3*, R. Lippmann, J. Moody, and D. Touretzky, eds., pp. 377-383 (Morgan Kaufmann, San Mateo CA).

[79] F. Rieke, D. Warland, and W. Bialek (1992). Coding efficiency and information capacity in sensory neurons, preprint.

[80] F. Rieke, D. Warland, R. R. de Ruyter van Steveninck, and W. Bialek (1992). Optimal real-time processing of visual movement signals: Theory and experiments in the blowfly, in preparation.

[81] F. Rieke, W. Yamada, K. Moortgat, E. R. Lewis, and W. Bialek (1992). Real-time coding of complex signals in the auditory nerve, in *Auditory Physiology and Perception: Proceedings of the 9th International Symposium on Hearing*, Y. Cazals, L. Demany, and K. Horner, eds., pp. 315-322 (Elsevier, Amsterdam).

[82] K. D. Roeder and R. S. Payne (1966). Acoustic orientation of a moth in flight by means of two sense cells, *Symp. Soc. Exp. Biol.* **20**, 251-272.

[83] A. Rose (1948). The sensitivity performance of the human eye on an absolute scale, *J. Opt. Soc. Am.* **38**, 196-208.

[84] R. R. de Ruyter van Steveninck (1986). *Real-time Performance of a Movement-Sensitive Neuron in the Blowfly Visual System* (Academisch Proefschrift, Rijksuniversiteit Groningen).

[85] R. de Ruyter van Steveninck and W. Bialek (1988). Real-time performance of a movement sensitive neuron in the blowfly visual system: Coding and information transfer in short spike sequences, *Proc. R. Soc. London Ser. B* **234**, 379-414.

[86] R. R. de Ruyter van Steveninck and W. Bialek (1992). Statistical reliability of a blowfly movement-sensitive neuron, in *Advances in Neural Information Processing 4*, J. Moody, S. J. Hanson, and R. Lippmann, eds., pp. 27-34 (Morgan Kaufmann, San Mateo CA).

[87] R. R. de Ruyter van Steveninck, W. Bialek, and W. H. Zaagman (1985). Vernier movement discrimination with three spikes from one neuron, *Perception* **13**, A47-48.

[88] R. R. de Ruyter van Steveninck, W. H. Zaagman, and H. Mastebroek (1986). Adaptation of transient responses of a movement-sensitive neuron in the visual system of the blowfly *Calliphora erythrocephela*, *Biol. Cybern.* **54**, 223-236.

[89] B. Sakitt (1972). Counting every quantum, *J. Physiol.* **223**, 131-150.

[90] B. Sakitt and H. B. Barlow (1982). A model for the economical encoding of the visual image in cerebral cortex, *Biol. Cybern.* **43**, 97-108.

[91] B. Sakmann and E. Neher, eds. (1983). *Single Channel Recording* (Plenum, New York).

[92] J. L. Schnapf, B. J. Nunn, M. Meister, and D. A. Baylor (1990). Visual transduction in cones of the monkey *Macaca fascicularis*, *J. Physiol.* **427**, 681-713.

[93] F. Schuling, H. Mastebroek, R. Bult, and B. Lenting (1989). Properties of elementary movement detectors in the fly *Calliphora erythrocphela*, *J. Comp. Physiol. A* **165**, 179-192.

[94] W. M. Siebert (1965). Some implications of the stochastic behavior of primary auditory neurons, *Kybernetik* **2**, 206-215.

[95] C. E. Shannon, (1949). Communication in the presence of noise, *Proc. I. R. E.* **37**, 10-21.

[96] J. A. Simmons (1989). A view of the world through the bat's ear: The formation of acoustic images in echolocation, *Cognition* **33**, 155-199.

[97] J. A. Simmons, M. Ferragamo, C. F. Moss, S. B. Stevenson, and R. A. Altes (1990). Discrimination of jittered sonar echoes by the echolocating bat, *Eptesicus fuscus*: The shape of target images in echolocation, *J. Comp. Physiol. A* **167**, 589-616.

[98] A. W. Snyder (1977). Acuity of compound eyes: Physical limitations and design, *J. Comp. Physiol. A* **116**, 161-182.

[99] A. W. Snyder, D. S. Stavenga, and S. B. Laughlin (1977). Information capacity of compound eyes, *J. Comp. Physiol.* **116**, 183-207.

[100] M. V. Srinivasan, S. B. Laughlin, and A. Dubs (1982). Predictive coding: A fresh view of inhibition in the retina *Proc. R. Soc. Lond. Ser. B* **216**, 427-459.

[101] R. B. Stein (1967). The information capacity of neurons using a frequency code, *Biophys. J.* **7**, 797-826.

[102] M. Tapiovaara (1990). Ideal observer and absolute efficiency of detecting mirror symmetry in random images, *J. Opt. Soc. Am. A* **7**, 2245-2253.

[103] M. C. Teich, L. Matin, B. I. Cantor (1978). *J. Opt. Soc. Am.* **8**, 386-402.

[104] M. C. Teich, P. R. Prucnal, G. Vannucci, M. E. Breton, and W. J. McGill (1982). Multiplication noise in the human visual system at threshold. I: Quantum fluctuations and the minimum detectable energy, *J. Opt. Soc. Am,* **72**, 419-431.

[105] M. C. Teich, P. R. Prucnal, G. Vannucci, M. E. Breton, and W. J. McGill (1982). Multiplication noise in the human visual system at threshold. III: The role of non-Poisson quantum fluctuations, *Biol. Cybern.* **44**, 157.

[106] F. Theunissen and J. P. Miller (1991). Representation of sensory information in the cricket cercal sensory system. II: Information theoretic calculation of system accuracy and optimal tuning curve widths of four primary interneurons, *J. Neurophys.* **66**, 1690-1703.

[107] H. de Vries (1943). The quantum character of light and its bearing upon threshold of vision, the differential sensitivity and visual acuity of the eye, *Physica* **10**, 553-564.

[108] H. A. van der Velden (1944). Over het aantal lichtquanta dat nodig is voor een lichtprikkel bij het menselijk oog, *Physica* **11**, 179-189.

[109] H. Wagner (1986). Flight performance and visual control of flight in the free-flying house fly *(Musca domestica L.).* I: Organization of the flight motor, *Phil. Trans. R. Soc. Lond. Ser. B* **312**, 527-551.

[110] H. Wagner (1986). Flight performance and visual control of flight in the free-flying house fly *(Musca domestica L.)*. II: Pursuit of targets, *Phil. Trans. R. Soc. Lond. Ser. B* **312,** 553-579.

[111] H. Wagner (1986). Flight performance and visual control of flight in the free-flying house fly *(Musca domestica L.)*. III: Interactions between angular movement induced by wide- and small-field stimuli, *Phil. Trans. R. Soc. Lond. Ser. B* **312,** 581-595.

[112] D. Warland (1991). *Reading Between the Spikes: Real-Time Processing in Neural Systems* (Dissertation, University of California at Berkeley).

[113] D. Warland, M. Landolfa, J. P. Miller, and W. Bialek (1991). Reading between the spikes in the cercal filiform hair receptors of the cricket, in *Analysis and Modeling of Neural Systems*, F. Eeckman, ed., pp. 327-333 (Kluwer Academic).

[114] N. Wiener (1949). *Extrapolation, Interpolation and Smoothing of Time Series.* (Wiley, New York).

[115] G. Westheimer (1981). Visual hyperacuity, *Prog. Sens. Physiol.* **1,** 1-30.

[116] G. Zweig (1975). Basilar membrane motion, *Cold Spring Harbor Symp. Quant. Biol.* **40,** 619-633.

[117] G. Zweig, R. Lipes, and J. R. Pierce (1976). The cochlear compromise, *J. Acoust. Soc. Am.* **59,** 975-982.